APOCALYPSE NEVER

APOCALYPSE NEVER

WHY ENVIRONMENTAL ALARMISM
HURTS US ALL

Michael Shellenberger

HARPER

An Imprint of HarperCollins*Publishers*
www.harpercollins.com

HarperCollins books may be purchased for educational, business, or sales promotional use. For information, please email the Special Markets Department at SPsales@harper collins.com.

FIRST EDITION

Library of Congress Cataloging-in-Publication Data

Names: Shellenberger, Michael, author.
Title: Apocalypse never : why environmental alarmism hurts us all / Michael
  Shellenberger.
Description: First edition | New York, NY : Harper, [2020] | Includes
  index.
Identifiers: LCCN 2020013145 (print) | LCCN 2020013146 (ebook) | ISBN
  9780063001695 (hardcover) | ISBN 9780063001701 (ebook)
Subjects: LCSH: Environmentalism. | Green movement. |
  Anti-environmentalism. | Nature—Effect of human beings on. | Human
  ecology.
Classification: LCC GE195 .S477 2020 (print) | LCC GE195 (ebook) | DDC
  304.2—dc23
LC record available at https://lccn.loc.gov/2020013145
LC ebook record available at https://lccn.loc.gov/2020013146

20 21 22 23 24   LSC   10 9 8 7 6 5 4 3

FOR JOAQUIN AND KESTREL

CONTENTS

INTRODUCTION

In early October 2019, a television journalist from Sky News in Britain interviewed two climate activists. Their group, Extinction Rebellion, was about to begin two weeks of civil disobedience in London and other cities around the world to protest lack of action on climate change.

A scientist and a professor had created Extinction Rebellion in spring 2018 and recruited environmentalists from across Britain to get arrested for the cause. In the fall of that year, more than six thousand Extinction Rebellion activists blocked the five main bridges that cross the River Thames, which flows through London, preventing people from getting to work or home.[1]

The organization's main spokesperson made alarming claims on national television. "Billions of people are going to die." "Life on Earth is dying." And, "Governments aren't addressing it."[2]

By 2019, Extinction Rebellion had attracted the support of leading celebrities, including actors Benedict Cumberbatch and Stephen Fry, pop stars Ellie Goulding and Thom Yorke, 2019 Oscar-winning actress Olivia Colman, Live Aid producer Bob Geldof, and Spice Girl Mel B.

While Extinction Rebellion may not have been representative of all environmentalists, nearly half of Britons surveyed told pollsters they supported the group.[3]

And the British were not alone. In September 2019, a survey of thirty thousand people around the world found that 48 percent believed climate change would make humanity extinct.[4]

But by the fall of that same year, public support for Extinction Rebellion, including the sympathy of journalists, rapidly declined after the organization shut down streets and public transit throughout London. "What about families?" the Sky News host asked the Extinction Rebellion spokespersons.

"I remember back in July, someone saying that he missed being at his father's bedside when he died in Bristol."[5]

"And that's really, really unfortunate," said Extinction Rebellion's Sarah Lunnon, putting her right hand over her heart, "and totally heartbreaking."

It was easy to see why Extinction Rebellion leaders chose Lunnon as their spokesperson. When I watched her apologize for the inconvenience, I didn't doubt she meant it.

"And when you think about it, it makes you feel absolutely dreadful," Lunnon told Sky News. She then pivoted to the topic at hand. "The pain and anguish that man suffered from being unable to say goodbye to his father is the pain and anguish we are suffering right now as we look at the future of our children, because it's very, very grave."

Three days before the Sky News interview, Extinction Rebellion had driven an old fire truck in front of the British Treasury in London and unfurled a banner that read Stop Funding Climate Death.

The Extinction Rebellion activists then opened up a fire hose and sprayed fake blood, which they had made from beet juice, onto the building. But they immediately lost control of the hose and ended up drenching the sidewalks and at least one bystander.[6]

Eleven days after the Sky News interview, Lunnon appeared on *This Morning*, one of Britain's most popular morning TV news shows.

By then, nearly two thousand Extinction Rebellion activists had been arrested; a few hours earlier, violence had erupted on the platform of a Tube station after Extinction Rebellion activists climbed on the roof of a train, forcing the conductor to hold the train in the station and evacuate the passengers.

"Why the Tube?" asked one of the irritated hosts of *This Morning*. "Why the *cleanest* way to travel across the capital?" The Tube is powered by electricity, which in Britain emits less than half the carbon now than it did in 2000.[7]

In the video, we see two Extinction Rebellion protesters climb on top of one of the train cars and unfurl a banner with white letters against a black background that read Business as Usual = DEATH.[8]

"One of the points of this particular action," said Lunnon, "is to identify the fragility of the systems that we're currently working with. The fragility of our transport systems—"

"But we all *know* that on a daily basis," interrupted the host. "If there's a power cut we know it's fragile. We *know* that. You don't need to *prove* that to us. What you've done is stop ordinary people going to work. Some of them are workers whose families will depend on an hourly rate by the money they make."

Video from the Tube protest showed hundreds of angry people on the platform, who had emptied out of the train cars, yelling at the Extinction Rebellion activists who stood defiantly on top of the train. The commuters shouted at the two young men to get down. "I'm just trying to get to work," one of the commuters said. "I'm just trying to feed my family."[9]

Things quickly descended into chaos. Some in the crowd threw cups of coffee and something made of glass, perhaps a bottle, which shattered. A woman started crying. People tried to find shelter from the chaos. "It was quite scary and there were some people who were quite frightened," recounted a reporter who was at the scene.[10]

A *This Morning* host said that 95 percent of people surveyed now said Extinction Rebellion was a hindrance to its cause. What was Extinction Rebellion thinking?[11]

In the video of the Tube protest, we see a commuter try to climb on top of the roof of the train to grab the Extinction Rebellion activist. The Extinction Rebellion activist responds by kicking the man in the face and chest. The man then grabs the Extinction Rebellion protester's legs and pulls him onto the ground. We see an angry mob of commuters start kicking him.

Back in the studio, Lunnon emphasized that the video showed the kind of disruption climate change would bring. "And not just transport," she said. "It's also power and it's also food. It's going to be empty supermarkets. It's going to be power systems turned off. And it's going to be the transport system disrupted."

Angry commuters at the Tube station descended into violence. In another video of the incident, we see a man knocking a man filming video of Extinction Rebellion action onto the floor and kicking him.[12] Later, outside the Tube station, a "man in a red jacket was punching the face of a woman," a man told a TV reporter, "who was calling on him to stop his violence."

Toward the end of *This Morning*, the cohosts did something odd: they appeared to agree with Extinction Rebellion's Sarah Lunnon about climate change.

"We are all hugely concerned and want to support you," said one of them. "Without question there is an enormous crisis," said the other.

Wait, what? I couldn't understand what they were saying. If the television hosts agreed that climate change was an enormous crisis, one in which "billions of people are going to die," how could they possibly be upset about commuters being late for work?

The Sky News host responded similarly. "I'm not trying to say that it's not deeply concerning," said the host. "The environment. But his very specific pain about not seeing his father. He might not think that's comparable."

But how could the disappointment of a single man possibly be comparable to "mass death, mass famine, and starvation"?

If "Life on Earth is dying," why did anybody care that somebody got splashed with a little beet juice?

Even if climate change were "only" going to kill *millions* of people, rather than *billions*, then the only reasonable conclusion to draw from Extinction Rebellion's tactics is that they weren't radical enough.

To be fair, the ITV and Sky News hosts didn't agree with Lunnon's extreme statements. They simply said they shared her concern about climate change.

But what, then, did they mean when they said climate change "is an enormous crisis"? If climate change isn't an existential crisis, meaning a threat to human existence, or at least to civilization, then what kind of a crisis is it, exactly?

At that moment, in the wake of a protest that could easily have resulted in the deaths of an Extinction Rebellion activist and videographer, it struck me that nobody was offering a particularly good answer to those questions.

I wrote *Apocalypse Never* because the conversation about climate change and the environment has, in the last few years, spiraled out of control, not unlike Extinction Rebellion's beet juice firehose.

I have been an environmental activist for thirty years and researched and written on environmental issues, including climate change, for twenty of them. I do this work because I care deeply about my mission to not only protect the natural environment but also to achieve the goal of universal prosperity for all people.

I also care about getting the facts and science right. I believe environmental scientists, journalists, and activists have an obligation to describe environmental problems honestly and accurately, even if they fear doing so will reduce their news value or salience with the public.

Much of what people are being told about the environment, including the climate, is wrong, and we desperately need to get it right. I decided to write *Apocalypse Never* after getting fed up with the exaggeration, alarmism, and extremism that are the enemy of a positive, humanistic, and rational environmentalism.

Every fact, claim, and argument in this book is based on the best-available science, including as assessed by the prestigious Intergovernmental Panel on Climate Change (IPCC), Food and Agriculture Organization of the United Nations (FAO), and other scientific bodies. *Apocalypse Never* defends mainstream science from those who deny it on the political Right and Left.

Apocalypse Never explores how and why so many of us came to see important but manageable environmental problems as the end of the world, and why the people who are the most apocalyptic about environmental problems tend to oppose the best and most obvious solutions to solving them.

Along the way, we will understand how humans save nature, not just destroy it. Through the stories of people around the world, and the species and environments they've saved, we will see how environmental, energetic, and economic progress constitute, in the real world, a single process.

Finally, *Apocalypse Never* offers a defense of what one might call mainstream ethics. It makes the moral case for humanism, of both secular and religious variants, against the anti-humanism of apocalyptic environmentalism.

My hope is that, amid the often chaotic and confusing debates about climate change and other environmental problems, there exists a hunger to separate scientific facts from science fiction, as well as to understand humankind's positive potential. I wrote *Apocalypse Never* to feed it.

APOCALYPSE
NEVER

1

IT'S NOT THE END OF THE WORLD

1. The End Is Nigh

If you scanned the websites of two of the world's most read newspapers on October 7, 2018, you might have feared the end of the world was near. A headline in *The New York Times* said: "Major Climate Report Describes a Strong Risk of Crisis as Early as 2040." Just below the bold headline was a photograph of a six-year-old boy playing with a dead animal's bones.[1] Said another headline in *The Washington Post* on the very same day: "The World Has Just Over a Decade to Get Climate Change Under Control, U.N. Scientists Say."[2]

Those stories in *The New York Times*, *The Washington Post*, and other media outlets around the world were based on a special report from the Intergovernmental Panel on Climate Change (IPCC), which is a United Nations body of 195 scientists and other members from around the globe responsible for assessing science related to climate change.

Two more IPCC reports would follow in 2019, both of which warned of similarly dire consequences: worsening natural disasters, sea-level rise, desertification, and land degradation. Moderate warming of 1.5 degrees Celsius would cause "long-lasting or irreversible" harm, they said, and climate change might devastate food production and landscapes. *The New York Times* reported that planetary warming threatens to worsen resource scarcity, and "floods, drought, storms and other types of extreme weather threaten to disrupt, and over time shrink, the global food supply."[3]

A NASA scientist predicted simultaneous collapses of food systems on multiple continents at once. "The potential risk of multi-breadbasket failure is increasing," she told *The New York Times*. "All of these things are happening at the same time."

An IPCC report on climate change and land in August 2019, prepared by more than a hundred experts from fifty-two countries, warned that "the window to address the threat is closing rapidly," and that "soil is being lost between ten and one hundred times faster than it is forming."[4]

Farmers will not be able to grow enough food to support the human population, scientists warned. "It's difficult to see how we could accommodate eight billion people or maybe even half of that," an agronomist said.[5]

"We can adapt to this problem up to a point," said Princeton University's Michael Oppenheimer, an IPCC contributor. "But that point is determined by how strongly we mitigate greenhouse-gas emissions." If emissions rise through 2050, then sea level rise will likely exceed 2 feet 9 inches by 2100, at which point "the job will be too big. . . . It will be an unmanageable problem."[6]

Too much warming could trigger a series of irreversible tipping points, experts said. For example, sea level rise could be slowing the circulation of water in the Atlantic Ocean, which could change surface temperatures.[7] Arctic permafrost covering an area nearly the size of Australia could thaw and release 1,400 gigatons of carbon into the atmosphere.[8] The glacier on the continent of Antarctica could collapse into the ocean. If that happens, sea level could rise thirteen feet.[9]

Rising atmospheric carbon dioxide levels are changing the chemistry of oceans in ways that scientists warn could harm marine life and even cause mass extinctions. A 2016 study published in *Nature* found that higher carbon dioxide levels were making coral reef fish species oblivious to predators.[10]

Many blamed climate change for wildfires that ravaged California. The death toll from fires skyrocketed from just one death from wildfires in 2013 to one hundred deaths in 2018. Of the twenty most destructive fires in California's history, half have occurred since 2015.[11] Today, California's fire season stretches two to three months longer than it was fifty years ago.[12] Climate change is increasing droughts and making trees vulnerable to disease and infestation.

"The reason these wildfires have worsened is because of climate change," said Leonardo DiCaprio.[13] "This is what climate change looks like," said

Rep. Alexandria Ocasio-Cortez.[14] "It's the end of California as we know it," concluded a columnist for *The New York Times*.[15]

In Australia, more than 135 bushfires burned in early 2020, claiming the lives of thirty-four people, killing an estimated one billion animals, and damaging or completely destroying nearly three thousand homes.[16]

David Wallace-Wells, author of *The Uninhabitable Earth*, warned that with a two degree increase, "the ice sheets will begin their collapse, 400 million more people will suffer from water scarcity, major cities in the equatorial band of the planet will become unlivable, and even in the northern latitudes heat waves will kill thousands each summer." [17]

"What we're playing for now is to see if we can limit climate change to the point where we don't wipe out civilizations," said environmental writer and climate activist Bill McKibben. "And at the moment we're headed in a direction where that won't happen." [18]

Said one IPCC contributor, "In some parts of the world, national borders will become irrelevant. . . . You can set up a wall to try to contain ten thousand and twenty thousand, one million people, but not ten million." [19]

"Around the year 2030, in ten years, 250 days, and ten hours, we will be in a position where we set off an irreversible chain reaction beyond human control that will most likely lead to the end of our civilisation as we know it," said student climate activist Greta Thunberg, in 2019. "I don't want you to be hopeful. I want you to panic." [20]

2. Resilience Rising

In early 2019, newly elected twenty-nine-year-old congresswoman Alexandria Ocasio-Cortez sat down for an interview with a correspondent for *The Atlantic*. AOC, as she is known, made the case for a Green New Deal, one that would address poverty and social inequality in addition to climate change. AOC pushed back against critics who claimed it would be too expensive. "The world is going to end in twelve years if we don't address climate change," she said, "and your biggest issue is how are we gonna pay for it?" [21]

The next day, a reporter for the news website Axios called several climate scientists to get their reactions to AOC's claim that the world was going to end in twelve years. "All the time-limited frames are bullshit," said Gavin Schmidt, a NASA climate scientist. "Nothing special happens when the 'carbon budget' runs out or we pass whatever temperature target you care about, instead the costs of emissions steadily rise."[22]

Andrea Dutton, a paleoclimate researcher at the University of Wisconsin–Madison, said, "For some reason the media latched onto the twelve years (2030), presumably because they thought that it helped to get across the message of how quickly we are approaching this and hence how urgently we need action. Unfortunately, this has led to a complete mischaracterization of what the report said."[23]

What the IPCC had actually written in its 2018 report and press release was that in order to have a good chance of limiting warming to 1.5 degrees Celsius from preindustrial times, carbon emissions needed to decline 45 percent by 2030. The IPCC did not say the world would end, nor that civilization would collapse, if temperatures rose above 1.5 degrees Celsius.[24]

Scientists had a similarly negative reaction to the extreme claims made by Extinction Rebellion. Stanford University atmospheric scientist Ken Caldeira, one of the first scientists to raise the alarm about ocean acidification, stressed that "while many species are threatened with extinction, climate change does not threaten human extinction."[25] MIT climate scientist Kerry Emanuel told me, "I don't have much patience for the apocalypse criers. I don't think it's helpful to describe it as an apocalypse."[26]

An AOC spokesperson told Axios, "We can quibble about the phraseology, whether it's existential or cataclysmic." But, he added, "We're seeing lots of [climate change–related] problems that are already impacting lives."[27]

But if that's the case, the impact is dwarfed by the 92 percent decline in the decadal death toll from natural disasters since its peak in the 1920s. In that decade, 5.4 million people died from natural disasters. In the 2010s, just 0.4 million did.[28] Moreover, that decline occurred during a period when the global population nearly quadrupled.

In fact, both rich and poor societies have become far less vulnerable to extreme weather events in recent decades. In 2019, the journal *Global En-*

vironmental Change published a major study that found death rates and economic damage dropped by 80 to 90 percent during the last four decades, from the 1980s to the present.[29]

While global sea levels rose 7.5 inches (0.19 meters) between 1901 and 2010,[30] the IPCC estimates sea levels will rise as much as 2.2 feet (0.66 meters) by 2100 in its medium scenario, and by 2.7 feet (0.83 meters) in its high-end scenario. Even if these predictions prove to be significant underestimates, the slow pace of sea level rise will likely allow societies ample time for adaptation.

We have good examples of successful adaptation to sea level rise. The Netherlands, for instance, became a wealthy nation despite having one-third of its landmass below sea level, including areas a full *seven* meters below sea level, as a result of the gradual sinking of its landscapes.[31]

And today, our capability for modifying environments is far greater than ever before. Dutch experts today are already working with the government of Bangladesh to prepare for rising sea levels.[32]

What about fires? Dr. Jon Keeley, a U.S. Geological Survey scientist in California who has researched the topic for forty years, told me, "We've looked at the history of climate and fire throughout the whole state, and through much of the state, particularly the western half of the state, we don't see any relationship between past climates and the amount of area burned in any given year."[33]

In 2017, Keeley and a team of scientists modeled thirty-seven different regions across the United States and found that "humans may not only influence fire regimes but their presence can actually override, or swamp out, the effects of climate." Keeley's team found that the only statistically significant factors for the frequency and severity of fires on an annual basis were population and proximity to development.[34]

As for the Amazon, *The New York Times* reported, correctly, that "[the 2019] fires were not caused by climate change."[35]

In early 2020, scientists challenged the notion that rising carbon dioxide levels in the ocean were making coral reef fish species oblivious to predators. The seven scientists who published their study in the journal *Nature* had, three years earlier, raised questions about the marine biologist who had made such claims in the journal *Science* in 2016. After an investigation, James Cook University in Australia concluded that the biologist had fabricated her data.[36]

When it comes to food production, the Food and Agriculture Organization of the United Nations (FAO) concludes that crop yields will increase significantly, under a wide range of climate change scenarios.[37] Humans today produce enough food for ten billion people, a 25 percent surplus, and experts believe we will produce even more despite climate change.[38]

Food production, the FAO finds, will depend more on access to tractors, irrigation, and fertilizer than on climate change, just as it did in the last century. The FAO projects that even farmers in the poorest regions today, like sub-Saharan Africa, may see 40 percent crop yield increases from technological improvements alone.[39]

In its fourth assessment report, the IPCC projected that by 2100, the global economy would be three to six times larger than it is today, and that the costs of adapting to a high (4 degrees Celsius) temperature rise would reduce gross domestic product (GDP) just 4.5 percent.[40]

Does any of that really sound like the end of the world?

3. The Apocalypse Now

Anyone interested in seeing the end of the world up close and in person could do little worse than to visit the Democratic Republic of the Congo in central Africa. The Congo[41] has a way of putting first-world prophecies of climate apocalypse into perspective. I traveled there in December 2014 to study the impact of widespread wood fuel use on people and wildlife, particularly on the fabled mountain gorillas.

Within minutes of crossing from the neighboring country of Rwanda into the Congolese city of Goma, I was taken aback by the extreme poverty and chaos: children as young as two years old perched on the handlebars of motorcycles flying past us on roads pock-marked with giant potholes; tin-roofed shanties as homes; people crammed like prisoners into tiny buses with bars over the windows; trash everywhere; giant mounds of cooled lava on the sides of the road, reminders of the volcanic anger just beneath the earth's surface.

In the 1990s and again in the early 2000s, Congo was the epicenter of the Great African War, the deadliest conflict since World War II, which involved nine African countries and resulted in the deaths of three to five million people, mostly because of disease and starvation. Another two million people were displaced from their homes or sought asylum in neighboring countries. Hundreds of thousands of people, women and men, adults and children, were raped, sometimes more than once, by different armed groups.[42]

During our time in the Congo, armed militias roaming the countryside had been killing villagers, including children, with machetes. Some blamed Al-Shabaab terrorists coming in from Uganda, but nobody took credit for the attacks. The violence appeared unconnected to any military or strategic objective. The national military, police, and United Nations Peacekeeping Forces, about six thousand soldiers, were either unable or unwilling to do anything about the terrorist attacks.

"Do not travel," the United States Department of State said, bluntly, of the Congo on its website. "Violent crime, such as armed robbery, armed home invasion, and assault, while rare compared to petty crime, is not uncommon, and local police lack the resources to respond effectively to serious crime. Assailants may pose as police or security agents."[43]

One reason I felt safe traveling to the eastern Congo and bringing my wife, Helen, was that the actor Ben Affleck had visited several times and even started a charity there to support economic development. If the eastern Congo was safe enough for a Hollywood celebrity, I reasoned, it would be safe enough for Helen and me.

To make sure, I hired Affleck's guide, translator, and "fixer," Caleb Kabanda, a Congolese man with a reputation for keeping his clients safe. We spoke on the telephone before I arrived. I told Caleb I wanted to study the relationship between energy scarcity and conservation. Referring to the North Kivu province capital of Goma, the sixth most populated city in the Congo, Caleb asked, "Can you imagine a city of nearly two million people relying on wood for energy? It's crazy!"

Ninety-eight percent of people in eastern Congo rely on wood and charcoal as their primary energy for cooking. In the Congo as a whole, nine out of ten of its nearly ninety-two million people do, while just one out of five has any access to electricity.[44, 45] The entire country relies on just 1,500 megawatts of electricity, which is about as much as a city of one million requires in developed nations.[46]

The main road Caleb and I used to travel from Goma to the communities around Virunga Park had recently been paved, but there was little else in the way of infrastructure. Most roads were dirt roads. When it rained, both the paved and unpaved roads and the surrounding homes were flooded because there was no flood control system. I was reminded of how much we take for granted in developed nations. We practically forget that the gutters, canals, and culverts, which capture and divert water away from our homes, even exist.

Is climate change playing a role in Congo's ongoing instability? If it is, it's outweighed by other factors. Climate change, noted a large team of researchers in 2019, "has affected organized armed conflict within countries. However, other drivers, such as low socioeconomic development and low capabilities of the state, are judged to be substantially more influential."[47]

There is only a barely functioning government in the Congo. When it comes to security and development, people are mostly on their own. Depending on the season, farmers suffer too much rain or not enough. Recently, there has been flooding once every two or three years. Floods regularly destroy homes and farms.

Researchers with the Peace Research Institute Oslo note, "Demographic and environmental variables have a very moderate effect on the risk of civil conflict."[48] The IPCC agrees. "There is robust evidence of disasters displacing people worldwide, but limited evidence that climate change or sea level rise is the direct cause."[49]

Lack of infrastructure plus scarcity of clean water brings disease. As a result, Congo suffers some of the highest rates of cholera, malaria, yellow fever, and other preventable diseases in the world.

"Lower levels of GDP are the most important predictor of armed conflict," write the Oslo researchers, who add, "Our results show that resource scarcity affects the risk of conflict less in low-income states than in wealthier states."[50]

If resources determined a nation's fate, then resource-scarce Japan would be poor and at war while the Congo would be rich and at peace. Congo is astonishingly rich when it comes to its lands, minerals, forests, oil, and gas.[51]

There are many reasons why the Congo is so dysfunctional. It is massive—it is the second largest African nation in area, behind only Algeria—and difficult to govern as a single country. It was colonized by the Belgians, who fled the country in the early 1960s without establishing strong government institutions, like an independent judiciary and a military.

Is it overpopulated? The population of Eastern Congo has doubled since the 1950s and 1960s. But the main factor is technological: the same area could produce much more food and support many more people if there were roads, fertilizers, and tractors.

The Congo is a victim of geography, colonialism, and terrible post-colonial governments. Its economy grew from $7.4 billion in 2001 to $38 billion in 2017,[52] but the annual per capita income of $561 is one of the lowest in the world,[53] leading many to conclude that much of the money that should flow to the people is being stolen.

For the last twenty years, the Rwandan government has been taking minerals from its neighbor and exporting them as its own. To protect and obscure its activities, Rwanda has financed and overseen the low-intensity conflict in Eastern Congo, according to experts.[54]

There were free elections in 2006 and optimism around the new president, Joseph Kabila, but he proved as corrupt as past leaders. After being re-elected in 2011, he stayed in power until 2018, when he installed a candidate who won just 19 percent of the vote as compared to the opposition candidate, who won 59 percent. As such, Kabila and his allies in the legislature appear to be governing behind the scenes.[55]

4. Billions Won't Die

On BBC Two's *Newsnight,* in October 2019, the journalist Emma Barnett asked Extinction Rebellion's sympathetic and empathic spokesperson, Sarah

Lunnon, how her organization could justify disrupting life in London the way it had.

"To be the cause of that happening is really very, very upsetting," said Lunnon, touching her heart, "and it makes me feel really bad to know that I'm disrupting people's lives. *And* it makes me really cross and angry that the lack of action over thirty years has meant that the only way I can get the climate on the agenda is to take actions such as this; if we don't act and protest in this way nobody takes any notice."[56]

Barnett turned to the man sitting next to Lunnon, Myles Allen, a climate scientist and IPCC report author.

"The name Extinction Rebellion is inherently pointing towards 'we're going to be extinct,' " said Barnett. "Roger Hallam, one of the three founders [of Extinction Rebellion], said in August . . . 'Slaughter, death and starvation of six billion people this century.' There's no science to back that up, is there?"

Said Allen, "There's a lot of science that backs up the very considerable risks we run if we carry on on a path to—"

"—but not six billion people. There's no science that calculates it to that level, is there?" asked Barnett.

Extinction Rebellion's Lunnon didn't let him answer.

"There are a number of scientists who've said if we get to four degrees of warming, which is where we're heading at the moment, they cannot see how the earth can support not one billion people, *a half* a billion people," she said. "That's six and a half billion people dying!"

Barnett appeared annoyed, and interrupted. "Sorry," she said, turning back to Myles. "So you're going to stand by, scientifically, a projection that says within this century we'll have the slaughter, death, and starvation of six billion people? It's just good for us to know."

"No," he said. "Because what we can do as scientists is tell you about the risks we face. The easy risks to predict, to be honest, are the ones that I do, how the climate system reacts to rising greenhouse gases. The harder risks are how people are going to respond to losing the weather they knew as kids. . . . So I imagine what they're talking about there is the risk of the human response to climate change as much as the risk of climate change itself."

"But I suppose the point is," pressed Barnett, "if there's no science that says that, do you understand why some people who *are* sympathetic to your cause also feel like you have fear-mongered? For instance, [Extinction Rebellion co-founder] Roger Hallam has also said our kids will be dead in ten to fifteen years."

"We are losing the weather we know!" Lunnon interrupted. "All of our agriculture and our food is based on weather that has been around for the last ten thousand years! If we don't have predictable weather, we don't have predictable food sources. We run the risk of multiple losses of harvest in the world's global breadbasket. That's no food!"

"Roger Hallam *did say,*" replied Barnett, "our kids would be dead in ten or fifteen years."

"There's a distinct possibility that we lose not only our food supplies but our energy supplies," said Lunnon. "In California, at the moment, millions of people do not have electricity."

In late November 2019, I interviewed Lunnon. We talked for an hour, and we exchanged emails where she clarified her views.

"I'm not saying billions of people are going to die," Lunnon told me. "It's not Sarah Lunnon saying billions of people are going to die. The science is saying we're headed to 4 degrees warming and people like Kevin Anderson of Tyndall Center and Johan Rockström from the Potsdam are saying that such a temperature rise is incompatible with civilized life. Johan said he could not see how an Earth at 4 degrees (Celsius) warming could support a billion or even half-billion people."[57]

Lunnon was referring to an article published in *The Guardian* in May 2019, which quoted Rockström saying, "It's difficult to see how we could accommodate a billion people or even half of that" at a four-degree temperature rise.[58] I pointed out that there is nothing in any of the IPCC reports that has ever suggested anything like what she is attributing to Anderson and Rockström.

And why should we rely on the speculations of two scientists over the

IPCC? "It's not about choosing science," said Lunnon, "it's about looking at the risk we're facing. And the IPCC report lays out the different trajectories from where we are and some of them are very, very bleak."[59]

To get to the bottom of the "billions will die" claim, I interviewed Rockström by phone. He said *The Guardian* reporter had misunderstood him. What he had actually said, he told me, was this: "It's difficult to see how we could accommodate eight billion people or even half of that," not "a billion people." Rockström said he had not seen the misquote until I emailed him and that he had requested a correction, which *The Guardian* made in late November 2019. Even so, Rockström was predicting four billion deaths.[60]

"I don't see scientific evidence that a four degree Celsius planet can host eight billion people," he said. "This is, in my assessment, a scientifically justified statement, as we don't have evidence that we can provide freshwater or feed or shelter today's world population of eight billion in a four degree world. My expert judgment, furthermore, is that it may even be doubtful if we can host half of that, meaning four billion."[61]

But is there IPCC science showing that food production would actually decline? "As far as I know they don't say anything about potential population that can be fed at different degrees of warming," he said.[62]

Has anyone done a study of food production at four degrees? I asked. "That's a good question. I must admit I have not seen a study," said Rockström, who is an agronomist. "It seems like such an interesting and important question."[63]

In fact, scientists have done that study, and two of them were Rockström's colleagues at the Potsdam Institute. It found that food production could increase even at four to five degrees Celsius warming above preindustrial levels.[64] And, again, technical improvements, such as fertilizer, irrigation, and mechanization, mattered more than climate change.

The report also found, intriguingly, that climate change *policies* were more likely to hurt food production and worsen rural poverty than climate change itself. The "climate policies" the authors refer to are ones that would make energy more expensive and result in more bioenergy use (the burning of biofuels and biomass), which in turn would increase land scarcity and drive up food costs. The IPCC comes to the same conclusion.[65]

Similarly, the UN Food and Agriculture Organization concludes that food production will rise 30 percent by 2050 except if a scenario it calls Sustainable Practices is adopted, in which case it would rise 20 percent.[66] Technological change significantly outweighs climate change in every single one of FAO's scenarios.

5. A Small Part of Big Conflicts

In 2006, a thirty-seven-year-old political science professor from the University of Colorado in Boulder organized a workshop for thirty-two of the world's leading experts to discuss whether human-caused climate change was making natural disasters worse, more frequent, or more costly. The professor, Roger Pielke, Jr., cohosted the workshop with a colleague, Peter Höppe, who at the time ran the Geo Risk division of Munich Reinsurance, which provides insurance to insurance companies and has a strong financial interest in knowing whether global warming will make natural disasters worse.

If there is a stereotype of an environmental sciences professor from Boulder, Colorado, Pielke fits it well. He wears hiking boots and plaid shirts. He is an avid hiker, skier, and soccer player. He is liberal, secular, and a Democrat. "I have written a book calling for a carbon tax," Pielke says. "I have publicly supported President Obama's proposed EPA carbon regulations, and I have just published another book strongly defending the scientific assessment of the IPCC with respect to disasters and climate change."[67]

The group met in Hohenkammer, Germany, outside of Munich. Pielke wasn't optimistic that the group would achieve consensus because the group included both environmental activists and climate skeptics. "But much to our surprise and delight," says Pielke, "all thirty-two people at the workshop—experts from academia, the private sector, and advocacy groups—reached a consensus on twenty statements on disasters and climate change."[68]

The experts agreed in their unanimous Hohenkammer Statement that climate change is real and humans are contributing to it significantly.[69] But they also agreed that more people and property in harm's way explained the rising cost of natural disasters, not worsening disasters.

When teaching his students, Pielke illustrates this point with a picture of Miami Beach in 1926 and in 2006. In 1926, Miami Beach had a single high-rise building vulnerable to hurricanes. By 2006, it had dozens of high-rise buildings in danger of having their windows blown out and flooded. Pielke shows the climbing, inflation-adjusted cost of hurricanes in the United States rising from near-zero in 1900 to more than $130 billion in 2005, when Hurricane Katrina hit New Orleans.[70]

Pielke then shows *normalized* hurricane losses for the same period. *Normalized* means that Pielke and his coauthors adjusted the damage data to account for the massive development of America's coastlines, like Miami's, since 1900. Once this is done there is no trend of rising costs.[71]

The lack of rising normalized costs matches the historical record of U.S. hurricane landfalls, which gave Pielke and his colleagues confidence in their results. Their results show a few big spikes in hurricane losses, including one rising to an inflation-adjusted and development-normalized $200 billion for the year 1926, when four hurricanes made landfall in the United States, exceeding the $145 billion of damage occurring in 2005.[72] While Florida experienced eighteen major hurricanes between 1900 and 1959, it experienced just eleven from 1960 to 2018.[73]

Is the United States unique? It's not. "Scholars have done similar analyses of normalized tropical cyclone losses in Latin America, the Caribbean, Australia, China, and the Andhra Pradesh region in India," Pielke notes. "In each case they have found no trend in normalized losses."[74]

And it's not just hurricanes. "There is scant evidence to indicate that hurricanes, floods, tornadoes or drought have become more frequent or intense in the U.S. or globally," he wrote later. "In fact we are in an era of good fortune when it comes to extreme weather."[75]

The IPCC says the same thing. "Long-term trends in economic disaster losses adjusted for wealth and population increases have not been attributed to climate change," notes a special IPCC report on extreme weather, "but a role for climate change has not been excluded."[76]

Pielke stresses that climate change may be contributing to some extreme weather events. "For instance," he notes, "some recent research is suggestive that regional warming in the western United States can be associated with increasing forest fires." [77]

But climate change so far has not resulted in increases in the frequency or intensity of many types of extreme weather. The IPCC "concluded that there's little evidence of a spike in the frequency or intensity of floods, droughts, hurricanes and tornadoes," explains Pielke. "There have been more heat waves and intense precipitation, but these phenomena are not significant drivers of disaster costs." [78]

What most determines how vulnerable various nations are to flooding depends centrally on whether they have modern water and flood control systems, like my home city of Berkeley, California, or not, like the Congo. [79]

When a hurricane hits Florida, it might kill no one, but when that same storm hits Haiti, thousands can die instantly through drowning and subsequently in disease epidemics like cholera. The difference is that Florida is in a wealthy nation with hardened buildings and roads, advanced weather forecasting, and emergency management. Haiti, by contrast, is a poor nation that lacks modern infrastructure and systems. [80]

"Consider that since 1940 in the United States 3,322 people have died in 118 hurricanes that made landfall," Pielke wrote. But when the "Boxing Day Tsunami struck Southeast Asia in 2004, more than 225,000 people died." [81]

Anyone who believes climate change could kill billions of people and cause civilizations to collapse might be surprised to discover that none of the IPCC reports contain a single apocalyptic scenario. Nowhere does the IPCC describe developed nations like the United States becoming a "climate hell" resembling the Congo. Our flood-control, electricity, and road systems will keep working even under the most dire potential levels of warming.

What about the claim IPCC contributor Michael Oppenheimer made that a 2-foot, 9-inch sea level rise would be "an unmanageable problem?" [82] To understand his reasoning, I interviewed him by phone.

"There was a mistake in the article by the reporter," he told me. "He had 2 feet, 9 inches. The actual number, which is based on the sea level rise amount in [IPCC Representative Concentration Pathway] 8.5 for its [Special Report on the Ocean and Cryosphere in a Changing Climate] report is 1.1 meters, which is 3 feet, 7 inches."[83]

I asked Oppenheimer why places like Bangladesh couldn't do what the Netherlands did. "The Netherlands spent a lot of time not improving its dikes due to two world wars and a depression," Oppenheimer said, "and didn't start modernizing them until the disastrous 1953 flood."[84]

The 1953 flood killed more than 2,500 people and motivated the Netherlands to rebuild its dikes and canals. "Most of humanity will not be able to avail itself of that luxury," said Oppenheimer. "So in most places, they will accommodate flooding by raising structures or floodable structures. Or you retreat."[85]

In 2012, said Oppenheimer, "People moved out of New York after Hurricane Sandy. I wouldn't call that unmanageable. Temporarily unmanageable. Meaning we wouldn't be able to maintain societal function around the world if sea level rise approaches those close to four feet. Bangladeshis might be leaving the coast and trying to get into India."[86]

But millions of small farmers, like the ones on Bangladesh's low-lying coasts, move to cities every year, I pointed out. Doesn't the word "unmanageable" suggest a permanent societal breakdown?

"When you have people making decisions they are essentially compelled to make," he said, "that's what I'm referring to as 'an unmanageable situation.' The kind of situation that leads to economic disruption, disruption of livelihoods, disruption of your ability to control your destiny, and people dying. You can argue that they get manageable. You recover from disasters. But the people who died didn't recover."[87]

In other words, the problems from sea level rise that Oppenheimer calls "unmanageable" are situations like the ones that already occur, from which societies recover, and to which they adapt.

6. Development > Climate

The Congo's underdevelopment is in part a consequence of having one of the most corrupt governments in the world.[88] Once, we were stopped by a police officer. I was in the back of the car and Caleb was in the front with the driver. As the police officer peered into the car, Caleb turned his head slightly toward the man and scowled. The officer checked the driver's papers and then waved us on.

"What was that all about?" I asked.

"He was trying to find something wrong so he could ask for a bribe," Caleb explained. "But I gave him my special stare."

Caleb confessed that he, like many other Congolese, loved watching the American TV series *24* (2001 to 2010) about a CIA agent who battles terrorists. "Everybody in Congo loves Jack Bauer!" Caleb said, referring to the CIA agent played by the American actor Kiefer Sutherland. I asked Caleb if people in the Congo love Sutherland as much as they love Ben Affleck, who was not only more famous than Sutherland but also was trying to help the Congo. Caleb paused for a moment to contemplate the question. "Not here!" he said. "Jack Bauer is more famous in the Congo. If Kiefer Sutherland came to the Congo and gave a press conference demanding that all the armed groups give up in twenty-four hours, all the fighting would come to a halt immediately!" Caleb laughed with delight at the thought of it.

We drove around the countryside and interviewed people at random. Caleb used his charm to reassure local villagers who were understandably suspicious about a foreigner asking them questions about their lives. Many people we interviewed were upset about baboons and elephants from nearby Virunga National Park, a protected area for wildlife, raiding their crops. Given the widespread hunger and poverty, losing your crops to wild animals is devastating. I was told that one woman was so upset about losing her crops to an elephant that she died of a heart attack the next day. And I was told that a chimpanzee had recently killed a two-year-old boy.

One man asked me to request Virunga Park officials install electric fences to keep animals out of their fields. Several people complained that when they approached park managers about the nuisance, they were told to capture the offending animals and bring them to the park, a request the villagers said was impossible and insulting.

A few weeks before I arrived, a group of young people organized a march to Virunga National Park headquarters to protest inaction on crop-raiding. In response, the park hired some of the youths to shoo away baboons.

Near the entrance to Virunga National Park, Caleb and I interviewed people from a local community. A crowd of about twenty or thirty people gathered around us, and many of them expressed outrage about the crop-raiding. "Can't you kill the baboons that are eating your crops?" I asked. Many people in the crowd let out a collective groan and said no, that they would go to prison, even though the animal was on their land and outside the park boundary.

There was a young mother with an infant on her breast in the crowd. I introduced myself and asked her name. It was Mamy Bernadette Semu-taga. She went by Bernadette. She was twenty-five years old. Her baby girl's name was Bibiche Sebiraro. She was Bernadette's seventh child.

Bernadette told us that baboons had eaten her sweet potatoes the night before. I asked her if she would take us to her plot of land so we could see for ourselves what had happened. She agreed. We talked in the car on the way there.

I asked Bernadette what her favorite memory was as a child. "When I was fourteen years old I visited my cousins in Goma and they bought new clothes for me," she said. "And when it was time to go back to my village, they paid for the ticket for me and gave me money to buy bread and cabbage to take home. I returned home very happy."

Much of the rest of Bernadette's life has been difficult. "I got married when I was fifteen years old," she said. "When I met my husband, he was an orphan. He had nothing. We've been always living with difficulties. I have never lived in happiness."

When we reached her small plot of land, Bernadette pointed to holes in the ground where sweet potatoes had been. I asked if I could take a photo. She said that was fine. In the photograph, she is frowning, but also looks proud. At least she had a plot of land to call her own.

Once we drove her back to the village, Caleb gave her some money, as a small token of our thanks, and to make up for the sweet potatoes.

We should be concerned about the impact of climate change on vulnerable populations, without question. There is nothing automatic about adaptation. And it's true that Bernadette is more vulnerable to climate change than Helen and I are.

But she is also more vulnerable to the weather and natural disasters *today*. Bernadette must farm to survive. She must spend several hours a day chopping wood, hauling wood, building fires, fanning smoky fires, and cooking over them. Wild animals eat her crops. She and her family lack basic medical care and her children often go hungry and get sick. Heavily armed militias roam the countryside robbing, raping, kidnapping, and murdering. Understandably, then, climate change is not on her list of things to worry about.

As such, it's misleading for environmental activists to invoke people like Bernadette, and the risks she faces from climate change, without acknowledging that economic development is overwhelmingly what will determine her standard of living, and the future of her children and grandchildren, not how much the climate changes.

What will determine whether or not Bernadette's home is flooded is whether the Congo builds a hydroelectric, irrigation, and rainwater system, not the specific change in precipitation patterns. What will determine whether Bernadette's home is secure or insecure is whether she has money to make it secure. And the only way she'll have money to make it secure is through economic growth and a higher income.

7. Exaggeration Rebellion

Economic development outweighs climate change in the rich world, too. Consider the case of California, the fifth largest economy in the world.

California suffers from two major kinds of fires. First, there are wind-driven fires on coastal shrubland, or chaparral, where most of the houses are built. Think Malibu and Oakland. Nineteen of the state's twenty most deadly and costly fires have taken place in chaparral.[89] The second type is forest fires in places like the Sierra Nevada where there are far fewer people.

Mountain and coastal ecosystems have opposite problems. There are too many fires in the shrublands and too few prescribed burns in the Sierras. Keeley refers to the Sierra fires as "fuel-dominated" and the shrubland fires as "wind-dominated."[90] The only solution to fires in the shrubland is to prevent them and/or harden homes and buildings to them.

Before Europeans arrived in the United States, fires burned up woody biomass in forests every 10 to 20 years, preventing the accumulation of wood fuel, and fires burned the shrublands every 50 to 120 years. But during the last 100 years, the United States Forest Service (USFS) and other agencies extinguished most fires, resulting in the accumulation of wood fuel.

Keeley published a paper in 2018 finding that all ignition sources of fires had declined in California except for electric power lines.[91] "Since the year 2000 there've been a half-million acres burned due to powerline-ignited fires, which is five times more than we saw in the previous twenty years," he said. "Some people would say, 'Well, that's associated with climate change.' But there's no relationship between climate and these big fire events."[92]

What then is driving the increase in fires? "If you recognize that 100 percent of these [shrubland] fires are started by people, and you add six million people [since 2000], that's a good explanation for why we're getting more and more of these fires," said Keeley.[93]

What about the Sierra? "If you look at the period from 1910 to 1960," said Keeley, "precipitation is the climate parameter most tied to fires. But since 1960, precipitation has been replaced by temperature, so in the last fifty years, spring and summer temperatures will explain 50 percent of the variation from one year to the next. So temperature is important."[94]

But isn't that also the period when the wood fuel was allowed to accumulate, I asked, due to suppression of forest fires? "Exactly," said Keeley. "Fuel is one of the confounding factors. It's the problem in some of the reports done by climatologists who understand climate but don't necessarily understand the subtleties related to fires."[95]

Would we be having such hot fires in the Sierra, I asked, had we not allowed wood fuel to build up over the last century? "That's a very good question," said Keeley. "Maybe you wouldn't." He said it was something he might look at. "We have some selected watersheds in the Sierra Nevadas where there have been regular fires. Maybe the next paper we'll pull out the watersheds that have not had fuel accumulation and look at the climate fire relationship and see if it changes."[96]

Fires in Australia are similar. Greater fire damage in Australia is, as in California, due in part to greater development in fire-prone areas, and in part to the accumulation of wood fuel. One scientist estimates that there is ten times more wood fuel in Australia's forests today that when Europeans arrived. The main reason is that the government of Australia, as in California, refused to do controlled burns, for both environmental and human health reasons. As such, the fires would have occurred even had Australia's climate not warmed.[97]

The news media depicted the 2019–2020 fire season as the worst in Australia's history but it wasn't. It ranked fifth in terms of area burned, with about half of the burned acreage as 2002, the fourth place year, and about a sixth of the burned acreage of the worst season in 1974–1975. The 2019–2020 fires ranked sixth in fatalities, about half as many as the fifth place year, 1926, and a fifth as many fatalities as the worst fire on record in 2009. While the 2019–2020 fires are second in number of houses destroyed, they razed about 50 percent less than the worst year, the 1938–39 fire season. The only metric by which this fire season appears to be the worst ever is in the number of non-home buildings damaged.[98]

Climate alarmism, animus among environmental journalists toward the current Australian government, and smoke that was unusually visible to densely populated areas, appear to be the reasons for exaggerated media coverage.

The bottom line is that other human activities have a greater impact on the frequency and severity of forest fires than the emission of greenhouse gases. And that's great news, because it gives Australia, California, and Brazil far greater control over their future than the apocalyptic news media suggested.

In July 2019, one of Lauren Jeffrey's science teachers made an offhand comment about how climate change could be apocalyptic. Jeffrey was seventeen years old and attended high school in Milton Keynes, a city of 230,000 people about fifty miles northwest of London.

"I did research on it and spent two months feeling quite anxious," she told me. "I would hear young people around me talk about it and they were convinced that the world was going to end and they were going to die."[99]

Studies find that climate alarmism is contributing to rising anxiety and depression, particularly among children.[100] In 2017, the American Psychological Association diagnosed rising eco-anxiety and called it "a chronic fear of environmental doom."[101] In September 2019, British psychologists warned of the impact on children of apocalyptic discussions of climate change. In 2020, a large national survey found that one out of five British children was having nightmares about climate change.[102]

"There is no doubt in my mind that they are being emotionally impacted," one expert said.[103]

"I found a lot of blogs and videos talking about how we're going extinct at various dates, 2030, 2035, from societal collapse," said Jeffrey. "That's when I started to get quite worried. I tried to forget it at first but it kept popping up in my mind.

"One of my friends was convinced there would be a collapse of society in 2030 and 'near term human extinction' in 2050," said Jeffrey. "She concluded that we've got ten years left to live."

Extinction Rebellion activists stoked those fears. Extinction Rebellion activists gave frightening and apocalyptic talks to schoolchildren across Britain. In one August talk, an Extinction Rebellion activist climbed atop a desk in the front of a classroom to give a terrifying talk to children, some of whom appear no older than ten years old.[104]

Some journalists pushed back against the group's alarmism. The BBC's Andrew Neil interviewed a visibly uncomfortable Extinction Rebellion spokesperson in her mid-thirties named Zion Lights.[105] "One of your founders, Roger Hallam, said in April, 'Our children are going to die in the next ten to twenty years,'" Neil says to Lights in the video. "What's the scientific basis for these claims?"

"These claims have been disputed, admittedly," Lights says. "There are some scientists who are agreeing and some who are saying that they're simply not true. But the overall issue is that these deaths are going to happen."

"But most scientists *don't* agree with this," says Neil. "I looked through [the Intergovernmental Panel on Climate Change's recent reports] and see no reference to billions of people going to die, or children going to die in under twenty years. . . . *How* would they die?"

Responds Lights, "Mass migration around the world is already taking place due to prolonged drought in countries, particularly in South Asia. There are wildfires in Indonesia, the Amazon rainforest, also Siberia, the Arctic."

"These are really important problems," Neil says, "and they can cause fatalities. But they don't cause *billions* of deaths. They don't mean that our young people will all be dead in twenty years."

"Perhaps not in twenty years," acknowledges Lights.

"I've seen young girls on television, part of your demonstration . . . *crying* because they think they're going to die in five or six years' time, crying because they don't think they'll ever see adulthood," says Neil. "And yet there's no scientific basis for the claims that your organization is making."

"I'm not saying that because I'm alarming children," replies Lights. "They're learning about the consequences."

Happily, not all of Britain's schoolchildren trusted Extinction Rebellion to honestly and accurately explain the consequences. "I did research and found there was a lot of misinformation on the denial side of things and also on the doomsayer side of things," Lauren Jeffrey told me.

In October and November 2019, she posted seven videos to YouTube and joined Twitter to promote them. "As important as your cause is," said Jeffrey in one of the videos, an open letter to Extinction Rebellion, "your persistent exaggeration of the facts has the potential to do more harm than good to the scientific credibility of your cause as well as to the psychological well-being of my generation."[106]

8. Apocalypse Never

In November and December 2019, I published two long articles criticizing climate alarmism and covering material similar to what I've written above. I did so in part because I wanted to give scientists and activists, including those whom I criticized, a chance to respond or correct any errors I might have made in my reporting before publishing this book. Both articles were widely read, and I made sure the scientists and activists I mentioned saw my article. Not a single person requested a correction. Instead, I received many emails from scientists and activists alike, thanking me for clarifying the science.

One of the main questions I received, including from a BBC reporter, was whether some alarmism was justified in order to achieve changes to policy. The question implied that the news media aren't already exaggerating.

But consider a June Associated Press article. It was headlined, "UN Predicts Disaster if Global Warming Not Checked." It was one of many apocalyptic articles that summer about climate change.

In the article, a "senior U.N. environmental official" claims that if global warming isn't reversed by 2030, then rising sea levels could wipe "entire nations . . . off the face of the Earth."

Crop failures coupled with coastal flooding, he said, could provoke "an exodus of 'eco-refugees,' " whose movements could wreak political chaos the world over. Unabated, the ice caps will melt away, the rainforests will burn, and the world will warm to unbearable temperatures.

Governments "have a ten-year window of opportunity to solve the greenhouse effects before it goes beyond human control," said the U.N. official.

Did the Associated Press publish that apocalyptic warning from the United Nations in June 2019? No, June *1989*. And, the cataclysmic events the U.N. official predicted were for the year 2000, not 2030.[107]

In early 2019, Roger Pielke reviewed the apocalyptic climate tract, *The Uninhabitable Earth*, for the *Financial Times*. In his review, Pielke described a filtering mechanism that results in journalists, like the one who wrote the book, getting the science so wrong.

"The scientific community produces carefully caveated scenarios of the future, ranging from the unrealistically optimistic to the highly pessimistic," Pielke wrote. By contrast, "Media coverage tends to emphasize the most pessimistic scenarios and in the process somehow converts them from worst-case scenarios to our most likely futures."

The author of *The Uninhabitable Earth*, like other activist journalists, simply exaggerated the exaggerations. He "assembled the best of this already selective science to paint a picture containing 'enough horror to induce a panic attack in even the most optimistic.' "[108]

What about so-called tipping points, like the rapid, accelerating, and simultaneous loss of Greenland or West Antarctic ice sheets, the drying out of and die-back of the Amazon, and a change of the Atlantic Ocean circulation? The high level of uncertainty on each, and a complexity that is greater than the sum of its parts, make many tipping point scenarios unscientific. That's not to say that a catastrophic tipping point scenario is impossible, only that there is no scientific evidence that one would be more probable or catastrophic than other potentially catastrophic scenarios, including an asteroid impact, super-volcanoes, or an unusually deadly influenza virus.

Consider the other threats humankind has recently been forced to cope with. In July 2019, NASA announced it had been caught by surprise when a "city-killer" asteroid passed by — just one-fifth of the distance between Earth and Moon.[109] In December 2019, a volcano unexpectedly erupted in New Zealand, killing twenty-one people.[110] And in early 2020, governments around the world scrambled to cope with an unusually deadly flu-like virus that experts say may kill millions of people.[111]

Have governments sufficiently invested to detect and prevent asteroids, super-volcanoes, and deadly flus? Perhaps, or perhaps not. While nations take reasonable actions to detect and avoid such disasters they generally don't take radical actions for the simple reason that doing so would make societies poorer and less capable of confronting all major challenges, including asteroids, super-volcanoes, and disease epidemics.

"Richer countries are more resilient," climate scientist Emanuel said, "so let's focus on making people richer and more resilient."

The risk of triggering tipping points increases at higher planetary temperatures, and thus our goal should be to reduce emissions and keep temperatures as low as possible without undermining economic development. Said Emanuel, "We've got to come up with some kind of middle ground. We shouldn't be forced to choose between growth and lifting people out of poverty and doing something for the climate."[112]

The new good news is that carbon emissions have been declining in developed nations for more than a decade. In Europe, emissions in 2018 were 23 percent below 1990 levels. In the U.S., emissions fell 15 percent from 2005 to 2016.[113]

The U.S. and Britain have seen their carbon emissions from electricity, specifically, decline by an astonishing 27 percent in the U.S. and 63 percent in the U.K., between 2007 and 2018.[114]

Most energy experts believe emissions in developing nations will peak and decline, just as they did in developed nations, once they achieve a similar level of prosperity.

As a result, global temperatures today appear much more likely to peak at between two to three degrees centigrade over preindustrial levels, not four, where the risks, including from tipping points, are significantly lower. The International Energy Agency (IEA) now forecasts carbon emissions in 2040 to be lower than in almost all of the IPCC scenarios.[115]

Can we credit thirty years of climate alarmism for these reductions in emissions? We can't. Total emissions from energy in Europe's largest countries, Germany, Britain, and France, peaked in the 1970s, thanks mostly to the switch from coal to natural gas and nuclear — technologies that McKibben, Thunberg, AOC, and many climate activists adamantly oppose.

2

EARTH'S LUNGS AREN'T BURNING

1. Earth's Lungs

In August 2019, Leonardo DiCaprio, Madonna, and soccer star Cristiano Ronaldo shared photographs of the green Amazon rainforest ablaze with smoke pouring out of it. On Instagram, DiCaprio wrote, "The lungs of the Earth are in flames." Ronaldo tweeted to his eighty-two million followers, "The Amazon Rainforest produces more than 20 percent of the world's oxygen." [1]

The New York Times explained, "The Amazon is often referred to as Earth's 'lungs,' because its vast forests release oxygen and store carbon dioxide, a heat-trapping gas that is a major cause of global warming." [2] The Amazon, which covers more than two million square miles of Brazil, Colombia, Peru, and other South American countries, could soon "self-destruct," the *Times* reported. It would be "a nightmare scenario that could see much of the world's largest rainforest erased from the earth. . . . Some scientists who study the Amazon ecosystem call it imminent." [3]

Wrote another *Times* reporter, "If enough [Amazon] rain forest is lost and can't be restored, the area will become savanna, which doesn't store as much carbon, meaning a reduction in the planet's 'lung capacity.' " [4]

Writers compared the Amazon fires to the detonation of nuclear weapons. "The destruction of the Amazon is arguably far more dangerous than the weapons of mass destruction that have triggered a robust response," wrote a reporter for *The Atlantic*. If another 20 percent of the Amazon is lost, wrote a reporter for *The Intercept*, it would release a "doomsday bomb of stored carbon." [5]

The news media, high-profile celebrities, and European leaders blamed Brazil's new president, Jair Bolsonaro. European leaders threatened not to ratify a major trade deal with Brazil. "Our house is burning—literally," tweeted French president Emmanuel Macron, days before he hosted a meeting of the seven largest economies, the G7, in France.[6]

Beyond the Amazon, reported the *Times*, "in central Africa, vast stretches of savanna are going up in flame. Arctic regions in Siberia are burning at a historic pace."[7]

One month later, Greta Thunberg and other student climate activists sued Brazil for not doing enough to stop climate change. "Brazil's rollbacks are already starting to have a damaging effect," the students' attorneys wrote. "As it stands, the Amazon acts as a large carbon sink, absorbing a quarter of the carbon taken up by forests around the world every year."[8]

Like many Generation Xers, my concern about rainforest destruction dates back to the late 1980s. In 1987, a San Francisco environmental group called Rainforest Action Network launched a consumer boycott against the fast-food giant Burger King, which was purchasing hamburger meat produced on land in Costa Rica that was formerly a rainforest.

In order to produce beef, farmers in Latin America and other countries clear rainforests in order to raise and graze cattle. I watched CNN and other TV news outlets show dramatic images of fires and indigenous people fleeing their ancestral homes.

Upset by the images of destruction, I held a backyard party on my sixteenth birthday to raise money for Rainforest Action Network. I charged people $5 to attend and raised about one hundred dollars.

Today, as back then, the use of land as pasture for beef production is humankind's single largest use of Earth's surface. We use twice as much land for beef and dairy production as for our second largest use of Earth, which is growing crops. Nearly half of Earth's total agricultural land area is required for ruminant livestock, which includes cows, sheep, goats, and buffalo.[9]

In the Amazon, the first people to exploit the forest are loggers, who extract valuable wood. They are followed by ranchers, who cut down the forest, burn it, and then graze cattle to establish ownership.

Because beef production was causing rainforest destruction, I stopped eating it and became fully vegetarian when I went to college in fall 1989.

For me, the nightmare of rainforest destruction was balanced by a feeling of success. By October 1987, Rainforest Action Network's Burger King boycott had succeeded. The fast-food chain announced it would stop importing beef from Costa Rica. In some small way, I felt I had helped save the rainforests.[10]

2. "There's No Science Behind That"

At fifteen years old, I started an Amnesty International chapter at my high school. A teacher asked my club supervisor, the school counselor, whether I was a communist. Two years later, I confirmed their suspicions by persuading my school principal to let me spend the fall semester of my senior year in Nicaragua to learn Spanish and witness the Sandinista socialist revolution. Afterward, I traveled throughout Central America, making relationships with small farmer cooperatives.

While attending college, I learned Portuguese so I could live in Brazil and work with the Landless Workers' Movement and the Workers Party in the semi-Amazonian state of Maranhão in Brazil. I returned there several times between 1992 and 1995. I loved Brazil and even imagined for a time that I would move there permanently to work with the Landless Workers' and Workers Party.

I attended the 1992 United Nations environment summit in Rio de Janeiro, where deforestation was a hot topic. The head of Rainforest Action Network, who five years earlier had forced Burger King to change its practices, staged a noisy protest. I was swept up in the excitement of a country emerging from several decades of military dictatorship.

I returned to Brazil several more times. I did field work in the semi-Amazon with small farmers defending their land from larger farmers seeking to take it over. I dated a Brazilian documentary filmmaker who was connected to the Workers Party and the left-wing non-governmental organization (NGO) scene in Rio de Janeiro. By 1995, I was interviewing the leading lights of Brazil's progressive movement, from the first Afro-Brazilian senator and *favelada*, Benedita da Silva, as well as Luiz Inácio "Lula" da Silva, who went on to be elected president in 2002.

I continued to write about the Amazon throughout the years and so, when the firestorm of publicity over the Amazon raged in the late summer of 2019, I decided to call Dan Nepstad, a lead author of a recent IPCC report on the Amazon. I asked him whether it was true that the Amazon was a major source of Earth's oxygen supply.

"It's bullshit," he told me. "There's no science behind that. The Amazon produces a lot of oxygen, but it uses the same amount of oxygen through respiration, so it's a wash." [11]

According to an Oxford University ecologist who studies them, Amazon plants consume about 60 percent of the oxygen they produce in respiration, the biochemical process whereby they obtain energy. Microbes, which break down rainforest biomass, consume the other 40 percent. "So, in all practical terms, the net contribution of the Amazon ECOSYSTEM (not just the plants alone) to the world's oxygen is effectively zero," the ecologist writes. "The same is pretty much true of any ecosystem on Earth, at least on the timescales that are relevant to humans (less than millions of years)." [12]

As for lungs, they absorb oxygen and emit carbon dioxide. By contrast, the Amazon, and all plant life, *store* carbon, though not 25 percent, as the student climate activists who sued Brazil claimed, but rather 5 percent. [13]

As for the photos that celebrities shared on social media, they weren't actually of the Amazon on fire. Many weren't even of the Amazon. [14] The photo Ronaldo shared was taken in southern Brazil, far from the Amazon—and it was taken in 2013, not 2019. [15] The photo Madonna shared was more than thirty years old. [16]

In reality, almost everything the news media reported in summer 2019 about the Amazon was either wrong or deeply misleading.

Deforestation had risen, but the increase had started in 2013, a full six years before President Bolsonaro took office. In 2019, the area of Amazon land deforested was just one-quarter of the amount of land that was deforested in 2004. [17] And while the number of fires in Brazil in 2019 was indeed 50 percent higher than the year before, it was just 2 percent higher than the average during the previous ten years. [18]

Against the horrifying picture painted of an Amazon forest on the verge of disappearing, a full 80 percent remains standing. Between 18 to 20 percent of the Amazon forest is still "up for grabs" (*terra devoluta*) and remains at risk of being deforested.[19]

It is true, however, that deforestation is fragmenting the Amazon and destroying the habitats of species of high conservation value. Big cats such as jaguars, pumas, and ocelots and other large mammal species need contiguous, unfragmented habitat to survive and thrive. Many tropical species, including ones in the Amazon, depend on "primary" old-growth forest. While mammals can reinhabit secondary forests, it often takes many decades or even centuries for forests to return to their original abundance.[20]

But rainforests in the Amazon and elsewhere in the world can only be saved if the need for economic development is accepted, respected, and embraced. By opposing many forms of economic development in the Amazon, particularly the most productive forms, many environmental NGOs, European governments, and philanthropies have made the situation worse.

3. Looking Down on the Poor

In 2016, the Brazilian model Gisele Bündchen flew over the Amazon forest with the head of Greenpeace Brazil as part of a National Geographic television series called *Years of Living Dangerously*. At first, they fly over an endless green forest. "The beauty seems to go on forever," Bündchen says in her voice-over, "but then [Greenpeace's Paulo] Adario tells me to brace myself." She is horrified by what comes next. Below her are fragments of forest next to cattle ranches. "All these large geometric shapes carved into the landscape are because of cattle?" she asks.

"Everything starts with small logging roads," Adario explains. "The road stays and then a cattle rancher comes and then he starts cutting the remaining trees."

"And cattle is not even natural of the Amazon!" Bündchen says. "It is not even supposed to be here!"

"No, definitely not," confirms Adario. "Imagine the destruction of this

beautiful forest to produce cattle. When you eat a burger you don't realize
your burger is coming from rainforest destruction." Bündchen starts to tear
up. "It's shocking isn't it?" asks Adario.[21]

But is it really so shocking? After all, agricultural expansion in Brazil is happening nearly identically to how it occurred in Europe hundreds of years ago.

Between 500 and 1350, forests went from covering 80 percent of western and central Europe to covering half of that. Historians estimate that the
forests of France were reduced from being thirty million hectares (about
seventy-four million acres) to thirteen million (about thirty-two million
acres) between 800 and 1300. Forests covered 70 percent of Germany in the
year 900 but just 25 percent by 1900.[22]

And yet developed nations, particularly European ones, which grew
wealthy thanks to deforestation and fossil fuels, are seeking to prevent Brazil
and other tropical nations, including the Congo, from developing the same
way. Most of them, including Germans, produce more carbon emissions per
capita, including by burning biomass, than do Brazilians even when taking
into account Amazon deforestation.[23]

The good news is that, globally, forests are returning, and fires are declining. There was a whopping 25 percent *decrease* in the annual area burned
globally from 1998 to 2015, thanks mainly to economic growth. That
growth created jobs in cities for people, allowing them to move away from
slash-and-burn farming. And economic growth allowed farmers to clear
forests for agriculture using machines, instead of fire.[24]

Globally, new tree growth exceeded tree loss for the last thirty-five years,
by an area the size of Texas and Alaska combined. An area of forest the size
of Belgium, Netherlands, Switzerland, and Denmark *combined* grew back
in Europe between 1995 and 2015.[25] And the amount of forests in Sweden,
Greta Thunberg's home nation, has doubled during the last century.[26]

Roughly 40 percent of the planet has seen "greening"—the production
of more forest and other biomass growth—between 1981 and 2016. Some of
this greening is due to a reversion of former agricultural lands to grasslands
and forests, and some of it is due to deliberate tree planting, particularly in
China.[27] This is even true in Brazil. While the world's attention has been
focused on the Amazon, forests are returning in the southeast, which is the

more economically developed part of Brazil. This is due to both rising agricultural productivity and environmental conservation.[28]

Part of the reason the planet is greening stems from greater carbon dioxide in the atmosphere, and greater planetary warming.[29] Scientists find that plants grow faster as a result of higher carbon dioxide concentrations. From 1981 to 2016, four times more carbon was captured by plants due to carbon-boosted growth than from biomass covering a larger surface of Earth.[30]

There is little evidence that forests around the world are already at their optimum temperature and carbon levels. Scientists find that higher levels of carbon dioxide in the atmosphere available for photosynthesis will likely offset declines in the productivity of photosynthesis from higher temperatures.[31] A major study of fifty-five temperate forests found higher growth than expected, due to higher temperatures resulting in a longer growing season, higher carbon dioxide, and other factors.[32] And faster growth means there will be a slower accumulation of carbon dioxide in the atmosphere.

None of this is to suggest that rising carbon emissions and climate change bring no risks. They do. But we have to understand that not all of their impacts will be bad for the natural environment and human societies.

Nor does any of this mean we shouldn't be concerned about the loss of primary old-growth forests in the Amazon and elsewhere in the world. We should be. Old-growth forests offer unique habitats to species. While the total amount of forest cover in Sweden has doubled during the last century, many of the new forests have been in the form of monocultural tree farms.[33] But if we are to protect the world's remaining old-growth forests, we're going to need to reject environmental colonialism and support nations in their aspirations to develop.

4. Romance and Reality

I am sensitive to the insensitive behaviors of developed-world environmentalists because I lived with the small farmers Bündchen looked down upon, and life was exceedingly difficult.

I grew up in middle-class comfort and was unprepared for the extreme poverty I experienced when I went to Nicaragua as a teenager. In lieu of hot showers and flush toilets, I poured bowls of cold water over my head, shivered, and used outhouses, like everyone else. Several times I became sick, likely from contaminated water.

The country was in its ninth year of civil war and people were increasingly desperate. One night, my Spanish teacher hosted her students for dinner. She lived in what can only be described as a shack, roughly thirty feet by ten feet. I helped make spaghetti. We drank beer and smoked cigarettes. I asked her, tactlessly, how much it would cost to buy a house like hers. She responded by offering to sell it to me for $100. I came home with intestinal parasites and a burning desire to help improve lives there.

Life in the Amazon was in many ways much harder than life in Central America because the communities are so much more remote. I lived in communities in Brazil that engaged in slash-and-burn agriculture. It starts with cutting trees in the forest, letting the wood and biomass dry, then burning it. The ash fertilizes the fields. Crops are then planted, returning a very small yield.

The people I worked with were too poor to have much livestock, though that was the next rung up the economic ladder. Slashing and burning was brutal work. The men drank large quantities of rum while doing it. We passed cooler and more pleasant afternoons fishing at the river.

The Amazon and semi-Amazonian northwest and central regions of Brazil are as hot as the Congo, with annual average temperatures near 90 degrees Fahrenheit. Higher temperatures reduce labor productivity, which helps explain why nations in tropical climates are less developed than nations in temperate ones. It is simply too hot to work for much of the day.[34]

In Brazil, as in Nicaragua, my enthusiasm for socialist cooperatives was often greater than of that the small farmers who were supposed to benefit from them. Most of the small farmers I interviewed wanted to work their own plot of land. They might be great friends with their neighbors and even be related to them by birth or marriage, but they didn't want to farm with them. They didn't want to be taken advantage of by somebody who didn't work as hard as them, they told me.

I can count on a single hand the number of young people who told me they wanted to remain on their family's farm and work their parents' land. The large majority of young people wanted to go to the city, get an education, and get a job. They wanted a better life than what low-yield peasant farming could provide. They wanted a life more like mine. And I knew, of course, that I didn't want to be a small farmer. Why did I ever think anyone else wanted to? The reality I lived, up close and in person, made it impossible for me to hold on to my romantic views.

In August 2019, the news media's portrayal of the burning rainforest as a result of greedy corporations, nature-hating farmers, and corrupt politicians annoyed me. I had understood for a quarter century that rising deforestation and fires are primarily the result of politicians responding to popular economic demands, not lack of concern for the natural environment.

The reason deforestation in Brazil rose again starting in 2013 was that of a severe economic recession and reduced law enforcement. The election of Bolsonaro in 2018 was as much an effect of rising demand for land as it was a cause of rising deforestation. Of Brazil's 210 million people, a full 55 million live in poverty. An additional 2 million Brazilians fell into poverty between 2016 and 2017.[35]

And the notion that the Amazon is populated mostly by indigenous people victimized by nonindigenous people is wrong. Just one million of the thirty million Brazilians who live in the Amazon region are indigenous, and some tribes control very large reserves.[36] There are 690 indigenous reserves covering an astonishing 13 percent of Brazil's landmass, almost all of them in the Amazon. Just nineteen thousand Yanomami Indians effectively own an area slightly larger than the size of Hungary.[37] Some engage in logging.[38]

Anyone looking to understand why Brazil cuts down its rainforests to produce soy and meat for export must start with the reality that it is trying to lift the last one-quarter of its population out of a poverty comparable to that of Bernadette in the Congo, of which environmentalists in Europe and North America are oblivious or, worse, unconcerned.

5. Fire and Food

Sometime between AD 900 and 950, Maori hunter-gatherers arrived by boat at what is today known as New Zealand, likely from other Pacific islands to the northeast. To their delight they found the island thick with moas, ostrich-like birds that stood an astonishing sixteen feet tall. Moas weren't able to fly. Nor did they have any other means to protect themselves from the Maori.[39]

To catch them, the Maori would set forest fires, which would push moas to the edges of the forest where they could be more easily slaughtered. The Maori came to rely so much on moas for food, as well as for tools and jewelry, that they called them their "primary source." During dry and windy seasons, the fires burned large landscapes, massively altered natural environments, and destroyed the habitats of other species.

In New Zealand, the conifer forests burned quickly during the hot and dry summer months and could not regenerate; they were replaced by bracken, fern, and scrub. But this hadn't put an end to the Maori practice of starting forest fires. "We saw either smoke by day and fires by night," Captain Cook wrote, "in all parts of it."[40]

Within three hundred years, half of New Zealand was deforested, moas teetered on the brink of extinction, and the Maori were facing rapid environmental and social change. By the time Cook arrived in the 1770s, the Maori had completely wiped out the moa, and they had been forced to take up slash-and-burn farming.

What happened in New Zealand was typical of what had happened around the world ten thousand years ago. A few million humans globally killed millions of large mammals each year, resulting in species extinctions.[41]

What we today view as a pleasing natural landscape—a grassy meadow surrounded by a forest and with a river running through it—is often a landscape created by humans to hunt game seeking out drinking water.[42] Using fire to create a meadow in which to slaughter animals is one of the most frequent mentions of the uses of fire by hunter-gatherers around the world. The meadows of the North American eastern forests would have disappeared had they not been burned annually by Indians for five thousand years. And in the Amazon, hunter-gatherers burned forests and introduced new species. Hunting by luring game is more energy efficient than chasing it. Over time, trapping wild animals within enclosed spaces evolved into the domestication of animals for livestock. [43]

Fire made communities more secure from human and nonhuman predators, allowed them to expand all over the world, and required new behaviors around eating, organizing societies, and procreating. Hunting with fire became a crucial milestone in the creation of both what we think of as nation-states and markets, through the demarcation of control by individuals and groups competing for food. Indeed, fire was used differently in different zones, for security, agriculture, and hunting. [44]

Fire allowed for the creation of sexually monogamous family units. And it allowed for the hearth as a place for reflection and discussion and widening social and group intelligence.

All over the planet, deforestation through fire gave rise to agriculture by fertilizing soils favoring blueberries, hazelnuts, grains, and other useful crops. Today, many tree species require fire for their seeds to grow into trees. Fire is also essential, as we saw with both California and Australia, for clearing woody biomass from the forest floor.

In short, fire and deforestation for meat production are major parts of what made us humans. [45] The only way Adario, Bündchen, and other environmentalists could find meat production in the Amazon so shocking is by knowing none of that history.

For twenty-first-century environmentalists, the word *wilderness* has positive connotations, but in the past it was a frightful "place of wild beasts." European farmers viewed forests as places of danger, which they often were, home to both dangerous animals like wolves and menacing humans like outlaw gangs. In the fairy tale "Hansel and Gretel," two children get lost in the forest and fall into the hands of a witch. In "Little Red Riding Hood," a little girl traveling through the forest is terrorized by a wolf. [46]

Thus, for early European Christians, removing the forest was good, not bad. Early Christian fathers, including Saint Augustine, taught that it was humankind's role to finalize God's creation on Earth and grow closer to

Him. Forests and wilderness areas were places of sin; clearing them to make farms and ranches was the Lord's work.

Europeans believed humans were blessed and distinct for their transformative powers. Monks tasked with the work of creating a clearing in the woods literally imagined themselves expelling the devil from Earth. They weren't trying to create Eden but rather a New Jerusalem: a civilization that mixed town and country, sacred and profane, commerce and faith.

It was only after humans started living in cities, and growing wealthier, that they started to worry about nature for nature's sake.[47] Europeans who, in the nineteenth century had viewed the Amazon as "jungle," a place of danger and disorder, came to see it in the late twentieth century as "rainforest," a place of harmony and enchantment.

6. Greenpeace Fragments the Forest

Insensitivity to Brazil's need for economic development led environmental groups, including Greenpeace, to advocate policies that contributed to the fragmentation of the rainforest and the unnecessary expansion of cattle ranching and farming. Environmental policies should have resulted in "intensification," growing more food on less land. Instead, they resulted in *ex*tensification and a political and grassroots backlash by farmers that resulted in rising deforestation.

"The mastermind of the soy moratorium was Paulo Adario of Greenpeace Brazil," said Nepstad. Adario is the man who made Bündchen cry. "It started with a Greenpeace campaign. People dressed up like chickens and walked through a number of McDonald's restaurants in Europe. It was a big international media moment."[48]

Greenpeace demanded a far stricter Forest Code than the one that had been imposed by the Brazilian government.[49] Greenpeace and other environmental NGOs insisted that landowners keep a large amount, 50 to 80 percent, of their property as forest through Brazil's Forest Code.

Nepstad said the stricter Forest Code cost farmers $10 billion in forgone profits and forest restoration. "There was an Amazon Fund set up in 2010 with $1 billion from Norwegian and German governments but none of it ever made its way to the large and medium-sized farmers," says Nepstad.

"Agribusiness is 25 percent of Brazil's GDP and it's what got the country through the recession," said Nepstad. "When soy farming comes into a landscape, the number of fires goes down. Little towns get money for schools, GDP rises, and inequality declines. This is not a sector to beat up on, it's one to find common ground with." [50]

Greenpeace sought stricter restrictions on farming in the savannah forest, known as the Cerrado, where much of the soy in Brazil is grown. "Farmers got nervous that there was going to be another moratorium by governments on Brazilian soy imports," explains Nepstad. "The Cerrado is 60 percent of the nation's soy crop. The Amazon is 10 percent. And so this was a much more serious matter." [51]

Greenpeace's campaign led journalists, policymakers, and the public to conflate the Cerrado savanna with the Amazon rainforest, and thus believe that the expansion of soy farming in the Cerrado was the same as logging the rainforest.

But there is far more economic and ecological justification for deforestation in the Cerrado, which is less biologically diverse and has soils more suited to soy farming, than in the rainforest. By conflating the two regions, Greenpeace and journalists exaggerated the problem and created the wrong impression that both places are of equal ecological and economic value.

Greenpeace wasn't the first organization that tried to prevent Brazil from modernizing and intensifying agriculture. In 2008, the World Bank published a report that "basically said that small is beautiful, that modern, technologically sophisticated agriculture (and especially the use of GMOs) was bad," wrote the World Bank's representative at the time to Brazil. The report said that "the path that should be followed was small and organic and local agriculture." [52]

The World Bank report enraged Brazil's agriculture minister, who called the Bank's representative and asked, "How can the World Bank produce such an absurd report. Following the 'wrong path' Brazil has become an agricultural superpower, producing three times the output we produced thirty years ago, with 90 percent of this coming from productivity gains!" [53]

The report added insult to injury. The World Bank had already cut 90 percent of its development aid for Brazil's agricultural research efforts as punishment because Brazil sought to grow food in the same ways that wealthy nations do.[54]

Brazil was able to make up for the aid that World Bank had denied it with its own resources. After it did so, Greenpeace pressured food companies in Europe to stop buying soy from Brazil.[55] "There's this exaggerated confidence, this hubris," said Nepstad, "that regulation upon regulation, without really thinking of the farmer's perspective."[56]

Much of the motivation to stop farming and ranching is ideological, Nepstad said. "It's really antidevelopment, you know, anti-capitalism. There's a lot of hatred of agribusiness. Or at least hatred of agribusiness in Brazil. The same standard doesn't seem to apply to agribusiness in France and Germany."[57]

The increase in deforestation in 2019 is to some extent Bolsonaro fulfilling a campaign promise to farmers who were "fatigued with violence, the recession, and this environmental agenda," Nepstad said. "They were all saying, 'You know, it's this forest agenda that will get this guy [Bolsonaro] elected. We're all going to vote for him.' And farmers voted for him in droves. I see what's happening now, and the election of Bolsonaro, as a reflection of major mistakes in [environmentalist] strategy."[58]

I asked Nepstad how much of the backlash was due to the government's enforcement of environmental laws and how much was due to NGOs like Greenpeace. "I think most of it was NGO dogmatism," he said. "We were in a really interesting space in 2012, '13, '14 because the farmers felt satisfied with the article of the Forest Code dedicated to compensating farmers, but it never happened."[59]

Brazil's soy farmers were willing to cooperate with reasonable environmental rules before Greenpeace started making more extreme demands. "What the farmers needed was basically amnesty on all of the illegal deforestation up through 2008," said Nepstad. "And winning that, they felt like, 'Okay, we could comply with this law.' I side with the farmers on this."[60]

What's happened in the Amazon is a reminder that concentrating farming in some areas allows governments to protect primary forest habitats so that they can remain relatively intact, wild, and biodiverse. The Greenpeace/ NGO strategy resulted in landowners clearing forest elsewhere, in order to expand their footprint. "I think the Forest Code has fostered fragmentation," Nepstad told me.[61]

Green NGOs have had a similar impact in other parts of the world. After environmentalists encouraged such fragmentation in palm oil plantations in Southeast Asia as a measure supposedly friendly to wildlife, scientists found a 60 percent reduction in the abundance of important bird species.[62]

7. "Take Your Dough and Reforest Germany"

Greenpeace's agenda fit neatly into the agenda of European farmers to exclude low-cost Brazilian food from the European Union. The two European nations that were the most critical of deforestation and fires in the Amazon also happened to be the two countries whose farmers most resisted the Mercosur free trade agreement with Brazil: France and Ireland.

"Brazilian farmers want to extend [the free trade agreement] EU-Mercosur," noted Nepstad, "but [French president Emmanuel] Macron is inclined to shut it down because the French farm sector doesn't want more Brazilian food products coming into the country."[63]

Indeed, it was President Macron who started the global news media fervor about Amazon deforestation just a few days before France hosted the G7 meeting. Macron said France would not ratify a major trade deal between Europe and Brazil so long as Brazil's president did nothing to reduce deforestation.

In Brussels, the capital of the European Commission, the attacks on Brazil by France and Ireland "raised eyebrows," noted *Forbes* business reporter Dave Keating. "These also happen to be the two countries who have been most vocally opposed to the Mercosur deal on protectionist grounds."[64]

According to Keating, "They are worried that their farmers will be overwhelmed by competition from South American beef, sugar, ethanol, and chicken. Beef, a staple of Argentinian and Brazilian agricultural exports, has been the most sensitive issue in these trade negotiations. Irish farmers in particular are expected to have a tough time competing with the influx."[65]

"I don't doubt the sincerity of Macron's wish to protect the Paris Agreement," a EU trade expert told Keating, "but it strikes me as suspicious that it's these two countries raising the objection. It makes you wonder whether the Amazon fires are being used as a smokescreen for protectionism."[66]

Macron's attacks enraged Brazil's president. "Few countries have the moral authority to talk about deforestation with Brazil," said President Bolsonaro. "I would like to give a message to the beloved [German Chancellor] Angela Merkel. Take your dough and reforest Germany, okay? It's much more needed there than here."[67]

There was nothing "right wing" about the anger of Brazil's president with foreign hypocrisy. Brazil's former socialist president grew just as angry at the hypocrisy and neo-imperialism of foreign governments more than a decade earlier. "The wealthy countries are very smart, approving protocols, holding big speeches on the need to avoid deforestation," said President Luiz Inácio "Lula" da Silva in 2007, "but they already deforested everything."[68]

8. After Amazon Alarmism

The increase in Amazon deforestation should lead the conservation community to repair its relationship with farmers and seek more pragmatic solutions. Farmers should be allowed to intensify production in some areas, particularly the Cerrado, to reduce pressure and fragmentation in other areas, particularly the rainforest.

Creating parks and protected areas goes hand-in-hand with agricultural intensification. Simply making farming and ranching more productive and profitable without protecting natural areas is insufficient. By protecting some areas and intensifying on already-existing farms and ranches, Brazilian farmers and ranches could grow more food on less land and protect the natural environment.[69]

Researchers have found that the production of beef in Brazil is at less than half of its potential, which means that the amount of land required to produce beef could be massively reduced. Brazil's lesser known Atlantic Forest, far more of which has been lost than the Amazon, could benefit enormously.

"There is enough land for a large-scale restoration of the Atlantic Forest, the 'hottest of hotspots,' " wrote a group of scientists, "where up to eighteen million hectares [an area twice the size of Portugal] could be restored without impeding national agricultural expansion. This would more than double the remaining area of this biome, slow the massive species extinctions, and sequester 7.5 billion tonnes of CO_2."[70]

Nepstad agrees. "There's a huge area of unproductive land that's growing fifty kilos of beef per hectare a year that should all go back to the forest."

In the Cerrado, the daily weight gain and milk production can be three times higher after simply switching to faster-growing and more nutritious grasses and using fertilizer. Doing so brings the added benefit of cutting emissions of methane, a greenhouse gas, per kilogram of meat in half, while reducing the amount of land required.[71]

"Let's get the agrarian reform reserves, which are huge and close to cities, to grow vegetables and fruits and staples for the Amazon cities instead of them importing tomatoes and carrots from São Paulo," said Nepstad.[72]

The World Bank and other agencies should support farmers seeking to intensify agricultural production. Research suggests that Brazilian farmers receiving technical assistance was the key factor in their adoption of methods proven to increase productivity.[73]

The determination by activist journalists and TV producers to paint deforestation in the Amazon as apocalyptic was inaccurate and unfair. Worse, it further polarized the situation in Brazil, making it harder to find pragmatic solutions between farmers and conservationists.

As for the myth that the Amazon provides "20 percent of the world's oxygen," it appears to have evolved out of a 1966 article by a Cornell University scientist. Four years later, a climatologist explained in the journal *Science* why there was nothing to be frightened of. "In almost all grocery lists of man's environmental problems is found an item regarding oxygen supply. Fortunately for mankind, the supply is not vanishing as some have predicted." [74]

Unfortunately, neither is the supply of environmental alarmism.

3

ENOUGH WITH THE PLASTIC STRAWS

1. The Final Straw

In summer 2015, a PhD student studying marine biology was in a boat off the coast of Costa Rica scraping parasites off a sea turtle's back when she noticed something stuck up its nose. Christine Figgener, then thirty-one years old, took out her video camera and asked a colleague to try to pull out the object. "Okay, I'm filming. You can go for it," she says. "He's going to be so happy." [1]

As her colleague uses a pair of pliers to pull on the object, the turtle sneezes. "Have you ever heard a turtle sneeze?" she asks.

"You know what this is?" the colleague asks.

"Brain?" another man answers.

"That's a worm," he replies.

"Oh, that is disgusting," says Figgener. "Oh my god."

The turtle writhes in pain as the researcher pulls on a thin gray object with a pair of pliers.

"What the fuck?" says Figgener.

Blood dribbles from the turtle's nostril.

"Oh, she's bleeding," says Figgener. "Is it a hookworm?"

"I think it's a tube worm," answers the man pulling on the object.

The turtle opens its mouth as though it wants to bite someone, and hisses. "I'm sorry, little one, but I think you'll like it better afterwards," Figgener says.

"I don't want to pull it too hard. I don't know what's attached to it," the man says.

"I understand that," Figgener says. "I mean it's bleeding already. Maybe it's in her brain already."

"Eso es un gusano," says one of the deckhands in Spanish. *That's a worm.* "Sí," says Figgener.

Later the man says, "Es una concha rara." *It's a strange kind of shell.*

After pulling out part of the object, they debate how hard to pull on the partially removed object as blood drips down the boat wall.

Finally the man says, "Es plastico." *It's plastic.*

"Is it a straw? Don't tell me it's a freaking straw," Figgener says. "Because in Germany we have those with a black stripe—"

"Es un pajilla," interrupts the man. *It's a plastic straw.*

"A straw! A plastic straw!" Figgener says.

"Ya lo mordí y es plastico," the man says. *I bit it and it's plastic.*

"Didn't we have this discussion the other day about how useless freaking straws are?" Figgener asks. "So this is the reason why we do not need plastic straws."

They return to pulling on the straw.

"I'm so sorry, baby," Figgener tells the turtle. "I just wonder how she can breathe properly with that shit inside her."

The turtle hisses and wriggles in pain. At eight minutes into the video we hear a suction sound as the researcher yanks the rest of the plastic straw free from the turtle's nose.

"Ah, man!" says Figgener. "Show me, please." In the final seconds of the video, blood drips from the turtle's nose.[2]

When Figgener returned home that night, she uploaded her video to YouTube.[3] Within two days, millions had seen it. By 2020, the video had over sixty million views.

Not long after Figgener's video went viral, the city of Seattle announced a ban on plastic straws, followed shortly afterward by Starbucks, American Airlines, and the city of San Francisco.[4]

In the months and years since, people told her that they had reduced their use of plastic, including straws. "I'm of course happy," she said later. "Everyone can do something at home, even if it's one thing."[5]

Perhaps. But when you consider that just 0.03 percent of the nine million tons of plastic waste that ends up in oceans every year is composed of straws, banning them seems like a profoundly small thing, indeed.[6]

2. The Persistence of Plastic

When I spoke to Figgener by phone in late 2019, she told me that bans on plastic straws were "a great first step and conversation starter, but it will not fix our problems. A lot of the stuff I find in the ocean are single-use plastics, Styrofoam, to-go cups, plastic bags."[7]

"The reason that I had this [turtle rescue] on camera was because I had been working with turtles for thirteen years, and plastic has been a constant byline to my work," said Figgener, who earned a PhD in marine biology at Texas A&M University in 2019.

Plastic waste can significantly increase sea turtles' mortality rates. Half of all sea turtles have eaten plastic waste and in some parts of the world, 80 to 100 percent have consumed plastic waste. Ingested plastic can kill turtles by reducing their ability to digest food as well as by rupturing their stomachs.[8]

"They ingest entire plastic bags," Figgener said, "but they also ingest smaller, five- to ten-centimeter pieces that [in] some of their stomachs can cause obstructions, perforations, and can lead to starvation and internal bleedings."[9]

Scientists in 2001 found that debris, mostly plastic, accounted for 13 percent of green turtle deaths that were studied off the coast of Brazil.[10] In 2017, scientists found that once a sea turtle had fourteen pieces of plastic in its gut, it had only a fifty-fifty chance of surviving.[11]

It's not just turtles. In spring 2019, a dead sperm whale was found in Italy with more than forty-eight pounds of plastic tubing, dishes, and bags in its stomach. Most of the plastic was undigested and intact. It may have killed her fetus, which experts said was "in an advanced state of decomposition."

One month earlier, scientists in the Philippines found a beached whale with eighty-eight pounds of plastic in its body. In 2018, scientists in Spain found sixty pounds of plastic trash in a dead sperm whale.[12] "For every pound of tuna we're taking out of the ocean, we're putting two pounds of plastic in the ocean," an ocean scientist reported. [13]

Seabird populations declined 70 percent between 1950 and 2010.[14] "Essentially seabirds are going extinct," a leading scientist said. "Maybe not to-morrow. But they're headed down sharply. Plastic is one of the threats they face."[15] Seabirds can eat up to 8 percent of their weight in plastic, which is the "equivalent to the average woman having the weight of two babies in her stomach," noted another scientist.[16]

The share of seabird species found to be ingesting plastic rose to an estimated 90 percent in 2015, with the scientists studying the issue pre-dicting an increase to 99 percent of species found with ingested plastic by 2050.[17]

Part of the reason we worry about plastics is that it seems to take so long for them to degrade. In 2018, the United Nations Environment Programme estimated that Styrofoam takes *thousands* of years to disintegrate.[18]

3. The Poverty of Waste

Consumption of plastics has skyrocketed during the last several decades. Americans use ten times as much, per capita, as we did in 1960.[19] Between 1950 and 2015, we went from producing 2 million tons of the stuff to nearly 400 million tons globally.[20] Scientists estimate that the amount of plastic waste could increase ten-fold between 2015 and 2025.[21]

One study found that just four developing nations, China, Indonesia, Philippines, and Vietnam, produce half of all mismanaged plastic waste at risk of entering the ocean. One-quarter came from China alone.[22]

The vast majority of plastic waste in the marine environment comes from land-based sources like littering, manufacturing materials, and waste asso-ciated with coastal recreational activities. The rest comes from ocean-based debris like fishing nets and lines.[23]

Fishing nets and lines account for half of all waste within the infamous Great Pacific Garbage Patch.[24] Figgener reports finding "ghost nets swim-ming in our oceans," rice bags, and other "big debris where turtles can be-come entangled."[25]

"Recycling isn't really working," Figgener explains. "We're not really re-cycling. When we do, it is kind of a downcycling, not an upcycling. Eventually, you know, recycling with plastics, unlike aluminum or glass, happens a few times, if at all, before it ends up in landfills."[26]

In the United States in 2017, nearly 3 million tons of plastic waste were recycled, 5.6 million tons were incinerated, and almost 27 million tons were sent to landfills.[27] Comparing 2017 to 1990, landfill and incineration amounts have doubled while plastic recycling amounts increased eight-fold. More than 25 million tons of plastic waste were produced in Europe in 2014, with 39 percent incinerated, 31 percent sent to landfills, and 30 percent recycled.[28]

"Even though you put it in your recycling bin, that doesn't mean it stays here in the U.S.," said Figgener. "It actually gets shipped to China and Asia, Indonesia, and Malaysia, other countries that do not have the infrastructure to deal with all that waste."[29]

In 2017, China abruptly announced that it would no longer accept large quantities of plastic waste from rich nations like the United States. At the time, China was importing $18 billion worth of solid waste. China's rejection of waste was part of a broader health and environmental effort.[30]

A few months later, Malaysia had replaced China as the world's solid waste dump, but its six-fold increase in solid waste imports in less than one year aroused domestic opposition. "Everybody knows those dumps are illegal," a Malaysian butcher told *The New York Times*. "We don't like them."[31]

Other countries seem less willing to accept waste. Vietnam announced it would stop importing plastic scrap waste by 2025. The Philippines refused to allow in a shipment of a plastic waste–based fuel from Australia in spring 2019 for reasons that included its terrible smell.[32]

That doesn't mean developed nations are off the hook. Even in fastidious Japan, where 70 to 80 percent of used plastic bottles, bags, and wrappers are collected and incinerated or recycled, twenty thousand to sixty thousand tons of plastic still ends up in the ocean.[33]

After two decades of growth in recycling, even in rich countries less than a third of plastic waste is recycled.[34] Figgener—who is from Germany, which incinerates much of its waste—notes that, "Germany still 'recycles' you know, in quotation marks, and we're still one of those that export our recycling to countries in Asia and Africa. We only incinerate those things that don't have a value anymore on the recycling market."[35]

The biggest factor in determining whether waste ends up in the ocean or not is whether a nation has a strong waste collection and management system. As such, if the concern is about plastic ending up in the ocean, nations will likely need to focus on either storing waste in landfills or incinerating it.

The experiences of American cities between the 1980s and 1990s starting up recycling systems tell us that the processing equipment and collection practices come at a massive premium over refuse collection—up to fourteen times the cost per ton.[36] Ultimately, it's just cheaper for plastic makers to simply produce new plastic resin from petroleum.[37]

For low income countries, which have collection rates of less than 50 percent, the transition from the open dumping toward efficient collection and sanitary landfilling should be the first step. A well-managed refuse and landfill system can cost ten times more than open dumping, yet will be necessary to avoid river and ocean pollution.[38]

Many experts thus believe that rich nations seeking to reduce plastic waste in the oceans should improve trash collection in poor ones. "Improving waste management infrastructure in developing countries is paramount," wrote the authors of a major study in 2015. Doing that would "require substantial infrastructure investment primarily in low- and middle-income countries."[39]

4. Things Fall Apart

Between 2007 and 2013, a team of nine scientists took twenty-four separate expeditions around the world to try to determine the total amount of plastic in the sea. They went to all five subtropical gyres, circular currents in the oceans that trap plastic waste. They towed nets behind boats 680 times to scoop up waste, which they separated from natural debris using microscopes, before counting and weighing to the nearest 0.01 milligram. They visually surveyed waste 891 times. And they even developed a model of how plastic waste spreads across the ocean, accounting for how the wind mixes plastic vertically.

The scientists seemed shocked by what they discovered: "The global weight of plastic pollution on the sea surface, from all size classes combined, is only 0.1 percent of the world annual production."[40] Even more astonishing, they found a *hundred times less* microplastic than they had been expecting to find.

So where are all the microplastics going? The scientists named several possibilities.

First, as large plastics get broken down into smaller and smaller particles, they start degrading ever more rapidly because the "volume relationship is increased dramatically and oxidation levels are higher, enhancing their biodegradation potential."[41]

Second, the sea life eating plastic waste appear to be "packaging microplastics into fecal pellets, thus enhancing sinking." While eating plastic can negatively impact the health of sea birds and mammals, it appears to also be contributing to "the removal of small microplastics from the sea surface."[42]

In the end, the scientists emphasized how much we still don't know. "The question 'where is all the plastic?' remains unanswered," they concluded, "highlighting the need to investigate the many processes that play a role in the dynamics of macro-, meso-, and microplastics in the world's oceans."[43]

Five years later, another group of scientists offered another possibility, at least for one of the most vexing forms of plastic waste: polystyrene, the plastic used in Styrofoam, plastic utensils, and endless other products.

In 2019, a team of scientists from Woods Hole Oceanographic Institution in Massachusetts and Massachusetts Institute of Technology announced that it had discovered that sunlight breaks down polystyrene in ocean water over a period as short as decades.[44]

It had long been known that sunlight breaks down plastics like polystyrene. "Just look at plastic playground toys, park benches, or lawn chairs," said one of the scientists, "which can rapidly become sun-bleached."[45]

But environmental groups have long considered polystyrene waste in the ocean to have a lifespan in the thousands of years, if not longer, because it can't be broken down by bacteria.

Therefore, even though polystyrene represents a small percentage of global plastic, its long life in nature was considered an environmental threat, one easily visualized as chunks of Styrofoam bobbing on waves and beaches.

In a lab, the scientists exposed five samples of polystyrene in seawater to light from a special lamp matching the sun's rays. What they discovered was that sunlight breaks down the polystyrene into organic carbon and carbon dioxide. The organic carbon dissolves in seawater, and the carbon dioxide enters the atmosphere. At the end of the process, the plastic is gone. "We used multiple methods, and they all pointed to the same outcome," said one of the researchers.

The same molecular property that made polystyrene molecules hard for microbes to eat was just the feature that made them vulnerable to having their bonds broken by sunlight. The scientists said their study was the first to provide direct proof of how, and how fast, sunlight breaks down polystyrene first into microplastics, then into individual molecules, and then into its elemental building blocks.[46]

One of the best pieces of good news to emerge from the study was that certain additives used to shape the flexibility, color, and other qualities of polystyrene can speed up or slow down its disintegration by sunlight in water. That discovery opens up the possibility of modifying how we make plastics to allow for a more rapid disintegration.[47]

5. The Elephant in the Room

For thousands of years, humans around the world made exquisite jewelry and other luxury items from the shells of hawksbill sea turtles, like the kind studied by Figgener and her crew in Costa Rica. Craftsmen heated the turtles over a fire, sometimes alive, so they could peel the misnamed "tortoiseshell" away from their skeletons. Sometimes the de-shelled sea turtles were returned to the sea.

Scientists estimate that since 1844, humans have killed nine million hawksbill turtles, or about sixty thousand each year. Humans killed so many hawksbill that the dramatic reduction in the species altered the function of coral reef and seagrass ecosystems around the globe.[48]

Around the world, artists and artisans used heat to flatten and mold tortoiseshell into various luxury items including eyeglasses, combs, lyres, jewelry, boxes, and, at least in Japan, penis rings, penis sheaths, and condoms.

Tortoiseshell was considered valuable in ancient Rome. Julius Caesar was thus overjoyed when, after invading Alexandria, Egypt, he discovered warehouses filled with the material. He went on to make the tortoiseshell a symbol of his victory.[49]

What was special about the shell of sea turtles wasn't just that it was smooth and beautiful but also that it was so *plastic*, which originally referred to things that were easily molded or shaped.

Tortoiseshell consists of keratin, a durable protein that protects cells from stress and damage. Keratin also comprises nails, horns, feathers, and hooves. Tortoiseshell is special in that it can be peeled into thin sheets to create veneer while remaining hard and water-resistant. When broken, it can even be repaired through the reapplication of heat and pressure.[50]

Like tortoiseshell, the ivory tusks of elephants were prized for their beauty and plasticity in making artistic and luxury items, including combs, piano keys, and billiard balls. The ancient Greek sculptor Phidias created a thirty-foot statue of Athena, daughter of Zeus and goddess of war, from gold and ivory. It was displayed for many years inside the Parthenon.[51]

In the Middle Ages, ivory was used to make caskets, goblets, and handles for swords and trumpets. Demand for ivory increased greatly in the nineteenth century as the material came to be used at an industrial scale. Americans, in particular, loved the material. From the 1830s to the 1980s, one of the largest ivory processing plants in the world was in Essex, Connecticut. It processed up to 90 percent of all ivory imported into the United States.[52]

Concerns about ivory shortages rose shortly after the American Civil War ended. "Dealers in ivory express considerable alarm lest the supply of elephants should run short in a few years," reported *The New York Times* in 1866, "and so throw them out of business." The reporter estimated twenty-two thousand elephants were being killed each year just "to supply the cutlery establishments of Sheffield, England, with handles for the knives and other cutlery made there."[53]

Already demand for ivory billiard balls had outpaced the available supply. "For some articles—billiard balls, for instance—made from ivory, no substitute for that material has been found," reported the *Times*. "A large dealer in billiard stock has offered a reward of several hundred dollars to any person who will produce a substance from which billiard balls can be made as durable and cheaper than ivory ones. As yet, no one has responded."[54]

Seven years later, in 1873, a reporter for the *Times* was despondent that a viable substitute for ivory would be found. "Think of the silence in the land, unless we could get ivory to make our piano-keys with!" The reporter estimated that U.S. ivory demand resulted in the killing of fifteen thousand elephants.[55] Later, a *Times* reporter estimated that the British imports resulted in the killing of eighty thousand elephants annually.[56]

Rising prices had encouraged entrepreneurs to look for alternatives. "The high price of ivory, together with its liability to warp and shrink have led to numerous efforts to find some good substitute for the article." Those included walrus and hippopotamus teeth, and the albumen of palm trees grown in the Andes, which was already being used to make rosaries, toys, and crucifixes.

In 1863, in upstate New York, a young man named John Wesley Hyatt learned about the billiard ball maker's offer of a $10,000 reward to anyone who could create a suitable substitute to ivory, and he started experimenting in his backyard shed with various materials. Six years later, he had invented celluloid from the cellulose in cotton.

By 1882, *The New York Times* warned of rising prices. "During the last quarter of a century ivory has been steadily increasing in price, until now it is selling at more than double its cost twenty years ago."[57] Europe and the United States were consuming two million pounds of ivory annually, which represented about 160,000 elephants.

"A prominent ivory merchant, who is a pessimist on the subject of the scarcity of ivory was recently heard to declare that it was his conviction that ivory would eventually grow to be so rare that in generations to come an ivory ring would be looked upon as one of the most costly gifts which a wealthy suitor could place upon the hand of his betrothed."[58]

A similar dynamic occurred with tortoiseshell. After Japan opened up to foreign trade in 1859, cheap, mass-manufactured products came into the country from Europe. "As Japan industrialized along Western patterns," notes a historian, "plastics replaced tortoiseshell for many of its uses, including the production of hair ornaments . . ."[59]

Combs were one of the first and most popular uses for celluloid. For thousands of years, humans had made combs of tortoiseshell, ivory, bone, rubber, iron, tin, gold, silver, lead, reeds, wood, glass, and porcelain. Celluloid replaced most of them.[60]

By the late 1970s, ivory was no longer used for piano keys. While some musicians claimed a fondness for ivory keys, most asserted the superiority of plastic. "I was glad to see it go," a quality control manager for a piano keyboard maker told *The New York Times* in 1977. "The tusks had to be handled very carefully to prevent disease. The plastic covering we use today is a far superior product in terms of its durability."

And it's not the case that plastic was uglier. "The best ivory has no grain and looks just like plastic."[61]

Celluloid had the advantage of being colored in ways to imitate the distinctive marbling of tortoiseshell combs. Hyatt produced a pamphlet boasting of the product's environmental benefits, claiming "it will no longer be necessary to ransack the earth in pursuit of substances which are constantly growing scarcer."[62]

In our conversation, after I told Figgener the history of how plastic helped save the hawksbill turtle, she laughed. "Plastic is a miracle product, you know? I mean, the advances in technology that also you know, help to develop. It wouldn't be possible without plastic. I mean, I don't want to lie about it. I'm not that hardline on it."[63]

6. The Real Killers

In September 2019, Helen and I traveled across the south island of New Zealand for vacation. We didn't have enough time to see both the glow worms and the island's rare penguins, so we opted for the penguins.

Before heading for the visitor's center, we stopped for lunch at the diner our guidebook had recommended. Fish and chips were on the menu. Americans don't know how to cook fish and chips well, and I had never enjoyed eating them until I had some in Britain a few years earlier. "I bet the fish and chips are good here," I said, checking Helen's face for agreement. She nodded, and we ordered it.

The fish, a blue cod, was delicious. It had been perfectly fried in a light batter. I devoured most of it, while Helen ate fish stew.

Within the hour we were at Penguin Place, a private farm that protects the nesting sites of yellow-eyed penguins. The farm's owner had created extensive trenches with blinds so tourists could spy on the penguins without scaring them. The trenches traversed roughly one kilometer over the hillside near the undeveloped coast. They were about five feet deep and were shielded with green blinds.

I hadn't read anything about the penguins in advance because I was on vacation and just wanted to enjoy the scenery. But before the tour, our guide started explaining how imperiled the species was. A chart plastered on the wall behind our guide showed the population of the island's yellow-eyed penguins: it hovered between three hundred and four hundred.

When he started talking about causes for the penguin's decline, the room of tourists fell quiet, and I started to fill with dread. There were several causes, he explained. Invasive species including the stoat, a kind of weasel, as well as dogs and cats, ate the penguins. But the biggest threat of late, he said, was that penguins were underweight. They weren't getting enough to eat.

Oh, no, I thought. *No, no, no.* I knew where this was headed. The big problem, said our guide—*Here it comes,* I thought—was from people over-fishing the areas where the penguins feed. And what kind of fish did the penguins prefer? I could have said the words before he did: blue cod. *How depressing,* I thought. We had literally eaten the poor penguins' lunch.

Penguin Place had started taking penguins into captivity for the sole purpose of fattening them up. "They can only stay here for three months," our guide said, "because if they stay longer they get sick and die."

"What exactly happens to them?" I asked. The guide said that they are greatly stressed by being around people and the stress appears to make them susceptible to falling ill from some kind of bacteria that is already in their bodies.

The International Union for Conservation of Nature's (IUCN) authoritative Red List classifies the yellow-eyed penguin as endangered, with its population declining. It estimates that between 2,528 to 3,480 birds, total, live in the wild.

Other major threats to the penguins are lack of habitat—most of the area where the penguins nest has been taken over by ranches and farms—invasive predators, and being caught by fishermen. "Its population has undergone extreme fluctuations," notes the IUCN, "and has potentially undergone a very rapid decline over the past three generations (twenty-one years) as a result of ongoing threats such as invasive predators and fisheries bycatch," which means being accidentally killed by fishermen.[64]

Warmer oceans from climate change might also be sending fish deeper in the ocean, requiring the penguins to dive deeper, expend more energy, and worsen their malnutrition.

The first yellow-eyed penguin we saw was in captivity. He was there getting fattened up with fish. Our guide told us to be absolutely quiet so as not to alarm him. He was resting on a piece of wood in a yard. He was distinctive and beautiful, with the yellow around his eyes forming a kind of mask. Our group of roughly thirty people peeked over the fence, our cameras making all kinds of noise. I don't know the first thing about penguins, but he looked stressed.

We boarded two school buses and drove to the entrance of the trenches. With an A-frame of green material shielding our heads, and walls of earth on both sides of us, it felt like we were entering the underworld. After walking a half kilometer or so through the trenches, our guide pointed to a lone penguin roughly two hundred yards away, just standing there. And roughly fifty yards in almost the opposite direction was a couple. The yellow-eyed penguins are not only deeply afraid of people, they fear each other.

The couple were protecting an egg, and barely moved. As we watched them, the guide explained that they had tagged and named all of the birds. On the walls inside the trench the scientists had posted laminated sheets describing each of the birds and photos to identify them.

Given how endangered they are, Penguin Place closely tracks their reproductive success. Tash, a fifteen-year-old female, had successfully raised
seven chicks. Jim, a twenty-five-year-old male, had raised twenty-one.
But Tosh, whose photo showed him looking down and to his left, perhaps
a bit dejected, had not raised a single chick, despite being sixteen years old.
There was a gay couple among the penguins, our guide told us. A scientist
had given them an egg, which they successfully hatched into a chick and
raised as their own.

Afterward, we went to the local visitor center to watch a video and read
the exhibits. On the wall was an image of a decaying body of a dead albatross seabird, its stomach filled with plastic garbage. But one of the exhibit's
videos indicated the major causes of albatross mortality were fishing boats
and invasive predators, not plastic.

The video was accurate. In the 1970s and 1980s, fishermen had used long
fishing lines with thousands of baited hooks. The albatrosses would eat the
bait and get snagged by the hook and die. Rabbits, cows, pigs, and cats have
negatively impacted the albatross population as well, and scientists believe
cats and pigs caused the local extinction of the southern royal albatross on
Auckland Island, while also preventing the species' return.[65]

As for climate change, if it were the only threat to the species, scientists
say, the penguins would likely be okay, while at least one species of albatross
does better in warmer water. "Unlike climate change, these factors could
be managed on a regional scale," one of the penguin scientists noted.[66] In
2017, scientists published research finding that "illegal fishing bycatch overshadows climate as a driver of [black-browed] albatross population decline."
And, in contrast to the sea turtles, the scientists found that warmer sea surface temperature "favors high breeding success."[67]

As many of the sea turtles scientists studied off the southern coast of Brazil were killed by fishing as were killed by plastic waste.[68] "We have massive losses of sea turtles to commercial fisheries and poaching," Figgener
explained.[69] "Over ten years a little more than half a million olive ridley turtles have died in fishing nets, and that's just in the economic zone, and so
we don't know about any of the things that happen in international waters.
Probably millions of turtles die every year in fisheries."

The habitat of olive ridleys, notes the IUCN, has been lost to coastal development, aquaculture ponds, and stress from a growing human population.[70]

As such, the intense media and public focus on plastic, like the intense focus on climate change, risks distracting us from other equally important—perhaps more important—threats to endangered sea life, which may be easier to address than climate change or plastic waste.

For example, overfishing, according to the Intergovernmental Panel on Climate Change, "is one of the most important non-climatic drivers affecting the sustainability of fisheries."[71]

The amount of fish and fish products for human consumption has increased from 11 percent in 1976 to 27 percent in 2016, and is projected to increase an additional 20 percent by 2030. According to the FAO, "Since 1961, the average annual increase in global apparent food fish consumption (3.2 percent) has outpaced population growth (1.6 percent) and exceeded consumption of meat from all terrestrial animals, combined (2.8 percent)."[72]

According to the IUCN, forty-two species of sharks are critically endangered, directly threatened by fishing. And apex predators like dolphins and sharks are slow to reproduce. Their populations cannot withstand such large losses.[73]

As for sea turtles, their direct killing by humans remains a threat. "There's still a lot of countries worldwide where turtle meat, shell, fat are still consumed," said Figgener. "There are beaches where they take literally 100 percent of every egg that is laid on that beach, preventing a new generation from being produced. They also take the nests."[74]

7. Plastic Is Progress

Today, Figgener worries that straws distract us from dealing with the root of the problem. "I don't want the corporations to feel like they're getting off easily just by eliminating plastic straws." She added, "I hope that in five years time, we don't even need to discuss plastic straws because there are too many alternatives." [75] Figgener said that in Germany they often use glass instead of plastic. [76]

But are the alternatives to fossil-based plastics really better for the environment?

Certainly not in terms of air pollution. In California, banning plastic bags resulted in more paper bags and other thicker bags being used, which increased carbon emissions due to the greater amount of energy needed to produce them. [77] Paper bags would need to be reused forty-three times to have a smaller impact on the environment. [78] And plastic bags constitute just 0.8 percent of plastic waste in the oceans. [79]

Glass bottles can be more pleasant to drink out of, but they also require more energy to manufacture and recycle. Glass bottles consume 170 to 250 percent more energy and emit 200 to 400 percent more carbon than plastic bottles, due mostly to the heat energy required in the manufacturing process. [80]

Of course, if the extra energy required by glass were produced from emissions-free sources, it wouldn't necessarily matter that glass bottles required more energy to make and move. "If the energy is nuclear power or renewables there should be less of an environmental impact," notes Figgener. [81]

As for bioplastics, they do not necessarily degrade faster than ordinary plastics made from fossil fuels. Some bioplastics, including cellulose, are just as durable as plastics made from petroleum products. While bioplastics biodegrade more quickly than fossil plastic, they are not reused as often as ordinary plastics, and they are more difficult to recycle. [82] The lack of reuse and recycling infrastructure reduces the resource-productivity of bioplastics, increasing both their environmental impact and economic cost. [83]

"People just assume that because it's 'bio' it means it's somehow better," said Figgener, "and it's just not. I mean, it depends also on where the raw materials come from. Just because it is made from cane sugar it is not necessarily biodegradable." [84]

A study of the life cycle of bioplastics made from sugar found higher negative respiratory health impacts, smog, acidification, carcinogens, and ozone depletion than from fossil plastics. When sugar-based bioplastics decompose, they emit more methane, a potent greenhouse gas, than fossil plastics. As a result, decomposing bioplastics often produce more air pollution than sending ordinary plastics to the landfill.[85]

And because bioplastics come from crops grown, rather than the resin waste product from the oil and gas industry, they have large land use impacts, just like biofuels have—from corn ethanol in the United States to palm oil in Indonesia and Malaysia, where they have destroyed the habitat of the endangered orangutan, one of the great apes.[86]

Plastics are made from a waste by-product of oil and gas production and thus require no additional land to be used. By contrast, switching from fossil plastics to bioplastics would require expanding farmland in the United States by 5 to 15 percent. To replace fossil plastic with corn-based bioplastic would require thirty to forty-five million acres of corn, which is equivalent to 40 percent of the entire U.S. corn harvest, or thirty million acres of switchgrass.[87]

Figgener told me she hopes that companies will develop better alternatives within five years. After I expressed skepticism, Figgener admitted, "The pace [the companies] are trying to change is just too slow for me and my turtles. I'm just a little impatient probably."[88]

8. Waste Not, Want Not

The plastics parable teaches us that we save nature by *not* using it, and we avoid using it by switching to artificial substitutes. This model of nature-saving is the opposite of the one promoted by most environmentalists, who focus on either using natural resources more sustainably, or moving toward biofuels and bioplastics.

We must overcome the instinct to see natural products as superior to artificial ones, if we are to save species like sea turtles and elephants. Consider how dangerous that instinct was in the case of tortoiseshell.

Rapid economic growth made the Japanese middle class rich by global and historic standards, and it increased their desire for luxury items, including natural tortoiseshell, much of which Japan sourced from Indonesia.

The Convention on International Trade in Endangered Species of Wild Fauna and Flora (CITES) finally banned the trade of hawksbill turtles in 1977.

Japan initially refused to join the ban and only relented in 1992.[89] Scientists estimate that during the 150 years of tortoiseshell trade, a shocking 75 percent of all tortoiseshell was traded during the single fifteen-year period between 1970 and 1985. The Japanese were responsible for a significant amount of that trade.[90]

As such, artificial substitutes are necessary but not sufficient to save wildlife like the hawksbill sea turtle and African elephants. We must also find a way to train ourselves to see the artificial product as superior to the natural one.

The good news is that, to some extent, this is already happening. In many developed nations, consumers condemn the consumption of natural products, like products made from ivory, fur, coral, and tortoiseshell.

Humankind is thus well-prepared to understand an important, paradoxical truth: it is only by embracing the artificial that we can save what's natural.

Toward the end of our conversation, Christine Figgener and I argued over her proposal that big companies like Coca-Cola provide the responsibility for waste management in poor nations like Nicaragua.

"If you have an unstable political situation," she asked me, "who is going to take care of the waste management?"

"Well, you have to have a functioning government, obviously," I said.

"Nicaragua is the best example," she said. "I mean, how many changes of government? African countries, how many changes of government? You always want to put it back on the government [but] poor countries often do not have a stable political situation."

"So are you going to have each company do it, rather than have a single waste management system?" I asked.

"In those countries who don't have that many choices, most of the things are produced by either Coca-Cola or PepsiCo, or maybe Nestlé. That's like two or three companies at most. So they would have to take responsibility for it. The first stepping stone might be to work together to circumvent the government, which often is corrupt."

"So we're gonna say that because your government is so messed up," I asked, "we're gonna make the companies—"

"Do you really believe that the burden should be on the government to pay for waste management that is created by companies?" she asked me.

"Everywhere in the world we do [waste collection] the same," I replied. "You're saying that to solve the plastic waste problem poor countries should do it differently. I'm not sure I understand why other than that you think the governments are corrupt."

"But it's still on the consumer," she said. "If you think about it, it's crazy. I mean, you're paying for trash that is produced by companies, and you can't even circumvent it because a lot of times there is no alternative."

"If you get Coca-Cola to pay for [waste collection] are they not going to pass that cost on to the consumer in the form of higher prices?" I asked.

"Yes! And then, what, people will consume less Coca-Cola? What would be so bad about that?" she asked.

"You want people to consume less Coca-Cola?" I asked. "Because I thought what you wanted was for them to have waste management."

"Well, that's reduction in a different kind of way," she said. "Because maybe it's not just gonna be about inconvenience or convenience, right?"

"I thought we were trying to solve the plastic waste problem," I said.

"I always said I want to reduce. And whatever's left over I want to have responsibly managed," she explained.

"But the big difference in terms of the problem we're all concerned about is whether or not you have a waste collection and management," I said. "It sounds to me like your impatience has led you to look for a solution that you think will be faster and easier."

"Countries in Africa, Central America, and Asia are not doing such a good job having to do with the level of poverty, corruption, and the level of instability of government," said Figgener. "So whatever works in Europe won't always work in those countries."[91]

Though we differed over solutions, I appreciated where Figgener was coming from. When I first went to Nicaragua in the late 1980s, I was horrified by the trash strewn everywhere. I remain bothered by the plastic waste I see when traveling in poor nations.

As a conservationist, there are few things more demoralizing than hiking or swimming to a place of great natural beauty only to discover plastic waste that has either been left behind by thoughtless people, or has migrated there through rivers and oceans.

But for the people who are often struggling to survive in poor and developing nations, there are many things more demoralizing than uncollected waste. In Delhi, India, in 2016, I visited a community right next to one of the city's main dumps. Even wearing a mask and goggles I could hardly tolerate the putrid smell. But the people I interviewed were understandably more focused on scavenging enough scrap metals and other materials in order to eat that night than they were on the smell.

Economic development brings waste management. In early 2020, China's top economic planning agency put out a five-year plan to reduce the production and use of plastic. By the end of 2020, supermarkets, malls, and food delivery services in China's largest cities will no longer use plastic bags. Notably, China is doing so long after having created a waste collection and management system.[92]

For poor nations, creating the infrastructure for modern energy, sewage, and floodwater management will be a higher priority than plastic waste, just as they were for the United States and China before them. The lack of a system to collect and manage *human* waste through pipes, sewers, and purification systems poses a far greater threat to human health. The lack of a floodwater management system is a far greater threat to homes, farms, and public health than the lack of a waste system, as we saw in the Congo. And, as the next chapter shows, the lack of a modern energy system poses one of the greatest threats to both people and endangered species in poor nations.

4

THE SIXTH EXTINCTION IS CANCELED

1. "We're Putting Our Own Survival in Danger"

When the more than six million visitors enter the American Museum of Natural History in New York each year, they are greeted by an imagined prehistoric encounter between predator and prey: an enormous barosaurus protecting its young from a menacing allosaurus.

There is also a more ominous message for visitors on a bronze plaque in the Theodore Roosevelt Rotunda, the museum's grand entrance hall. "Five major worldwide extinction events have struck biodiversity since the origin of complex animal life some 535 million years ago," the plaque reads. "Global climate changes and other causes, probably including collisions between the Earth and extraterrestrial objects, were responsible for the mass extinctions of the past. Right now, we are in the midst of the Sixth Extinction, this time caused solely by humanity's transformation of the ecological landscape."[1]

One million animal and plant species are at risk of becoming extinct because of humans, according to a 2019 report from something called the Intergovernmental Science-Policy Platform on Biodiversity and Ecosystem Services (IPBES). The rate of species extinction "is already at least tens to hundreds of times higher than it has averaged over the past 10 million years," said the IPBES summary.[2] Earth could lose 40 percent of all amphibians, 30 percent of marine mammals, 25 percent of mammals, and 20 percent of reptiles, IPBES warned.[3]

The 1,500-page report was prepared by 150 leading international experts on behalf of 50 governments. It was the most comprehensive review of the decline in biodiversity globally, and the threat it poses to humans, to date.[4]

"The loss of species, ecosystems and genetic diversity is already a global and generational threat to human well-being," said the IPBES chairman.

The ultimate victims, many warn, will be us. Elizabeth Kolbert, author of the 2014 book *The Sixth Extinction: An Unnatural History*, writes, "By disrupting these systems—cutting down tropical rainforests, altering the composition of the atmosphere, acidifying the oceans—we're putting our own survival in danger." Said the anthropologist Richard Leakey, coauthor of the 1995 book *The Sixth Extinction: Patterns of Life and the Future of Humankind*, "Homo *sapiens* might not only be the agent of the sixth extinction but also risks being one of its victims." [5]

Claims that the extinction rate is accelerating and that "half a million terrestrial species . . . may already be doomed to extinction" rest upon something called *species area model*. Conservation biologists Robert H. MacArthur and E. O. Wilson created the model in 1967. This model rests on the assumption that the number of new species that migrate to an island would decline over time. The idea was that as more species competed for declining resources, fewer would survive. [6]

Fortunately, the assumptions of the species area model proved to be wrong. In 2011, the British scientific journal *Nature* published an article titled "Species-Area Relationships Always Overestimate Extinction Rates from Habitat Loss." It showed that extinctions "require greater loss of habitat than previously thought." [7]

Around the world, the biodiversity of islands has actually doubled on average, thanks to the migration of "invasive species." The introduction of new plant species has outnumbered plant extinctions one hundred fold. [8] The "invaders" didn't crowd out "natives," as Wilson and MacArthur feared.

"More new plant species have come into existence in Europe over the past three centuries than have been documented as becoming extinct over the same period," noted a British biologist. [9]

Kolbert acknowledges the failure of the species area model. "Twenty-five years later it's now generally agreed that Wilson's figures . . . don't match observation," she writes. [10] Kolbert says the model's failure "should be chastening to science writers perhaps even more than to scientists." [11] And yet it was not chastening enough for her to modify her book's title claim.

In truth, nobody needed to know the model's fine workings to know it was wrong. If the species area model were true, then half of the world's species should have gone extinct during the last two hundred years, notes an environmental scholar.[12]

2. Exaggerating Extinction

It turns out that IPBES is not the principal scientific organization studying species, extinctions, and biodiversity. That status belongs to the International Union for Conservation of Nature (IUCN), and it says 6 percent are critically endangered, 9 percent are endangered, and 12 percent are vulnerable to becoming endangered.[13]

The IUCN has estimated that 0.8 percent of the 112,432 plant, animal, and insect species within its data have gone extinct since 1500. That's a rate of fewer than two species lost every year, for an annual extinction rate of 0.001 percent.[14]

The huge increase in biodiversity during the last 100 million years massively outweighs the species lost in past mass extinctions. The number of genera, a measure of biodiversity more powerful than species count alone, has nearly tripled over the course of this time period.[15] After each of these past five mass extinctions, the biodiversity in the fossil record dips between 15 to 20 percent, but each extinction is followed by much larger growth.[16]

Some say the erroneous claims of a sixth mass extinction undermine conservation efforts. "To a certain extent they're claiming it as a way of frightening people into action when in fact, if it's actually true that we're in a sixth mass extinction, then there's no point in conservation biology," noted one scientist. "People who claim we're in the sixth mass extinction don't understand enough about mass extinctions to understand the logical flaw in their argument."[17]

Conservationists, it turns out, are skilled at maintaining small populations of animals, from yellow-eyed penguins of New Zealand to mountain gorillas of central Africa. The real challenge is expanding the size of their populations.

It's not the case that humankind has failed to conserve habitat. By 2019, an area of Earth larger than the whole of Africa was protected, an area that is equivalent to 15 percent of Earth's land surface.[18] The number of designated protected areas in the world has grown from 9,214 in 1962 to 102,102 in 2003 to 244,869 in 2020.[19]

The same is true in a part of Congo, Uganda, and Rwanda known as the Albertine Rift. The area under protection there rose from 49 percent to 60 percent between 2000 and 2016.[20]

The real problem is not extinction but rather the decline in animal populations and their overall habitat. The populations of wild mammal, bird, fish, reptile, and amphibian species declined by roughly half between 1970 and 2010. The worst impacts were in Latin America, which saw an 83 percent decline in wild animal populations, and in South and Southeast Asia, which saw a 67 percent decline.[21]

This reality has led some environmentalists to claim species are threatened by fossil fuels and economic growth. In 2014, the Oscar-nominated documentary film *Virunga* depicted the possibility of oil-drilling in Virunga Park as a major threat to the mountain gorilla, and thus mountain gorilla tourism.

But *Virunga* was misleading. "The areas where the gorillas are have never had prospects of oil," primatologist Alastair McNeilage, of Wildlife Conservation Society, told me. McNeilage first went to Uganda in 1987 to study butterflies.

"The gorillas are on an escarpment so there is no danger or desire of drilling or disturbing the area where the gorillas live," he said. "And nobody makes that clear. A lot have fought oil companies in the name of the gorillas, but the oil companies just aren't interested in those areas."

The real threat to the gorillas and other wildlife isn't economic growth and fossil fuels, I learned during my visit in December 2014, but rather poverty and wood fuels. In the Congo, wood and charcoal constitute more than 90 percent of residential primary energy. "The place where gorillas are located," noted Caleb on our phone call, "is near villages that need charcoal for cooking."[22]

Indeed, when Helen and I arrived at the lodge in Virunga Park, we could see from a distance the smoke from several fires burning inside the park.

3. Wood Kills

Nobody knows for certain if Senkwekwe, a 500-pound silverback gorilla, smelled, heard, or saw the men who killed him and the four females in his group in July 2008. Even if Senkwekwe had, there was no reason he would have been alarmed. After all, he and the others in his twelve-member family had been habituated to the smell of conservation biologists, park rangers, and tourists.

Virunga Park rangers discovered their bodies the next day. The rangers quickly surmised that the killers were not seeking gorilla body parts. After all, they had not cut off their hands or heads. Nor were the men seeking gorilla babies for a foreign zoo; they found a traumatized infant gorilla left behind, cowering in the jungle. The gorillas appeared to have been assassinated by the mafia.

It wasn't the first such killing. The month before, park rangers found a female gorilla who had been shot in the head with her infant, still alive, clinging to her breast. Another female gorilla went missing and was presumed dead. Men had killed seven mountain gorillas in all.

Local villagers carried the gorillas, including the giant silverback Senkwekwe, on a homespun stretcher. Some wept.[23]

A few months later the director of Virunga Park was charged with taking bribes to turn a blind eye to charcoal production in the park. The killings appeared to be retaliation by the charcoal mafia after the park director, under pressure from European conservationists, stepped up efforts to stop charcoal production.[24]

People prefer charcoal for cooking because it is lighter, burns cleaner, and does not become infested, like wood, with insects. And charcoal is labor-saving: you can put a pot of beans onto a charcoal fire and go do something else. Unlike with a wood fire, you don't have to constantly blow on or fan the flames.

Whole parts of Virunga Park had been taken over by the charcoal makers, who produce charcoal for the two million people in the city of Goma.
To make charcoal, you slow-roast wood underground for three days. At the
time of the gorilla killings, the charcoal trade was worth $30 million an-
nually, while gorilla tourism brought in just $300,000. By the early 2000s,
25 percent of the old-growth, hardwood forest in the southern half of
Virunga National Park had been lost to charcoal production.[25] By 2016, the
value of the charcoal trade had risen to $35 million annually.[26]

Ninety percent of the wood harvested from the Congo Basin is used for
fuel. "Under a 'business as usual' scenario," concluded researchers in 2013,
"charcoal supply could represent the single biggest threat to Congo Basin
forests in the coming decades."[27]

Caleb agreed. "The only place people have to get wood from is Virunga
National Park," he told me in 2014. "In such a situation you cannot expect
the gorillas will remain safe."

The Congolese government named Emmanuel de Merode the new director of Virunga Park a few months after the charcoal mafia had killed Senkwekwe. Merode is a Belgian primatologist in his late thirties. He got the
job by offering the Congolese government a vision of economic development for communities around the park, financed by European governments
and American philanthropist Howard Buffett, son of the legendary investor
Warren Buffett.[28]

Merode's vision centered on building a small hydroelectric dam, schools,
and a facility to make soap from palm oil. The European Union, Buffett Foundation, and other donors contributed more than $40 million to
Merode's efforts between 2010 and 2015.[29]

"What Emmanuel is doing is impressive and admirable," said Michael
Kavanagh, who has lived in and reported on the Congo for years. "A
four-megawatt dam doesn't sound like much, but for this world, it's huge.
Emmanuel will always say that what we're doing now in Congo is harvesting palm oil and sending it to Uganda to process and ship back, and that's
insane. If Congo had power, it could build those factories and provide those
jobs."[30]

One of the benefits of the dam was that it would reduce the government's economic incentive to drill for oil in Virunga Park. "Congo is run by a small group of elites," said Kavanagh, "and if you can incentivize them to not explore oil, then it would be possible for them not to."

One day, Caleb and I visited the hydroelectric dam under construction near the town of Matebe. We met the twenty-nine-year-old Spanish engineer, Daniel, who was overseeing the dam construction. Caleb became boyish in his enthusiasm. "Once this project is built," Caleb said, "Buffett will be like Jesus." As we walked from Daniel's office to the dam, Caleb held Daniel's hand, which is something Congolese men do when they are friends.

I asked Daniel if he was married. "Yes, to my work!" he laughed. "This construction is my wife and mistress and children!" Daniel said that he was bringing the project in on time and within budget.

The people we interviewed around Virunga Park knew about Merode's plans to build a dam and were excited to get electricity, which they said they would use for lighting, cell phone charging, ironing, and electric stoves.

And yet the threat of violence remained ever-present, including when Helen and I visited in 2014. Earlier that year, Merode had left the courthouse in Goma and was driving back to Virunga Park in his Land Rover. He was all alone except for the AK-47 at his side. About halfway to park headquarters, Merode turned a corner and saw a gunman 200 meters away. "As I approached, I saw him raise his rifle, and I saw two other men crouched in the forest," Merode told a reporter. "At that point, the bullets started hitting the vehicle so I ducked."[31]

Merode was hit, as was the engine or circuit board of his Land Rover, which stalled. Merode grabbed his AK-47 and tumbled out the door. Once in the bushes, he started firing wildly. After the would-be assassins were out of sight, Merode stumbled onto the road, and waved for help. Despite his being covered in blood, aid agency cars sped past him.

Two farmers on a motorcycle finally stopped, chucked their crops to the side, and strapped him to the back. "The really painful part was being carried on the back of the motorbike on the Congolese roads, which are extremely bumpy," Merode recalled.[32]

The farmers took him to an army checkpoint and loaded him onto a truck. But then the truck ran out of gas. "I had to reach into my pocket and give them twenty bucks," said Merode, which was covered in blood.[33] Then the truck broke down. He was transferred to another army vehicle, which took him to a park vehicle. Finally, he arrived at the hospital. The forty-four-year-old Merode had emergency surgery and somehow survived.

All of which raises a question: why, if Merode was doing so much for the people around Virunga Park, had somebody tried to kill him?

4. Colonial Conservation

In 470 BCE, Hanno the Navigator, an explorer from Carthage, the North African city near modern day Tunisia in North Africa, mistook a group of gorillas for humans. In the foothills of the mountains of what is today Sierra Leone, Hanno's local guides led him to an island in the middle of a lake "inhabited by a rude description of people," wrote Hanno. "The females were much more numerous than the males and had rough skins. Our interpreters called them Gorillae."

Hanno decided he needed some as specimens and so he and his men gave chase.

> We pursued but could take none of the males; they all escaped to the top of precipices, which they mounted with ease, and threw down stones; we took three of the females, but they made such violent struggles, biting and tearing their captors, that we killed them, and stripped off the skins, which we carried to Carthage: being out of provisions we could go no further.[34]

Hanno's fellow Carthaginians must have been as fascinated by the creatures as he was, because when the Romans invaded Carthage 300 years later, the skins were still on display.

Fascination with mountain gorillas grew among European colonizers in the nineteenth and twentieth centuries. The mountain gorillas of central Africa got their species name (*Gorilla gorilla Berengei*) from a German officer who killed two of them in 1902. In 1921, a Swedish prince killed fourteen mountain gorillas. Between 1922 and 1924, an American killed and captured nine. Throughout the next twenty-five years, hunters killed about two gorillas annually, ostensibly for scientific research.[35]

A turning point came when a naturalist working for the American Museum of Natural History killed five gorillas in 1921 and was consumed with regret. "As he lay at the base of the tree," wrote Carl Akeley, "it took all of one's scientific ardor to keep from feeling like a murderer. He was a magnificent creature with the face of an amiable giant who would do no harm except perhaps in self-defense or in defense of his friends."

Akeley traveled to Belgium and met with King Albert. Coincidentally, the king had been to the newly opened Yellowstone National Park in the United States and was inspired by Akeley to protect the gorillas by giving them their own park. He called it Virunga National Park.[36]

But the hunting didn't stop. Zoos in Europe, the United States, and other nations wanted gorillas to display to their publics. But Akeley was right about gorillas: they give their lives to defend children and family members. In 1948 alone, sixty gorillas were killed trying to prevent eleven of their infants from being taken for sale to foreign zoos.

Even though it was foreigners, not local people, who killed mountain gorillas, European colonizers sought to expel the locals from the areas they had designated for parks. American conservationists including Sierra Club founder John Muir successfully advocated for governments to evict indigenous people from Yellowstone and Yosemite parks in the 1860s and 1890s. King Albert of Belgium brought that same model to the eponymously named Albertine Rift, where many people already lived and, indeed, humankind had been born, 200,000 years ago.

The Albertine Rift is spectacularly beautiful and biodiverse, with forests, volcanoes, swamps, erosion valleys, and mountains with glaciers. It is home to 1,757 terrestrial vertebrate species, half of all of Africa's birds, and 40 percent of its mammals.[37] Mountain gorillas today can be found in Rwanda's Volcanoes National Park, Congo's Virunga Park, and Uganda's Bwindi Impenetrable National Park.

But creating those parks involved the eviction of local communities, which has led to disputes and violence. "Virunga Park was created during colonial times," noted Helga Rainer, a conservationist with the Great Ape Program. "Land is the resource at the heart of the conflict, and it was European colonialists who changed or confused land tenure systems."[38]

Scientists estimate that between five and "tens of millions" of people have been displaced from their homes by conservationists since the creation of Yosemite National Park in California in 1864. A Cornell University sociologist estimated that Europeans created at least fourteen million conservation refugees in Africa alone.[39]

Displacing people from their lands wasn't incidental to conservation but rather central to it. "The displacement of people who herded, gathered forest products, or cultivated land was a central feature of twentieth century nature conservation in southern and eastern Africa and India," noted two scholars.[40]

The Ugandan government and conservationists expelled the Batwa from Uganda's Bwindi Park in the early 1990s. Their poaching of wild animals for meat threatened the gorillas, conservationists believed.[41] "Whole societies like the Batwa of Uganda," writes Mark Dowie, the author of the 2009 book *Conservation Refugees*, have been "transformed from independent and self-sustaining to deeply dependent and poor communities."[42]

Conservation refugees can suffer from very high stress and poor health. Scientists took saliva samples from eight thousand indigenous people in India who had been evicted from their villages by the government to create a lion sanctuary. The scientists discovered that the people suffered from shortened telomeres, a sign of premature aging from stress, despite having been compensated and given new homes.[43]

Something similar happened to Uganda's Batwa. Since they had depended on harvesting meat, honey, and fruit from within Bwindi Park for centuries, they did not know how to create a new life as farmers. "Consequently," noted researchers ten years later, "the other community members have taken advantage of the Batwa's deprivation to exploit them."[44]

On my return to the United States from the Congo, I interviewed Francine Madden, a conservationist who worked to reduce human–wildlife conflicts in Uganda in the early 2000s. I told her about how the baboons had eaten Bernadette's sweet potatoes, and about the widespread complaints I had heard of crop-raiding.

"People go to parks asking for compensation for their animals coming out and destroying their crops," said Madden, "which in many ways is quite reasonable. If your neighbor's cows came and ate your crops, you'd want to be compensated. But few parks can set up compensation systems [that are] manageable."

Other conservationists agreed that crop-raiding was a major problem. "In Uganda, crop-raiding was one of the biggest issues conservationists had to deal with," McNeilage said, "and was the biggest source of conflict with community, along with access to resources of one kind or another. So it's not surprising that that's what people were telling you."[45]

In 2004, researchers with Wildlife Conservation Society found that two-thirds of the people they surveyed around Virunga Park reported having their crops eaten by baboons once a week. Significant percentages reported crop-raiding by gorillas, elephants, and buffalo.[46]

Another primatologist, Sarah Sawyer, studied gorillas both in Uganda and in Cameroon, which, like Congo, was too poor to benefit from ecotourism. "The idea of conservation that local people [in Cameroon] held was that you get kicked off your land and you get no money. While I got used to being called 'white man,' both in Cameroon and other countries, at my field site we were called 'conservation,' instead, in a very derogatory way. And it hurt. 'We don't want conservation here,' they would say.[47]

"Talking about conservation there felt totally wrong because the local people felt like conservation was simply a way to rob them of their resources. Talking to them about conserving gorillas was not talking to them in their own language. This reminded me of what I had read in grad school about conservation as neocolonialism."[48]

5. "Fighting the Locals Is a Losing Battle"

By the 1990s, it appeared most conservationists had internalized the lesson that "fighting the locals is a losing battle." Conservation non-governmental organizations (NGOs) made strong statements declaring their support for protecting local people. They worked to improve their treatment of the people who live in or near protected areas. International development agencies like the United States Agency for International Development (USAID) have spent millions seeking to protect impacted indigenous people and others living near parks and protected areas.

In 1999, the International Union for Conservation of Nature (IUCN) officially recognized indigenous people's right to the "sustainable traditional use" of their land. In 2003, the World Parks Congress passed a "do no harm" principle and promised to financially compensate poor and developing nations protecting natural and wild areas like Virunga. In 2007, the United Nations approved a strong statement of support for indigenous rights impacted by conservation efforts.

Today, conservationists point to efforts by NGOs to promote alternatives to charcoal and to the success of gorilla tourism as proof that conservation can pay for itself and reduce pressure on habitats.[49]

But, says primatologist Sawyer, "there are few species like the mountain gorilla that relatively wealthy foreigners will spend thousands of dollars to see." In Rwanda today, it costs $1,500 to view gorillas for one hour.[50] "And even with those species, the ecotourism opportunity is lost when there is no infrastructure, security, and economic development."[51]

Wildlife Conservation Society's Andrew Plumptre and his colleagues conducted interviews with people representing 3,907 households around Albertine Rift parks in the early 2000s and found that few benefited from tourism. "When people were asked about the benefits of the forest for themselves, or their communities, tourism ranked very low," they wrote. "Tourism was only perceived as being useful to the country [as a whole]."[52]

Meanwhile, NGO efforts to promote alternatives to charcoal, such as through wood pellets and special stoves, have failed. "The pellets haven't been a great success anywhere," said McNeilage. "I don't know anywhere where they've cottoned on and become popular."[53]

Plumptre agreed. "The World Wildlife Fund has long had a program of planting trees for sustainable use, but it has not had a great impact on the push for charcoal out of the park," he told me. The reason is that the locals used the plantation to make more lucrative poles for building, rather than wood fuel to use as charcoal.[54]

"The need to move to modern fuels is a bone of contention of mine," said Dr. Helga Rainer of the Great Ape Program. "That we still talk about energy-saving stoves is disappointing."[55]

The situation is increasingly desperate. In Plumptre's study of the people around Virunga Park, he and his colleagues found that half reported not having sufficient access to firewood.[56]

The experience of working in places where conservationists aren't welcome made conservation scientist Sawyer question whether her efforts were worth it. "When I first got in [conservation biology] I thought 'find the most vulnerable areas and protect them,' " she said. "Now I see that it might not be a lost cause, but in many places, conservation is the last priority. You have to be careful what situations you try to conserve.

"The places where you can get the biggest bang for the buck aren't always the places where things work. I thought, 'You can't lose species from the planet,' but then you have to ask, 'Is it worth saving this species at the cost of the social, political, and economic cost on the human side?' Or do we say, 'We hope the species makes it, but there are other priorities in this area right now'?"[57]

Plumptre worries that the management of Virunga Park by foreigners will weaken local support. "My problem is when it's seen as outsiders coming in and taking care of everything, you are at risk of the park being seen as an ex-pat's playground. And so when threats come, there may be little support for them."

Merode, the Virunga National Park director, is himself a Belgian prince who married into conservation royalty. His wife, Louise, is the daughter of Richard Leakey, who warned of a sixth extinction, and granddaughter of Louis Leakey, the primatologist who uncovered humankind's evolutionary origins in the Albertine Rift.[58]

Merode's first major act upon becoming director was to crack down on the small farmers planting within Virunga Park. Recall that, due to lack of fer-

tilizer, roads, and irrigation, people in the area are desperate for land. "He
reduced the ranger force from 650 to 150 and then tried to go to war with
people at gunpoint, and force them out," Plumptre said.[59]

Merode's hard line surprised the conservation community. "He did his
whole PhD on community conservation and so it was quite a shock to us
when he militarized and forced people out of the park at gunpoint," said
Plumptre. "Historically he had worked with communities."[60]

Merode's crackdown backfired. "They ended up with many more people in
the park afterwards," Plumptre said. "That was a big mistake. The idea was to
use the same funds to pay a higher salary, but they took the number [of park
rangers] down to such a low number that they couldn't control the park."[61]

Madden, the scientist who studied crop-raiding, said aggressive approaches to conservation had, in the past, resulted in local people killing
wildlife. "If people don't feel respected and recognized, they will retaliate,
and may retaliate in a disproportionate way," she said. Journalists and scientists have documented such behavior for decades.[62]

Plumptre said the animosity Merode created among local residents may
have resulted in the deaths of as many as 250 elephants. "We did a census
and found just thirty-five elephants left in Virunga Park, down from 300
elephants in 2010," Plumptre said. "Some may have moved to Queen Elizabeth Park in Uganda, but we haven't seen 240 or 250 new elephants in
Uganda, and we did surveys on both sides of the border."[63]

"It's impossible for a person to retaliate against the government," Madden said. "But you can retaliate against the wildlife that the government is
trying to protect. It is a symbolic retaliation. It's a 'Screw you' psychological
retaliation."

I asked Plumptre why Merode had opted for such a heavy hand against
locals. "He felt there were a lot of people living in the park who shouldn't
be there, and one of the things he had to do was try to stamp control on the
park," he said. "I think he didn't realize how much he needed the support of
the local traditional chiefs. They were involved in illegal activities, and he
probably thought if he could get away with not having to deal with them,
that would be much easier."[64]

Madden believes that the personalities of many conservation scientists undermine their relationships with locals. Conservation scientists "are highly introverted and analytical," she said. "They want to make big decisions by themselves in a corner with people who think like them, and then give it to people who experience it as an imposition. It's not like they try to be assholes. They want to get it right. They don't have the same values, and it looks disrespectful and blows up."

Between 2015 and 2019, crop-raiding by Virunga Park animals had worsened. In late 2019, a local farmer told Caleb, "We need the park to protect our farms by building the electric fence to prevent animals from coming to raid our crops."

"The park has to take responsibility to keep the animals out of their farms," said Madden. "There was a full-scale insurrection against the crop-raiding when I was there. Gorillas were getting killed and sent over to Congo for barbeque. People were getting hurt, with half of their thigh ripped out. It was on the verge of anarchy."

Cameroon, too, stands as a warning to Congo that things can go from bad to worse, says primatologist Sarah Sawyer. "When I arrived at the reserve [in Cameroon] it was being discussed as deserving extra protection," she said. "When I left, it was being discussed as a potential logging concession."

Anyone exposed to books like *The Sixth Extinction*, reports like IPBES's 2019 Platform, and films like *Virunga*, along with the publicity they generate, might reasonably think that protecting wildlife requires restricting economic growth, strictly enforcing park boundaries, and fighting oil companies. Worse, they might give developed-world audiences the impression that African wildlife parks are best run by Europeans.

"When Virunga Park was run by the Congolese," said Plumptre, "there were more large mammals, less political problems, and fewer people invading the park, even though they didn't have anywhere near the money coming in. Today, there's infrastructure, but the mammal numbers have crashed and there is a lot of cultivation in the park that wasn't there five or six years ago."

6. The 800-Pound Gorilla

In 2018, Virunga Park closed after an armed group murdered a twenty-five-year-old park ranger and kidnapped three people, two of whom were British tourists. The kidnappers released the tourists and their Congolese driver two days later. The victim was one of only twenty-six female rangers working at the park at the time.[65]

After closing for eight months, Virunga Park reopened to tourists in early 2019, but only a few weeks later, a local armed group killed another park ranger.[66]

As for Bernadette, armed men appeared with the apparent intention to kidnap or kill her husband, and they were forced to flee. I found this out after I hired Caleb in late 2019 to do a follow-up interview with her. After some searching, Caleb was able to track her down.

They met in person, and Bernadette told him that her husband had been targeted for political reasons. "My husband is the grandson to the chief," she said, "whom bandits took into the bush and killed." His son took over but the bandits killed him too. The brother to the older assassinated chief took over but the bandits killed him too. The next in the line of succession fled to Goma. It was then that the bandits came for Bernadette's husband.[67]

"We awoke to people trying to break in. My husband was scared and I was still deep asleep. He got up slowly and went to our daughter's room. I awoke and called out to him. 'Jackson's dad! Jackson's dad!' Nobody answered. I grew scared.

"Then, I heard them trying to open the door, and I said to myself, 'There is another woman who died at the place of her husband when he was being hunted. I may also get shot in the head.'

"And so I got up slowly and took my baby, who began crying. They heard how I was moving in my daughter's bedroom. I found my husband standing there. I stood up next to him. I don't know why but God inspired him to take the telephone and check what time it was. When they heard the phone ringing, they were afraid and left the door.

"They left the front door and went behind the house and we heard them calling each other. We stayed quiet, no shouting. We stayed sitting the whole night until morning because we were still afraid.

"Three days later, we awoke to the guys again trying to force open the door. We got up and went to stand up in our daughter's bedroom because we thought they might shoot. My husband took out his telephone and started pretending to be talking to somebody and we heard them leave."

The next day, Bernadette's mother in-law urged them to flee. "My husband said, 'No. My father has passed away! How am I going to leave you alone? Who am I going to leave you with?' His mother said, 'We have picked up too much of your brothers' brains already and that is enough. Just go. God will take care of me.' "

Bernadette left seven of their children with her mother in-law, and she and her husband and three children went to work on other people's farms.

When Caleb interviewed her, Bernadette sounded depressed. "In order to survive here you have to work for someone," she told Caleb. "You have to clear or cultivate to get food to eat. I'm suffering with life. We are not overcoming."

When I asked Caleb what could happen to Bernadette if she were kidnapped, Caleb replied, "When they kidnap the women they also rape them. The men they beat. They call your relatives and put you on speaker phone, so your family can hear you being tortured."

But how could such poor people possibly afford to pay a ransom? He said it was by selling their land and borrowing from relatives.

7. Why Congo Needs Fossil Fuels

Ultimately, for people to stop using wood and charcoal as fuel, they will need access to liquefied petroleum gas, LPG, which is made from oil, and cheap electricity. Researchers in India proved that subsidizing rural villagers in the Himalayas with LPG reduced deforestation and allowed the forest ecosystem to recover.[68]

Some conservationists, including the Wildlife Conservation Society's McNeilage, believe it is inevitable that the Congolese government will one

day drill for oil in Virunga National Park, and that the results could be positive. "I think the chances they'll leave the oil underground indefinitely is
pretty minimal. The oil price has gone way down and maybe that's why
they're not moving as fast as they were in Uganda. The estimates were of
two to three billion barrels. Not all of them would be recoverable. But there
could be 30,000 to 60,000 barrels per day and even 250,000 barrels per day.
It could make a big difference to regional fuel needs." [69]

McNeilage's colleague, Plumptre, agrees. "If they had hydro and oil, and
if it can be done in a clean way, to generate electricity and gas to use instead
of charcoal, that would be good for the environment."

McNeilage recognizes that his point of view remains controversial among
conservationists. "There's two lines of thought," he said. "One is that it is
illegal. It's a [United Nations] World Heritage site. In an ideal world, doing
what we can to stop it is the right thing to do.[70]

"The other line of thought—and this is me going out on my own—is
that if it goes ahead anyway, it would be better that [British Petroleum] or
the most responsible oil company do the best job they can with the minimum
impact and maximum benefit for local people and transparency and the rest
of it.[71]

"If you succeed in scaring away Soco," a London-based oil company that
sought to drill inside Virunga Park, "does that mean a worst-case scenario
will come along and you'll have, I don't know what, a company from Asia
somewhere? A less responsible company takes the oil out and doesn't give
two hoots what the world thinks? I think often you get someone worse, who
cares less about the danger." [72]

McNeilage had helped the French oil company Total SA drill sustainably
for oil in Uganda's Murchison Falls National Park, northeast of Virunga
Park, a major park in the Albertine Rift. "I've always come from a background of wanting to build the capacity for the countries I'm working in
to do this work themselves, and understand the importance of it," he said.
"That's fundamentally the right approach." [73]

McNeilage's motivation to work with the oil companies is based on a mix
of realism and humanism. "I realized that we were in crisis in many ways
more than I thought we were. The oil industry has the potential, and the

scale of the income is more than enough to pay for the upkeep of the park. Potentially, they can generate a lot of income if the oil companies involved are committed to protecting the areas they are exploiting. If they do it right, it can do good for publicity, but also harm for publicity, if they make a mess somewhere." [74]

Oil drilling has occurred with surprisingly little impact. "You can find a level of disturbance," McNeilage said. "Animals will tend to move away from the immediate area of any areas of operations. But it's not a large disturbance at the level of the park as a whole if the company takes precautions to minimize impact. And there is a lot they have been able to do to keep impact to a minimum. The impacts we saw on animals would have been transitory." [75]

That said, McNeilage says it hasn't been easy. "The actual practicalities of getting the companies to do what they need has been difficult. It has been frustrating to work with Total because it's a huge bureaucracy and it was sometimes hard to get them to do things that they should have done."

I asked McNeilage how the experience had changed him since he first arrived in the Congo more than a quarter century earlier to study butterflies. "I don't think I'm a fundamentally different person," he said, "just a lot less naïve."

8. Power for Progress

As for the Virunga dam built by Merode with Howard Buffett and EU money, most experts believe it cannot be easily scaled up. "That project is not a sustainable project and can't be copied throughout the country," said Kavanagh, the reporter. "You have a wealthy donor putting up the money. You have a particular situation because of the gorillas and the park and because of Emmanuel. Virunga is practically a fiefdom of ICCN run by Emmanuel. They can get around the things the average company can't get around." [76]

Eventually the dam will provide electricity to twenty thousand people, but that is a drop in the bucket when considering that there are two million people in Goma alone.[77] And the high cost of electricity from the Virunga Park dam means that only the relatively wealthy have been able to afford it. The upfront cost of $292 to connect to the grid is prohibitive for most.[78] Recall that the average annual income is $561.[79]

"It's helping people open factories, garages, and grinding machines, but lower-class people can't afford it," one man told Caleb. "And so it's not stopping the exploitation of the park for charcoal. It's still coming out of the park. The objective was to stop the cutting of trees in the park for making charcoal. But now the electricity is very expensive. As long as it is, people will continue making charcoal from trees within the park."

Experts agree that the easiest and cheapest way for Congo to produce abundant supplies of cheap electricity is by building the long-planned Grand Inga Dam on the Congo River. "You have 100,000-megawatt potential through Inga," Kavanagh said. "You can provide all of Africa with that power."

The Inga would be fifty times larger than the Hoover Dam, which serves eight million people in California, Arizona, and Nevada.[80]

But for cheap electricity and LPG to pay for themselves, and not depend on charitable donations from European governments and American philanthropists, the Congo needs security, peace, and industrialization of the kind that has lifted so many nations out of poverty in the past.

5

SWEATSHOPS SAVE THE PLANET

1. War on Fashion

In fall 2019, during London Fashion Week, hundreds of Extinction Rebellion activists protested the industry's impact on the climate. Some covered themselves in fake blood and lay down in the street. In front of Victoria Beckham's fashion show they held up signs that read, "Fashion = Ecocide" and "Business as Usual Equals the End of Life on Earth."[1] An Extinction Rebellion spokesperson said, "As a race, we are—quite literally—facing the End of Days."[2]

About two hundred Extinction Rebellion protesters staged a fake funeral march complete with a giant coffin and marching band. They walked down the Strand thoroughfare in central London and blocked traffic. They chanted and handed out pamphlets. The activists pointed to estimates suggesting that the fashion industry was responsible for 10 percent of all carbon emissions.[3]

More than a half-dozen protesters in London, wearing blood-red dresses and white face paint, protested in front of H&M, a retailer that sells inexpensive "fast-fashion."[4] A post on Extinction Rebellion's Facebook page read, "The industry still adheres to an archaic system of seasonal fashions, adding pressure to relentlessly create new fashion from new materials."[5]

On its website, Extinction Rebellion wrote, "Globally we produce up to one hundred billion pieces of clothing a year, taking a terrible toll on the planet and people who make them. What's worse, new reports predict the apparel and footwear industry will grow by 81 percent by 2030, putting an unprecedented strain on already devastated planetary resources."

The protests appear to have had an impact. A recent survey found that more than 33 percent of customers switched to brands they believe are more sustainable, and 75 percent indicated environmental sustainability is important to them when shopping for clothes.[6]

Others argue that fashion and other consumer product industries are inherently unsustainable, focused as they are on increasing consumption. "Stopping people consuming is really the only way of having any impact at this point, which is a difficult message for many people to take on board," said an Extinction Rebellion activist. "As a communication tool, fashion is so influential," said another activist. "We all have to put clothes on and that has power."[7]

Environmentalists target a range of consumer products, not just shoes and clothing. In 2011, Greenpeace activists protested Barbie doll maker Mattel in California. In a statement, Greenpeace told *The Washington Post* and other journalists that it targeted Mattel because the entertainment and toy manufacturer sourced from Asia Pulp and Paper, which Greenpeace said bought pulp made from wood taken from Indonesian rainforests.[8]

During the protest, a Greenpeace activist dressed like Barbie drove a pink bulldozer. "Do you think they will let me park this at the mall?" she asked onlookers. An activist climbed onto the roof of Mattel's headquarters in El Segundo, California, and unfurled a banner with the face of a despondent-looking Ken doll that read: Barbie, It's Over. I Don't Date Girls That Are into Deforestation.[9]

But Mattel is hardly a major agent of deforestation. Compared to daily newspapers, Mattel's consumption of paper is minimal. The reason Greenpeace targeted Mattel wasn't because it was a major paper user, but because Barbie is such a recognized brand. Attacking the toy company was sure to attract media attention.

The truth about clothing and other consumer items made in factories in poor and developing countries is actually the opposite of what Extinction Rebellion and Greenpeace claim. Rather than being the main culprit in the destruction of forests, factories have been, and remain, an engine for saving them.

2. Leaving the Farm

In 1996, I left graduate school in Santa Cruz, California, and returned to San Francisco to work on activist campaigns with Global Exchange, Rainforest Action Network, and other progressive and environmental organizations.

At the time, concern over the labor and environmental impact of manufacturing clothing and other products, from Barbie dolls to chocolates, was growing. So we decided to launch what we labeled a "corporate campaign" against one of the largest and most profitable multinational companies in the world. Our model was Rainforest Action Network's boycott against Burger King, which I had raised money for in high school. Our strategy was to target a big, well-recognized brand. We quickly settled on Nike.

At the time, Nike had just started promoting its shoes by linking them with women's empowerment through sports for women and girls. My colleagues at Global Exchange and I decided to focus on women's rights. Already, Global Exchange had brought a Nike factory worker from Indonesia to tour the United States and talk publicly about what life was like in the apparel company's factories, generating extensive national publicity.

We drafted an open letter to Phil Knight, the founder and then-chairman of Nike. We circulated the letter among feminist leaders and provided a copy to the *New York Times*. In the letter, we asked that Nike allow local independent monitors to inspect its factories in Asia and increase wages. For example, its factory workers in Vietnam were making only $1.60 per day at the time.

In fall 1997, the *Times* ran a story with the headline, "Nike Supports Women in Its Ads but Not Its Factories, Groups Say." The reporter wrote, "A coalition of women's groups have attacked Nike as hypocritical for its new television commercials that feature female athletes, asserting that something is wrong when the company calls for empowering American women but pays its largely female overseas work force poorly."[10]

Our campaign seemed like a success. We had generated so much negative publicity that we damaged Nike's brand. As importantly, we sent a message to other corporations that they would be held accountable for conditions in factories they contracted with abroad. "I go back to 1997 to find the first clearly [corporate social responsibility] related event I can recall—the boycott of Nike—that had a real impact on the company," said Geoffrey Heal, a Columbia University Business School professor.[11]

Not everybody agrees that the Nike campaign was a success. Some, like Jeff Ballinger, whose work with Indonesian factory workers dates back to 1988, believes Nike hyped "environmental sustainability" as a public-relations tool to overshadow continued human exploitation. "The grinding outsource paradigm still rules in most low-skill manufacturing," wrote Ballinger[12]

Meanwhile, environmental experts and activists say consumer product companies haven't done anything significant to improve their environmental practices. "The mission of sustainable fashion has been an utter failure, and all small and incremental changes have been drowned by an explosive economy of extraction, consumption, waste and continuous labour abuse," said activists in 2019.[13]

In June 2015, a few months after visiting Congo, Rwanda, and Uganda, I decided to go to Indonesia and see for myself what the situation was like for factory workers there. I hired a twenty-four-year-old Indonesian journalist, Syarifah Nur Aida, who called herself Ipeh. She had reported on labor issues at the factories and had recently uncovered military corruption.[14]

"I was beaten up last year after I covered how military officers had bought up land for a low price," she told me. "It was very scary for my parents, but they never insisted I quit."

Ipeh had arranged interviews with several factory workers for me, one of whom was a twenty-five-year-old named Suparti, who had come from a small village on the coast. Her first job was at a Barbie factory, and her second at a chocolate factory. We met twice, first at Suparti's union office, and again at her home. She wore a bright pink hijab, which she fastened with a large brooch.

"Every Sunday we spent playing in the water, but I never learned how to swim and never fully went in," she said. "I lived in a strict Islamic community where we couldn't even go to social gatherings if men were there."

Trips to the beach were infrequent. "We rarely went to the beach because there was always so much work to do."[15]

After school, Suparti worked alongside her parents and siblings in the fields. "We were poor relative to the other houses in the community. Our house had four rooms and was made of bamboo. We had no electricity or TV. We cooked with rice husks."

Her family grew rice and a little bit of eggplant, chiles, and green beans. They rotated the rice crops with soy to fertilize the soil. She helped her parents tie spinach into bundles to sell at the local market.

Some of the greatest threats Suparti and her family faced were wild animals, disease, and natural disasters. One time, wild dogs got loose in the village. "My parents worried they would eat our rabbits," she explained. "During the avian bird flu epidemic, my parents worried about their chickens but they turned out to be okay.

"Everyone was scared of tsunamis and earthquakes because we lived so close to the sea," said Suparti. "Some people were so scared they moved their stuff to the mountains. But then the volcano Merapi erupted. The rich people who had moved their stuff up into the hillside lost their belongings to the lava. We felt very vulnerable to nature."

Eventually Suparti was drawn to the city. "As a child, I heard from my aunt about what it was like to work in a factory, and I imagined I would work in one. My parents didn't want me to. 'Stay here, do your chores, and wait for a good man to marry you,' they said. My mother in particular didn't want me to go. I explained to them that I wanted to send money back to them."

And so, after she turned seventeen years old, Suparti left home.[16]

3. Manufacturing Progress

When young people like Suparti leave the farm for the city, they must buy food rather than grow it. As a consequence, the declining number of farmers in poor and developing countries must produce even more food.

In Uganda, I had a conversation with a middle-aged woman who worked at our ecolodge, where we went to see gorillas for a second time. I told her that just two out of every hundred Americans are involved in farming, whereas two out of every three Ugandans are farmers.

"How can you grow enough food?" she asked.

"With very large machines," I replied.

For more than 250 years, the combination of manufacturing and the rising productivity of farming have been the engine of economic growth for nations around the world. Factory workers like Suparti spend their money buying food, clothing, and other consumer products and services, resulting in a workforce and society that is wealthier and engaged in a greater variety of jobs. The declining number of workers required for food and energy production, thanks to the use of modern energy and machinery, increases productivity, grows the economy, and diversifies the workforce.

While a few oil-rich nations like Saudi Arabia have achieved very high standards of living without ever having embraced manufacturing, almost every other developed country in the world, from Britain and the United States to Japan to South Korea and China, has transformed its economy with factories.

Increased wealth from manufacturing is what allows nations to build the roads, power plants, electricity grids, flood control, sanitation, and waste management systems that distinguish poor nations like Congo from rich nations like the United States.

Cities, meanwhile, concentrate human populations and leave more of the countryside to wildlife. Cities cover just more than half a percent of the ice-free surface of the earth.[17] Less than half a percent of Earth is covered by pavement or buildings.[18]

As farms become more productive, grasslands, forests, and wildlife are returning. Globally, the rate of reforestation is catching up to a slowing rate of deforestation.[19]

Humankind's use of wood has peaked and could soon decline significantly.[20] And humankind's use of land for agriculture is likely near its peak and capable of declining soon.[21] All of this is wonderful news for everyone who cares about achieving universal prosperity and environmental protection.

The key is producing more food on less land. While the amount of land used for agriculture has increased by 8 percent since 1961, the amount of food produced has grown by an astonishing 300 percent.[22]

Though pastureland and cropland expanded 5 and 16 percent, between 1961 and 2017, the maximum extent of total agriculture land occurred in the 1990s, and declined significantly since then, led by a 4.5 percent drop in pastureland since 2000.[23] Between 2000 and 2017, the production of beef and cow's milk increased by 19 and 38 percent, respectively, even as total land used globally for pasture shrank.[24]

The replacement of farm animals with machines massively reduced land required for food production. By moving from horses and mules to tractors and combine harvesters, the United States slashed the amount of land required to produce animal feed by an area the size of California. That land savings constituted an astonishing one-quarter of total U.S. land used for agriculture.[25]

Today, hundreds of millions of horses, cattle, oxen, and other animals are still being used as draft animals for farming in Asia, Africa, and Latin America. Not having to grow food to feed them could free up significant amounts of land for endangered species, just as it did in Europe and North America.

As technology becomes more available, crop yields will continue to rise, even under higher temperatures. Modernized agricultural techniques and inputs could increase rice, wheat, and corn yields five-fold in sub-Saharan Africa, India, and developing nations.[26] Experts say sub-Saharan African farms can increase yields by nearly 100 percent by 2050 simply through access to fertilizer, irrigation, and farm machinery.[27]

If every nation raised its agricultural productivity to the levels of its most successful farmers, global food yields would rise as much as 70 percent.[28] If every nation increased the number of crops per year to its full potential, food crop yields could rise another 50 percent.[29]

Things are headed in the right direction regarding other environmental measures. Water pollution is declining in relative terms, per unit of production, and in absolute terms in some nations. The use of water per unit of agricultural production has been declining as farmers have become more precise in irrigation methods.

High-yield farming produces far less nitrogen pollution run-off than low-yield farming. While rich nations produce 70 percent higher yields than poor nations, they use just 54 percent more nitrogen.[30] Nations get better at using nitrogen fertilizer over time. Since the early 1960s, the Netherlands has doubled its yields while using the same amount of fertilizer.[31]

High-yield farming is also better for soils. Eighty percent of all degraded soils are in poor and developing nations of Asia, Latin America, and Africa. The rate of soil loss is twice as high in developing nations as in developed ones. Thanks to the use of fertilizer, wealthy European nations and the United States have adopted soil conservation and no-till methods, which prevent erosion. In the United States, soil erosion declined 40 percent in just fifteen years, between 1982 and 1997, while yields rose.[32]

As such, buying cheap clothing, and thus increasing agricultural productivity, is one of the most important things we can do to help people like Suparti in Indonesia and Bernadette in Congo, while also creating the conditions for the return and protection of natural environments, including rainforests.

4. The Great Escape

Suparti was happy and anxious to leave her home in the country for the big city. "I can remember the excitement I felt on taking a bus by myself," she recounted. "We left at 5 p.m. and arrived at 8 a.m. the next day. I only got two hours of sleep because I was so excited. I was met by my aunt and sister, who lived thirty minutes away and worked in a factory.

"My first interview was just two hours later at 10 a.m., and I screwed it up because they asked me my address and I didn't know it! I spent a week at my aunt's house and then I went to an interview at the Mattel factory."

Hundreds of young women had started lining up at 5 a.m. just to get a ticket to enter a lottery, from which candidates would be selected for interviews. Suparti arrived a couple of hours late, after the tickets were already gone, but was able to sneak into the factory with a friend when a security guard wasn't looking.

"Part of the interview was to put clothes and accessories on a Barbie doll," she said. "We were judged on speed. As a girl, I had played with fake Barbie dolls. Plus, I knew this would be the test and was mentally prepared. The other tests included making a ponytail and putting shoes on the Barbie."

The tests started at 10 a.m. Five hours later, they announced who was being hired. "I was faster than others and got the job," Suparti said. "I was happy but not surprised because I thought I could do it."

But the job—as well as the culture inside the Mattel factory—wasn't what Suparti imagined it would be. "There was never any physical abuse, but there was constant yelling," she said, "and I had never in my life been yelled at.

"Javanese people speak slowly and quietly. Sumatrans speak loudly. They don't mean to yell, but that's how they talk. I couldn't bear it. I was used to going to bed at 7 p.m., but now I worked late. One time, I fell asleep on the line, and a manager came by and lifted up my chair saying, 'Wake up!' I cried every day after work.

"My family said, 'It's okay. There are many kinds of people. Be patient in finding a new job,' " she said. "They never said, 'We told you so.' "

Shortly after Suparti turned eighteen, she found employment at a chocolate factory. Her first job was to pour liquid chocolate into moldings and package them. Then she was promoted to deliver chocolate and supplies to other parts of the factory by trolley, and finally she worked behind a desk, printing product labels, plastic wrap and expiration dates, and barcodes for retailers.[33]

Around the world, for hundreds of years, young women have been voting with their feet. They have moved to cities from the countryside not because the urban areas are utopian but because they offer many more opportunities for a better life.

Urbanization, industrialization, and energy consumption have been over-whelmingly positive for human beings as a whole. From preindustrial times to today, life expectancy extended from thirty to seventy-three years.[34] Infant mortality declined from 43 to 4 percent.[35]

Before 1800, notes Harvard University's Steven Pinker, most people were desperately poor. "The average income was equivalent to that in the poorest countries in Africa today (about $500 a year in international dollars)," he writes, "and almost 95 percent of the world lived in what counts today as 'extreme poverty' (less than $1.90 a day)." The Industrial Revolution constituted what Pinker calls the "Great Escape" from poverty.[36]

The Great Escape continues today. From 1981 to 2015, the population of humans living in extreme poverty plummeted from 44 percent to 10 percent.[37]

Our prosperity is made possible by using energy and machines so fewer and fewer of us have to produce food, energy, and consumer products, and more and more of us can do work that requires greater use of our minds and that even offers meaning and purpose to our lives.

Moving to the city gives women more freedom in who they marry. "My parents encourage the *ta'aruf*, Muslim way of marriage," said Suparti. "This is where you explain yourself to a religious teacher or preacher, and they will introduce you to someone they think is a good match. But it's still up to you. I'm still rebellious and thus would like to get to know the man before marriage."

Cities and manufacturing bring other positive benefits. The human population growth rate peaked in the early 1960s alongside rising life expectancy and declining infant mortality.[38] Total population will peak soon.[39] And thanks to rising agricultural productivity, the share of humans who are malnourished declined from 20 percent in 1990 to 11 percent today, about 820 million people.[40]

Consider the differences between Bernadette, Suparti, and Helen. Bernadette must produce all of her food and energy whereas Suparti and Helen can purchase theirs. And while Suparti prepares most of her own food, Helen, as a working professional, is able to buy prepared food and even have it delivered to our home.

While Bernadette must farm to live, Helen is wealthy enough that she can garden for pleasure. And while Helen must also ward off wildlife, namely moles, our food supply and lives are not at risk if the rodents eat her plants, as Bernadette's is when a baboon does the same.

Machines liberate women from drudgery. I once asked my mother's oldest sister what her happiest childhood memories were while growing up on their farm in Indiana. She recalled the day when a clothes wringer her mother had ordered from Sears arrived. It was little more than two rollers and a hand crank, but it spared my grandmother's hands the hard labor of pulling, squeezing, and twisting clothes. Later came the electric-powered washing machines and dryers that fully liberated women from having to wash, wring, and hang out to dry the family's clothes.[41]

Scholars including Harvard's Benjamin Friedman and Steven Pinker find that rising prosperity is strongly correlated with rising freedom among, reduced violence against, and greater tolerance for, women, racial and religious minorities, and gays and lesbians. Such was the case in Indonesia.[42]

"My favorite singer is Morrissey," Ipeh told me. "I went to his show last year and was nervous that [Islamic] extremists would threaten to bomb it, and the show would be canceled. That's what happened with Lady Gaga, who had to cancel her show."

"Why didn't they?" I asked.

"Maybe they didn't realize that Morrissey is gay?" she said.

I asked Ipeh if she disapproved of homosexuality. "Do I think it's a sin? Yes. You can tell when someone is gay because they wear super tight clothes. But I'm cool with people being gay. It's not going to affect me."[43]

5. The Power of Wealth

In the eighteenth century, Britain pioneered the factory system, which combined people, machines, and energy in a way that made the manufacturing of clothing, shoes, and consumer products far faster and cheaper than the cottage-level, home-based crafts system that predated it. In the opening chapter of his 1776 book, *The Wealth of Nations*, Adam Smith explains that an individual worker is fifty times more productive when he is focused on a single task in making a pin than if he made the entire pin himself.[44]

Factories required people to harness energy to power machines. Owners built factories next to rivers so they could use water wheels. Over time, they transitioned to using steam from coal to power factories, and then to electricity.[45]

Today, economists point to three reasons why manufacturing, as opposed to other sectors of the economy, has allowed poor nations to develop into rich ones.

First, poor nations can become as efficient as rich nations in making things, and even surpass them. It is relatively easy for poor nations to steal manufacturing secrets from rich ones. Americans stole factory "know-how" from Britain in the eighteenth and nineteenth centuries, just as China has stolen intellectual property from the United States and other nations in recent years.[46]

Second, goods made in factories are easy to sell to other countries. This allows developing nations to make things they cannot yet afford to buy, and to buy things that they themselves cannot yet make. Even when the rest of the economy isn't working because of high levels of corruption in government or other factors, history shows that factories can continue to be productive and, indeed, drive economic growth.

Finally, factories are labor-intensive, which allows them to absorb large numbers of unskilled, small farmers. Former farmers like Suparti don't need to learn a new language or special skills to work in factories. "It is comparatively easy to turn a rice farmer into a garment factory worker," notes Harvard University economist Dani Rodrik.[47]

During the last 200 years, poor nations found that they didn't need to end corruption or educate everyone to develop. As long as factories were allowed to operate freely, and the politicians didn't steal too much from their owners, manufacturing could drive economic development. And, over time, as nations became richer, many of them, including the U.S., became less corrupt.

"You could start with very poor initial conditions, get a few things right to stimulate the domestic production of a narrow range of labor-intensive manufactures—and voilà! You had a growth engine going," Rodrik says.[48]

Such was the case with Indonesia. In the 1960s, it was as poor as many sub-Saharan African nations are today. A civil war and mass killings from

1965 to 1966 led to the deaths of more than one million people, or maybe more, according to recent estimates.[49] The government was and remains famously corrupt.[50] Recall that Ipeh was beaten up after she uncovered Indonesian soldiers involved in a crooked real estate deal.

And yet, for as dysfunctional and corrupt as Indonesia was, and in many ways still is, it has been able to attract enough manufacturing to drive development; per capita annual incomes rose from $54 to $3,800 between 1967 and 2017.[51]

For Suparti, what that has meant is that her wages have more than tripled since she first started working in the city. As a factory worker, she was able to purchase a flat-screen TV, a motor scooter, and even a home by the age of twenty-five.

In the early 2000s, a young Dutch economist named Arthur van Benthem was working for Shell Oil Company in the Netherlands to develop scenarios to predict future energy supply and demand.

In the 1960s, Shell pioneered scenario planning, which is essentially creating plausible, but distinct, stories about what the future might bring. Such stories helped Shell anticipate oil price increases in the 1970s and reductions in the 1980s, and thus hedge its bets. To forecast market collapses, Shell's scenario planning depended on thinking in counterintuitive, contrarian ways, and continually seeking new evidence, rather than relying on assumptions.[52]

Like many energy analysts at the time, van Benthem assumed that more energy-efficient lightbulbs, refrigerators, computers, and virtually every other technology meant that poor nations could get rich using far less energy than rich economies. "Because all these energy-efficient technologies are also available in China and Asia today," said van Benthem, "you might expect to find lower growth in energy consumption than you had in the United States or Europe when they were at the same GDP levels."[53]

Van Benthem decided to try to determine whether such energy "leapfrogging," as it is called, had actually occurred. He created a database from data on GDP, energy prices, and energy consumption from seventy-six countries. He crunched the numbers. He found no evidence of leapfrogging.

"If anything," he told me, "I found it slightly the other way around, that developing countries exhibit more energy-intensive growth at the same levels of GDP than developed countries did."[54]

Thanks to energy efficiency, things like lighting, electricity, and air conditioning are a lot cheaper. But that has just meant that people use them more, which reduces the energy savings that would have occurred had consumption levels not risen.[55]

The same has occurred for big, expensive, and energy-intensive products like cars. "If you look at India today, one high-selling car [the Suzuki Maruti Alto] is a $3,500, very energy-efficient vehicle that gets 40 [miles per gallon]," van Benthem explained. "That's way more efficient than cars used in the United States a century ago."[56]

Since 1800, lighting has become five thousand times cheaper. As a result, we use much more of it in our homes, at work, and outdoors. Cheap light-emitting diodes (LEDs) allow Suparti to consume much more lighting than our grandparents could when they were at similar income levels.[57]

And by making cars cheaper, more people can buy them, increasing energy consumption. "The Suzuki Maruti Alto's much lower price and its higher efficiency allows a greater number of poorer Indians to use it," notes van Benthem.[58]

Van Benthem's finding wasn't particularly new. The fact that energy efficiency, a form of resource productivity, lowers prices, which increases demand, is basic economics. And economists demonstrated that cheaper lighting led to greater consumption in 1996 and again in 2006.[59]

How wealthy we are is thus reflected in the amount of energy we consume. The average Congolese person consumes the energy equivalent of 1.1 kilograms of oil per day (kg/day). The average Indonesian consumes the energy equivalent of 2.5 kg/day. The average U.S. citizen consumes 19 kg/day.[60]

But these numbers obscure huge differences in the quality of energy. Almost all of the average Congolese person's energy consumption is in the form of burning wood and other biomass, where just 24 percent of the average Indonesian's is, and nearly none of the average American's is. In the seven years since we bought our home, Helen and I have not lit a fire in our fireplace once, and do not intend to do so in the future, because of the pollution it would create.

Countries with high population density, like Germany, Britain, and Japan, consume less energy per person than population-diffuse places like California, thanks to reduced car usage, but nowhere near the levels of people like Suparti, much less Bernadette.

As nations like Indonesia industrialize, they at first require more energy per unit of economic growth, but as they deindustrialize, like the United States, they require less.

Globally, the history of human evolution and development is one of converting ever-larger amounts of energy into wealth and power in ways that allow human societies to grow more complex.

6. Energy Density Matters

When you interview women who are small farmers about what it's like to cook with wood you might assume they would complain about the toxic smoke they must breathe. After all, such indoor air pollution shortens the lives of four million people per year, according to the World Health Organization.[61] But, around the world, what they complain about more often is how much time it takes to chop wood, haul wood, start fires, and maintain them.

After Suparti moved to the city, she was able to use liquefied petroleum gas as cooking fuel instead of rice husks. Doing so produces far less pollution, including one-third fewer carbon emissions.[62] But more importantly LPG saves her time she can spend doing other things.

When coal is burned in a power plant miles away, smoke can be removed from the home entirely, even when cooking and heating with natural gas. But even burning coal indoors, with the right fireplace, produces less indoor pollution than burning wood.[63]

Humans have been moving away from wood to fossil fuels for hundreds of years. Globally, wood went from providing nearly all primary energy in 1850 to 50 percent in 1920 to just 7 percent today.[64]

As we stop using wood for fuel, we allow grasslands and forests to grow back and wildlife to return. In the late 1700s, the use of wood as fuel for cooking and heating was a leading cause of deforestation in Britain. In the United States, per capita consumption of wood for fuel peaked in the 1840s. It was used at a per capita rate that was fourteen times higher than today.

Fossil fuels were thus key to saving forests in the United States and Europe in the eighteenth and nineteenth centuries. Wood went from being 80 percent of all primary energy in the United States in the 1860s to 20 percent in 1900, before reaching 7.5 percent in 1920.[65]

The environmental and economic benefits of fossil fuels are that they are more energy-dense and abundant. A kilogram of coal has almost twice as much energy as a kilogram of wood, while a kilogram of Suparti's liquefied petroleum gas has three times the energy as the rice husk biomass she cooked with back on the farm.[66]

Centralizing energy production has been essential to leaving more of planet Earth for natural landscapes with wild animals. Today, all hydroelectric dams, all fossil fuel production, and all nuclear plants require less than 0.2 percent of the Earth's ice-free land. The earth's food production takes 200 times more land than this.[67]

While the energy density of coal is twice as high as the energy density of wood, the *power* density of coal *mines* is up to twenty-five thousand times greater than forests.[68] Even eighteenth-century coal mines were four thousand times more power-dense than English forests and sixteen thousand times more power-dense than crop residues, like the kind Suparti's family used.[69]

The more people and wealth an area has, the higher the power density. Manhattan has a power density twenty times higher than New York City's outer boroughs, and the wealthy island nation of Singapore has a power density seven times higher than that of the global urban average.[70]

Thanks to fertilizers, irrigation, petroleum-powered tractors, and other farm machines, the power densities of farms rise ten-fold as they evolve from the labor-intensive techniques used by Suparti's parents to the energy-intensive practices used on California's rice farms.[71]

Power-dense factories and cities require energy-dense fuels because they are easier to transport and store, and they pollute less. Horse-drawn carriages made New York City unlivable in the years before the introduction of the automobile. The streets were dirty and dusty and stank of urine and feces, which brought flies and disease. Petroleum-powered vehicles allowed for much higher power densities with much less pollution.[72]

During the last 250 years, the power density of factories has increased dramatically. By the 1920s, Henry Ford's River Rouge Complex factory in Detroit had a power density fifty times higher than that of America's first large, integrated clothing factory, the Merrimack Manufacturing Company, one hundred years earlier.[73]

That fifty-fold increase in power density between Merrimack and River Rouge was made possible through electricity, which is the flow of electrons, the subatomic particles that are technically matter but act as a kind of pure, matter-less energy. Electricity is, technically, an "energy carrier," not a fuel or primary energy. Nevertheless, the increase shows the power of humankind's evolution away from *matter*-dense fuels to *energy*-dense ones.

People are often willing to accept even high levels of air pollution to enjoy the benefits of electricity. In 2016, I interviewed people living around an old and dirty coal power plant in India. The plant provided them with free electricity, but also sometimes emitted toxic ash, which they said irritated and burned their skin. However much they hated the pollution, none said they would be willing to give up the free, dirty electricity for cleaner electricity at a cost.

Even coal-burning has become dramatically cleaner over the last 200 years. A simple technical fix added to coal plants in developed nations after 1950 reduced dangerous particulate matter by 99 percent. High-temperature coal plants are nearly as clean as natural gas plants, save for their higher carbon emissions. Natural gas is still, as a rule, superior to coal, for inherently physical reasons. But on the question of air pollution, the extent to which coal plants have become far cleaner is remarkable.[74]

None of this is to say that burning coal is "good," only that it is, on most human and environmental measures, better than burning wood. As we will see, natural gas is similarly better than coal on most measures. People burn wood not coal, and coal not natural gas, when those fuels are all they can afford, not because those are the fuels they would prefer.[75]

As a result of cleaner-burning coal, the transition to natural gas, cleaner vehicles, and other technological changes, developed nations have seen major improvements in air quality. Between 1980 and 2018, U.S. carbon monoxide levels decreased by 83 percent, lead by 99 percent, nitrogen dioxide by 61 percent, ozone by 31 percent, and sulfur dioxide by 91 percent. While death rates from air pollution can rise with industrialization, they decline with higher incomes, better access to health care, and reductions in air pollution.[76]

Despite this positive trend, the shift from biomass to fossil fuels is far from complete. Humans today use more wood for fuel than at any other time in history, even as it constitutes a lower share of total energy. Ending the use of wood for fuel should thus be one of the highest priorities for people and institutions seeking both universal prosperity and environmental progress.[77]

7. The Manufacturing Ladder

The real risk to forests comes not from the expansion of energy-intensive factories in poor nations, as Greenpeace and Extinction Rebellion claim, but rather from the declining need for them.

On one hand, Africa has made real progress during the last half century. Agricultural productivity has risen, but manufacturing was a higher percentage of the economy in the mid-1970s than it is today. "Most countries of Africa are too poor to be experiencing de-industrialisation," writes Rodrik, "but that is precisely what seems to be taking place."[78]

One exception is Ethiopia, which has attracted Calvin Klein, Tommy Hilfiger, and fast-fashion leader H&M,[79] both because of its low wages compared to places like China and Indonesia, where they have risen, as well as to its investments in hydroelectric dams, the electricity grid, and roads.[80] "Ethiopia has experienced GDP growth of more than 10 percent per annum over the last decade," notes Rodrik, "due in large part to the increase in public investment, from 5 percent to 19 percent of GDP."[81]

Ethiopia had to end and recover from a bloody seventeen-year civil war, which resulted in at least 1.4 million deaths, including one million from famine, before its government could invest in infrastructure. "The resources spent on investment—in basic infrastructure such as roads and hydroelectricity—appear to have been well spent," says Rodrik. The infrastructure investments "have raised the overall productivity of the economy and reduced rural poverty." [82]

Leadership matters. "To be successful, industrialization has to come from the very top," Hinh Dinh, a former World Bank economist who advised the Ethiopian government, told me. For over two decades Dinh has researched and written on how poor nations can attract manufacturing. "Ethiopia got good results because of the now-deceased prime minister [Meles Zenawi] who went to China to get garment and shoe factories." [83]

I asked Dinh if he shared Rodrik's view that poor nations might need to find a path to development other than manufacturing. "In the U.S., manufacturing employment peaked at twenty million people in 1978," said Dinh, "and since then, it has shed its low-end industries to focus on higher, more specialized manufacturing. That's different from Nigeria de-industrializing at 7 or 8 percent (share of manufacturing in GDP) before its manufacturing reached the maturity stage."

Dinh added, "In a lot of developing countries, the de-industrialization is due to poor policy, bad governance, or neglect, not because of some natural peak as the case of the U.S. or Europe."

Dinh dismissed the notion that China's high productivity made the expansion of manufacturing in African nations irrelevant. "In the developing world, everyone, rich or poor, needs simple things like chairs or shoes," he said. "But when I was in Zambia I went to buy shoes, but not a single pair was made in Zambia!

"In any country, poor or rich, you have a lot of consumer goods that have to be produced," he explained. "The number of goods produced in the household only increases over time with income and never stops. I'm referring to the basic necessities of clothing, shoes, and household items."

Dinh observed the same phenomenon as Van Benthem. "What our grandparents had is completely different from what we have, and we will produce a lot more. I have no fear that at one point the industrialization will stop because demand will be saturated."

I asked Dinh if poor nations could become rich ones through agriculture, as Brazil has been attempting.

"There is nothing wrong with growing through agriculture," said Dinh. "But historically, nations did not do it that way because the scope for innovations is fairly limited. We're better at producing a bushel of wheat today than we were fifty years ago. But the wheat is pretty much the same. By contrast, a TV today and a TV thirty years ago are two completely different products."

He pointed to the difference between South Korea, whose per capita GDP is $30,000, and Argentina, where it is $14,000. "Argentina in 1920 had a higher per capita income than Italy—and higher than Korea. While there are many factors involved, one cannot help but observe that Korea's development path has been based on manufacturing and Argentina on other things, especially agriculture."

If Congo ever got its act together, I asked Dinh, what should it do? "I was asked to advise Osun State [in Nigeria]," he said. "I advised them to open up to foreign direct investment and try to get as many jobs created as possible. For now, forget about who owns them. Just bring them over. Get in touch with the Chinese or Vietnamese or Malaysians and ask them to bring factories over."

There is no shortcut to success, Dinh emphasizes.

"I gave a lecture at the Harvard Africa [Business] Club and someone said, 'We don't want to produce clothing and start with cheap products like China. We want to go right to the higher value added.' But you can't go directly from making bicycles to making a satellite. First you make bicycles and that allows you to make motorcycles. From there you can go to automobiles. From automobiles you can start thinking about satellites.

"The goal in Ethiopia is to have as many jobs as possible, and have the education system turning out the factory workers that you need. That's why I push for light manufacturing. It's not just the skills but also the discipline instilled in people. Later on, when the country reaches the second stage, the education system should produce more skilled workers capable of producing medium tech products, and so on."

Governments should train small farmers how to be factory workers, argues Dinh. "When industrialization first started in Vietnam in the early 1990s, you would see women and girls in the countryside going from the fields to informal shops in the village to do some sewing. They would make clothing that would then be sold domestically and for export. There was a culture of when your clothes were torn you sewed them up with needle and thread. That really helped."

8. Fast-Fashion for Africa

Contrary to what I and others have long believed, the positive impacts of manufacturing outweigh the negative ones. We should thus feel pride, not guilt, when buying products made by people like Suparti. And environmentalists and the news media should stop suggesting that fast-fashion brands like H&M are behaving unethically for contracting with factories in poor nations.

That doesn't mean we shouldn't want companies like Mattel, Nike, and H&M to improve factory conditions. Consumers can play a positive role in pressuring companies to do the right thing. But that depends on them continuing to buy cheap products produced in developing nations in the first place.

Many demographers believe that how quickly the human population peaks and starts to decline, globally, depends on how quickly sub-Saharan nations like the Congo industrialize and people like Bernadette move to the city, get jobs in factories, earn money, and choose to have fewer kids.

Understanding this process leads to an apparently counterintuitive conclusion. "If you want to minimize carbon dioxide in the atmosphere in 2070, you might want to accelerate the burning of coal in India today," said MIT climate scientist Kerry Emanuel. "It doesn't sound like it makes sense. Coal is terrible for carbon. But it's by burning a lot of coal they make themselves wealthier, and by making themselves wealthier they have less children. The population doesn't grow, and you don't have as many people burning carbon. You might be better off in 2070." [84]

Late economic developers like the Congo have a much harder time competing in international markets than did early economic developers like the United States and Europe. That means early developers, today's rich nations, should do everything they can to help poor nations industrialize. Instead, as we will see, many of them are doing something closer to the opposite: seeking to make poverty sustainable rather than to make poverty history.

Before we left, I asked Suparti what she felt she had accomplished as a labor union organizer.

"My proudest accomplishment was to win menstrual leave," she said, "so that when you are on your period you can have two days off. This was good because we had coworkers who were in so much pain from their periods that they cried and one even fainted."

I asked her whether she was lonely, and thought about returning to the village. "I really miss home," she confessed, "especially the nice chats with my mom and her cooking. But I don't have any desire to go back. I'm grateful to be doing what I'm doing."

I asked Suparti if she was worried about her parents. "I'm not worried yet about their retirement, but I am, however, saving up money to be able to send them to Mecca as a gift."

Suparti told me that once she's married, she would prefer to stay home and be a housewife. She'd like to have four children. "I used to want to have two kids, but now I think the house will be too quiet, and I don't want to be alone."

Before we left, I asked Suparti if I could take a photograph of her. "Where?" she asked.

I told her to choose her favorite part of the house. She decided to stand next to a sewing machine she rarely used. In front of and on top of the sewing machine's arm, Suparti had placed framed photographs of family and friends, plastic flowers, and tiny toy electric guitars.

In the photo I took, she rests her left arm on the helmet of her motor scooter. Above her is a Muslim prayer rug. Suparti smiles and looks proud.[85]

6

GREED SAVED THE WHALES, NOT GREENPEACE

1. Greenpeace and the Whales

If anything qualifies as a miracle of nature, it's the blue whale. As a baby, it gains ten pounds *an hour* drinking its mother's milk. It takes one ten years to achieve maturity. Full-grown, the blue whale is the largest known creature to have graced Earth—nearly three times bigger than the biggest dinosaur.[1] A single blue whale can stretch ten stories long and weigh as much as the National Football League.[2]

While we know much more about them than we did fifty years ago, whales remain mysterious and mystical. We know humpbacks work together to blow air bubbles to trap schools of herring and other fish in a sphere before lunging upward, mouths open, to gorge themselves. But we don't know whether they use acoustics, the stars, Earth's magnetic fields, or some other means of navigating from Hawaii to Alaska every year.[3]

Indigenous people reportedly treated whales with reverence. In Vietnam, those who set off to sea prayed to the whale, addressing the creature as "ngai," or "lord," a tradition that continues today.[4] Fishermen accord an elaborate funeral upon the beached whale, equal to one they would give a king.[5] In one version of the Inuit creation tale, a fisherman who finds a beached whale is told by the Great Spirit to eat magic mushrooms to gain the strength needed to return the whale to the sea and restore order to the world.

But by the mid-twentieth century, with the rise of massive industrial ships, humans nearly hunted whales to extinction.

Scientists raised the alarm about declining whale stocks, and a small group of committed young activists set out to save them. The activists started documenting the brutality of whaling, and the humanity of whales.

Things came to a boil in the summer of 1975. A small group of anti-whaling activists left the port of Vancouver on a sixty-six-foot halibut fishing boat. They headed to a whaling region in the North Pacific, where they intended to confront Russian whalers.

Once there, the activists boarded their Zodiac, a high-speed inflatable boat, and drove between a Soviet catcher vessel, the *Vlastny*, and a pod of sperm whales. An activist gripped his Super 8 camera with anticipation. The *Vlastny*'s cannon boomed and a 250-pound grenade harpoon whizzed past the bearded activists. It slammed into the back of a small female humpback.

One of the young men involved in the 1975 Vancouver incident later described the confrontation with the whalers.

> The whale wavered and towered motionless above us. I looked up past the daggered six-inch teeth into a massive eye, an eye the size of my fist, an eye that reflected back an intelligence, an eye that spoke wordlessly of compassion, an eye that communicated that this whale could discriminate and understood what we had tried to do . . .
>
> On that day, I knew emotionally and spiritually that my allegiance lay with the whales first and foremost over the interests of the humans that would kill them.[6]

A few nights later, Walter Cronkite aired the Zodiac crew member's Super 8 footage on *CBS Evening News*, and millions of people would learn the name of the new organization: Greenpeace.

After another seven years of media publicity, grassroots organizing, and political pressure, in 1982 environmental activists successfully inspired the world to impose a complete ban on commercial whaling. Today, whales of all species, including the mighty blue, are rebounding in numbers.

2. "Grand Ball Given by the Whales"

The story of a small band of committed nature lovers saving the environment appeals to us. It is the story we learn from TV and movie documen-
taries, books, and news reports. It is an exciting drama with obvious heroes
and villains. On one side there are greedy, cowardly people destroying nature for profit, and on the other side there are idealistic, brave youths. It is a
story that has inspired millions to take action.

The only problem with it as a guide for protecting the environment is that
nearly everything about it is wrong.

Whatever reverence some traditions have for whales, humans around the
world have mostly treated them as prey and sought to eat them, not worship
them. The Inuits may have freed beached whales but they also survived by
hunting them.

In the early 1600s, an English explorer observed Native Americans whaling around what is now Cape Cod, Massachusetts. "They go in company
of their King," he wrote, "with a multitude of their boats, and strike him
with a bone made in fashion of a harping iron fastened to a rope" (harpoon).
After shooting the whale with arrows—and then drowning it, or bleeding
it out—the Indians would return to shore with it and "sing a song of joy."[7]

A Jesuit explorer from Spain described a daring team of indigenous war-
riors in what is now Florida. They paddled in their canoes right up next to a
whale. Then, one commando leaped atop the mammal and thrust his spear
into its blowhole. At that moment, the Jesuit claims, the animal plunged into
the sea. Upon surfacing, the man was hanging on for dear life, stabbing the
creature to death.[8]

Organized whaling is at least as old as the eighth century. It was then
that the Basque people in what is now Spain built towers to spot them and
hunt them.[9] In seventeenth-century Japan, during the same decade, six
companies formed a consortium to hunt whales. Ten to twelve boats would
form a semicircle and drop a net to trap whales close to shore. "The climax
was reached when one of the men administered the *coup de grâce* with a long

sword," writes a historian. "As in bullfighting, these whaling toreadors were celebrated as national heroes."[10]

In eighteenth- and nineteenth-century United States, whales were hunted in open sail ships. Once a whale was spotted, two groups of six men, each in a small boat, were lowered into the sea. They would quietly row next to their target. When the small rowboats were rubbing right up next to their prey, "wood to blackskin," one of the men hurled a harpoon. Typically, the whale would bolt ahead, dragging the men along for the ride of their lives. Eventually, the whale would tire, at which point the men would pull next to it, drive a sharp steel lance into its lungs, and then twist it. Other times, whales plunged into the abyss, dragging their human predators to their death.[11]

Sometimes men killed whales easily and other times "she will hold the Whale-men in Play near half a Day together, with their Lances," a naturalist in 1725 wrote, "and sometimes will get away after they have been lanced and spouted Blood." A geyser of blood from the whale's blowhole would excite the men. "Chimney's afire!" men would shout.[12]

By 1830, the United States was the global whaling leader.[13] Whale oil was a luxury commodity because it burned brighter than candles and cleaner than wood fires. Whales provided much else: food, soap, machine lubricants, the base oil for perfume, and from their baleen, corsets, umbrellas, and fishing rods.[14]

Rising demand for whale oil led entrepreneurs to look for alternatives.[15] One of them was named Samuel Kier. In 1849, a doctor prescribed Kier's wife "American Medicinal Oil," petroleum, to treat an illness. His idea wasn't new. The Iroquois had used petroleum as an insect repellent, salve, and tonic for hundreds of years.[16]

When his wife felt better, Kier recognized a business opportunity. He launched his own brand, "Kier's Petroleum, or Rock Oil," and sold bottles for fifty cents through a sales force traveling through the region by wagon.

Kier was ambitious and sought other uses for his product. A chemist recommended distilling it and using it as lighting fluid. Kier's contribution to the emerging petroleum revolution was the creation of the first industrial-scale refinery in downtown Pittsburgh.[17]

A group of New York investors believed Kier had created a major business opportunity. They hired an itinerant and disabled engineer with good expertise in salt drilling to poke around in Pennsylvania for petroleum. In 1859, the man, Edwin Drake, drilled and hit a gusher of oil near Titusville, Pennsylvania.

The discovery of the Drake Well led to widespread production of petroleum-based kerosene, which rapidly took over the market for lighting fluids in the United States, thus saving whales, which were no longer needed for their oil. At its peak, whaling produced 600,000 barrels of whale oil annually.[18] The petroleum industry achieved that level less than three years after Drake's oil strike.[19] In a single day, one Pennsylvania well produced as much oil as it took a whaling voyage three or four years to obtain, a dramatic example of petrolem's high power density.[20]

In 1861, two years after Drake's oil strike, *Vanity Fair* magazine published a cartoon showing sperm whales standing on their fins and dressed in tuxedos and ball gowns, toasting one another with champagne at a fine celebration. The caption read, "Grand ball given by the whales to celebrate the discovery of the oil wells in Pennsylvania."[21]

While whalers over-hunted whales, historians conclude that "there is no evidence that American whaling contracted because of a serious shortage of whales." The creation of a substitute with a much higher power density was sufficient. This is an important lesson since it means we need not wait for inferior products, environmentally and otherwise, to run out before replacing them.[22]

3. How Congo Saved the Whales

As fate would have it, capitalism would save the whales not once, but twice.

By 1900, whaling looked like a dying industry. U.S. whaling output was less than 10 percent of its peak.[23] The only reason whaling didn't disappear altogether is that Norwegians were able to continue whaling certain species at low costs, and because of continuing demand for "whalebone," or baleen.[24] People hadn't yet invented petroleum-based plastics to substitute for baleen, the material taken from the underside of whale mouths, which people prized for its plasticity.

But then, whaling came back, and in a big way. Between 1904 and 1978, whalers killed one million whales, nearly three times more than had been harvested during the nineteenth century.[25]

A series of breakthroughs made whale oil newly useful for different products. In 1905, European chemists invented a way to turn liquid oil into solid fat for making soap. The process was called hydrogenation because it involved blowing hydrogen gas over nickel fillings into the oil.[26] Then, in 1918, chemists discovered how to solidify whale oil while eliminating the smell and taste, allowing it to be used for the first time as margarine.[27]

But then, industrial chemists succeeded in making margarine almost entirely from palm oil, eliminating the need for whale oil. By 1940, palm oil, much of it coming from the Congo, had become cheaper than whale oil. Between 1938 and 1951, the use of vegetable oils used for margarine quadrupled, while the use of whale and fish oil declined by two-thirds. The share of whale oil as an ingredient in soap fell from 13 percent to just 1 percent.[28] Whale oil as a share of global trade in fats declined from 9.4 percent in the 1930s to 1.7 percent in 1958, resulting in declining whale oil prices in the late 1950s.[29]

Journalists realized what was going on. In 1959, *The New York Times* reported that "the growing output of vegetable oils . . . has forced down the market value of whale oil and may, in the end, save the whales."[30] By 1968, Norwegian whalers were reduced to selling whale meat to pet food manufacturers. The *Times* reported that "the market for once-prized whale oil has slipped from $238 a ton in 1966 to $101.50. It has lost out to Peruvian fish oil and African vegetable oils."[31]

This time, rising scarcity of whales *did* incentivize their replacement with vegetable oil. A group of economists concluded that "economic growth brought with it a declining demand for whale products, whilst decreasing stock levels fed back into more and more expensive harvesting effort . . ."[32]

Whaling peaked in 1962, a full thirteen years before Greenpeace's heavily publicized action in Vancouver, and declined dramatically during the next decade. The United Nations called for a ten-year moratorium in 1972, and the United States banned whaling under the Marine Mammal Protection Act. By 1975, the year of Greenpeace's celebrated Vancouver action, an international agreement between forty-six nations, which prohibited all hunting of the humpback, the blue whale, the gray whale, and some species of right, fin, and sei whales, was already in place.[33]

It was vegetable oil, not an international treaty, that saved the whales. Ninety-nine percent of all whales killed in the twentieth century had occurred by the time the International Whaling Commission (IWC) got around to imposing a moratorium in 1982.[34] The Commission's moratorium on whaling in the 1980s, according to the economists who did the most careful study, was a "rubber stamp" on a "situation that had already emerged. . . . Regulation was not important in stabilizing populations."

The International Whaling Commission set whaling quotas, but they weren't low enough to prevent over-whaling. "In theory, the IWC was meant to regulate the killing of whales; in practice, the IWC functioned more like an international hunting club." Concludes the leading historian of the period, "The thirty years of work by the IWC have proven a fiasco."[35]

Those nations that thundered the loudest against whaling after the Greenpeace action didn't themselves hunt whales. "Strong anti-whaling positions became a rather convenient way of portraying a green image as virtually no material costs were involved for nations without whaling interests."[36]

Rising prosperity and wealth created the demand for the substitutes that saved the whales. People saved the whales by no longer needing them, and they no longer needed them because they had created more abundant, cheaper, and better alternatives.

Today, the populations of blue whales, humpback whales, and bowheads, three species for which there is great concern, are all recovering, albeit slowly, as is to be expected due to their large size and thus slow rate of reproduction.[37] Not a single whale species is at risk of extinction. Nations harvest fewer than two thousand whales annually, an amount that is 97 percent less than the nearly seventy-five thousand whales killed in 1960.[38]

The moral of the story, for the economists who studied how vegetable oil saved the whales, was that, "to some extent, economies can 'outgrow' severe environmental exploitation."[39]

4. A System Without a Schedule

While consulting for General Electric in the early 1970s, a playful, forty-something-year-old Italian nuclear physicist named Cesare Marchetti became friendly with one of GE's in-house economists. The man had recently coauthored a paper, "A Simple Substitution Model of Technological Change."[40] The model calculated how quickly new products become suitable replacements for older ones in order to predict how quickly new products would saturate the market. For a company like GE, with many different technology products, having such a model could be valuable.

Marchetti did not ordinarily think much of economic modeling. "As an old-time physicist," he wrote, "I always had the tendency to tease my economist friends for their supreme ability to construct beautifully structured models that will never be used in practice, and will never be splattered with the mud of this low world."[41]

This time was different. The GE economists were plugging actual data into the model to see if it worked. "I was impressed by the fact that the model was sloshing joyfully in the mud," Marchetti wrote.[42] He was so taken with the model that he took it with him when he left GE to work at the International Institute for Applied Systems Analysis (IIASA), a rare research collaboration funded by the Organisation for Economic Co-operation and Development (OECD), in 1974.

The United States and Soviet Union created IIASA (pronounced "yah-saw") as a means for scientific cooperation to bring the communist East and capitalist West closer together. It was focused on breaking down barriers, not just between nations but between disciplines. IIASA pioneered an interdisciplinary approach to *systems analysis*, a version of which would later be adopted by the Intergovernmental Panel on Climate Change.

The picture of evolution as a series of replacements inspired Marchetti. For much of his life he has collected typewriters, from some of the first models to some of the last, an extension of his fascination with technological innovation as a kind of Darwinian evolution.

At IIASA, Marchetti tested the replacement, or substitution, model on primary sources of energy, which he treated like "commodities competing for a market."[43] By *primary energy,* Marchetti meant those natural resources, or fuels, that can be used in a variety of ways, namely wood, coal, petroleum, natural gas, and uranium. (*Secondary energies,* by contrast, are things like electricity, kerosene, hydrogen, LPG, and gasoline, which must be made from primary energies.)

Throughout the next summer, Marchetti and a colleague inputted data from three hundred cases of energy transitions from around the world. The transitions were from wood to coal, whale oil to petroleum, coal to oil, and many other combinations. "I could not believe my eyes," he wrote, "but it worked."[44] He added, "The whole destiny of an energy source seems to be completely predetermined in the first childhood."[45] The study of what we today call *energy transitions* was born.

Wars, big changes in energy prices, and even depressions, Marchetti found, had no effect on the rate of energy transition. "It is as though the system had a schedule, a will, and a clock," he wrote.[46]

Older histories emphasized the role of scarcity in raising prices and stimulating innovation, such as how Europeans had to import wood from increasingly distant forests, making it more expensive, and the newcomer fuel, coal, relatively cheaper.[47] But Marchetti found that "the market regularly moved away from a certain primary energy source, long before it was exhausted, at least at world level."[48]

While scarcity helps incentivize entrepreneurs like Drake's investors to create alternatives, it is often rising economic growth and rising demand for a specific energy service, like lighting, transportation, heat, or industry, that allows fossil fuels to replace renewables, and oil and gas to replace coal.

That's what happened with whales. Other substitutes, principally hog fat and ethanol, emerged before the discovery of oil fields in Pennsylvania and the distillation of petroleum into kerosene. It was petroleum's abundance and superior power density that ultimately led to its triumph over biofuels.[49]

Coal started declining as a share of energy around World War I, even though "coal reserves were in a sense infinite" as oil and natural gas started to replace it.[50]

Energy transitions have occurred in the way that Marchetti predicted, from more energy-dilute and carbon-dense fuels toward more energy-dense and hydrogen-dense ones. Just as coal is twice as energy-dense as wood, petroleum is more energy-dense than coal, as is natural gas, when converted to liquid form.[51]

The chemistry is simple to understand. Coal is comprised of roughly one carbon atom for every hydrogen atom. Petroleum is comprised of one carbon atom for every two hydrogen atoms. And natural gas, or rather, its main component, methane, has four hydrogen atoms to one carbon atom, hence its molecular expression as CH_4.[52]

As a consequence of these energy transitions, the carbon-intensity of energy has declined for more than 150 years. Between 1860 and the mid-1990s, the carbon intensity of primary global energy declined about 0.3 percent per year.[53]

Marchetti was right that human societies tend to move from energy-dilute to energy-dense fuels, but wrong that "the system had a schedule . . . and a clock." While the direction of energy transitions he predicted was broadly correct, Marchetti's timing was off. For example, in the United States, the share of electricity coming from coal declined from more than 45 percent in 2010 to just less than 25 percent in 2019.[54] Europe saw similarly large declines in electricity from coal and increases from natural gas during the last two decades.[55] Marchetti predicted the coal-to-gas transition would occur in the 1980s and 1990s and was thus two decades off. And he optimistically predicted that few humans would, by today, still be burning wood and other forms of biomass where in reality more than 2.5 billion people still do.[56]

What determines the rate of those transitions is politics. And, as we will see, sometimes politics can move societies away from energy-dense fuels and back toward more energy-dilute ones.

5. The Gasland Deception

In spring 2010, a documentary filmmaker released the trailer to his new film, *Gasland,* about the natural gas boom in the United States. The background music, similar to what we hear in trailers for horror-fantasy movies, grows in volume and speed. We hear people say that hydraulic fracturing, or fracking, of shale, an underground rock formation, is poisoning their water and causing neurological diseases and brain lesions. It shows documents describing lung disease and cancer.

Three-quarters into the trailer, we hear an ominous chorus typical of what you might expect to hear when dragons take flight. A man stands by his sink with a hand-written sign above it reading, Do Not Drink This Water. We then see a congressman with a southern accent saying, with frustration in his voice, "What we're doing is searching for a problem that does not exist!"

The trailer then cuts back to the man at the sink. He is holding a lit cigarette lighter near the faucet's tap, which he turns on, igniting huge flames that force him to jump backward.[57]

The New York Times and other national media picked up on the story and depicted fracking for natural gas as a significant threat to America's natural environment, helping spawn a grassroots movement to end the practice.[58]

But the film's depiction of the flammable water was deceptive. In 2008 and 2009, the man from the film and two other Colorado residents filed formal complaints to Colorado's main oil and gas regulator, the Colorado Oil and Gas Conservation Commission. The commission took water samples from the three homes and sent them to a private laboratory. The laboratory found that the gas from the man's faucet and one other home was 100 percent "biogenic," or natural, and something people have safely dealt with for decades. It was created not by frackers but by Mother Nature. The third home had a mixture of biogenic and thermogenic methane; the owner and operator reached a settlement in the case.[59]

The independent regulator of Colorado's oil and gas industry took sharp objection to *Gasland*, noting that it informed producer Josh Fox of the facts of the cases well before he produced his movie, and he chose not to include them.[60]

People have documented water catching fire naturally for centuries.

There are reports of water on fire dating back to the ancient Greeks, Indians, and Persians. We now know that they were naturally occurring methane seeps. In 1889, a driller burned his beard after lighting water from a well he drilled in Colfax, Louisiana. There's a historical marker at the site of the well, which was featured in Ripley's Believe It or Not.[61]

An Irish documentary filmmaker named Phelim McAleer called out Fox for his mischaracterization of fracking at a 2011 *Gasland* screening.

MCALEER: There's a [flaming water] report from 1976 . . .

FOX: Well, I don't care about the report from 1976. There were reports from 1936 that people say they can light their water on fire in New York State.

MCALEER: I'm curious why you didn't include relevant reports from 1976 or from 1936 in the documentary? Most people watching your film would think that lighting your water started with fracking. You have said yourself people lit their water long before fracking started. Isn't that correct?

FOX: Yes, but it's not relevant.[62]

The Irish filmmaker posted the exchange on YouTube. Fox alleged copyright infringement. At first YouTube obeyed Fox's demand and removed the video, before eventually restoring it.[63]

6. Fracking the Climate

For nearly a decade, climate activists led by Bill McKibben of 350.org have claimed that natural gas is worse for the climate than coal.[64]

And yet, on virtually every metric, natural gas is cleaner than coal. Natural gas emits 17 to 40 times less sulfur dioxide, a fraction of the nitrous oxide that coal emits, and almost no mercury.[65] Natural gas is one-eighth as deadly as coal, counting both accidents and air pollution.[66] And burning gas rather than coal for electricity requires 25 to 50 times less water.[67]

The technological revolution allowing for firms to extract far more natural gas from shale and the ocean floor is the main reason that U.S. carbon emissions from energy declined 13 percent between 2005 and 2018, and a big part of the reason why global temperatures are unlikely to rise more than 3 degrees centigrade above pre-industrial levels.[68]

McKibben makes his claim that coal is better than natural gas by using an inappropriately short timeframe for global warming of just twenty years. The United States government and most experts agree that the appropriate timeframe to use is one hundred years. His timeframe thus exaggerates the impact of natural gas as a heat-trapping gas.[69]

Despite a nearly 40 percent increase in natural gas production since 1990, the EPA reported a 20 percent decrease in methane emissions in 2013, in part because of improved gaskets, monitoring, and maintenance.[70]

Natural gas fracking also resulted in the 62 percent decline in the mountaintop mining for coal between 2008 and 2014.[71]

Where fracking for natural gas cracks shale below the Earth's surface, imposing very small impacts aboveground, coal mining devastates mountain ecosystems. More than 500 mountains, covering more than one million acres, have been destroyed in central and southern Appalachia by mountaintop removal.[72] When mining companies demolish mountains with explosives to harvest coal, they dump millions of tons of crushed rock into nearby valleys, destroying forests and headway streams. Exposed rock leeches heavy metals and other toxins, which hurt wildlife, insects, and humans. Dust that blows into the air from such operations can harm miners and people who live in nearby communities.[73]

No energy transition occurs without human and environmental impacts. Fracking brings pipelines, rigs, and trucks, which can disrupt peaceful landscapes that people rightly care about. Frackers have created small earthquakes and improperly disposed of fracking wastewater. These problems are serious and should be addressed, but they are nowhere as bad as coal mining, which has in many ways become worse throughout the decades, not better, culminating in mountaintop removal and the destruction of river ecosystems.[74]

What explains the lower environmental impact of natural gas fracking as compared to coal mining is power density. A natural gas field in the Netherlands is three times more power-dense than the world's most productive coal mines.[75]

Today, many if not most scientists and environmentalists support natural gas as a substitute for coal. "People are placing too much emphasis on methane," climate scientist Ray Pierrehumbert told *The Washington Post.* "People should prove that we can actually get the CO_2 emissions down first, before worrying about whether we are doing enough to get methane emissions down."[76]

Pollution regulations helped make coal plants more expensive to build and operate. But, as Marchetti predicted, and similar to what we saw with whales, what mattered most was the creation of a more power-dense, abundant, and cheaper alternative. What Marchetti didn't foresee was how powerful and important opposition to the new technology, particularly from upper classes of society, could be in the case of energy transitions.

7. Fish Go Wild

In late 2015, the U.S. Food and Drug Administration approved a genetically modified salmon, one that delivered major environmental benefits over existing farmed salmon. Critics loved it. "The flesh is exquisite," wrote one food writer. "Buttery, light, juicy. Just as Atlantic salmon should be."[77]

The AquAdvantage salmon, developed by AquaBounty Technologies in 1989, grows twice as fast and needs 20 percent less feed than Atlantic salmon. While eight pounds of feed is needed to harvest one pound of beef, only one pound of feed is required for one pound of AquAdvantage salmon.

Unlike the majority of farmed salmon, which is produced in floating sea cages in coastal areas, AquAdvantage is produced in hatcheries and facilities in warehouses on land. It thus minimizes the impact of aquaculture on natural ocean environments and prevents harmful interactions with wild species, which can result in disease. And AquaBounty estimates it will produce 23 to 25 percent fewer carbon emissions than traditionally farmed salmon.[78]

Atlantic salmon is already one of the world's healthiest foods. It is low in calories and saturated and trans fats, and a good source of protein and omega-3 polyunsaturated fatty acids. By genetically altering the salmon, AquaBounty also eliminated the need for antibiotics, which public health officials warn can contribute to antibiotic resistance. "AquAdvantage salmon is as safe to eat as any non-genetically engineered Atlantic salmon, and also as nutritious," the FDA says.[79]

Fish farming is critical for saving wild fish and other marine species, such as the yellow-eyed penguin and the albatross. That's because the total population of ocean fish that humans hunt and eat for food has declined by nearly 40 percent since 1970. Overfishing has resulted in many local extinctions, including of sharks.

Today, 90 percent of the world's fish stocks are either overfished or at capacity, meaning they are close to or just barely above the maximum they can be harvested before seeing their populations collapse entirely.[80] Where 15 percent of Earth's land surface is protected, just less than 8 percent of the world's oceans are.[81]

Since 1974, humankind has tripled the share of fish stocks being harvested at unsustainable levels.[82] And the pressure on wild fish continues to rise: between now and 2050, thanks to rising wealth and a growing number of people, global demand for fish is expected to double.[83]

The good news is that fish farming, or aquaculture, is developing rapidly. Aquaculture output doubled between 2000 and 2014, and today it produces half of all fish for human consumption.[84] The Food and Agriculture Organization of the United Nations (FAO) reported in 2018 that aquaculture "continues to grow faster than other major food production sector," and that by 2030, "the world will eat 20 percent more fish than in 2016."[85]

A big environmental benefit from aquaculture comes from moving fish farms from oceans to land. Doing so reduces their impact on marine environments and allows for closed or near-closed systems where water is constantly being cleaned and recycled.[86]

The technologies pioneered to create genetically engineered fish bring side benefits. Scientists say such genetic modification offers the possibility of eliminating the deadly avian flu virus.[87]

And yet, the most outspoken critics of replacing wild fish consumption with farmed fish are environmental groups, including the Natural Resources Defense Council (NRDC) and Sierra Club, which claimed AquAdvantage salmon might contaminate populations of wild salmon.[88] After the FDA approved AquAdvantage, the head of the Center for Food Safety, another environmental group, announced it was filing a lawsuit to stop "the introduction of this dangerous contaminant."[89]

In response, several large supermarket chains, including Trader Joe's and Whole Foods, announced they would not carry the AquaBounty fish, even though spokespersons from both chains admitted the stores offer other foods produced with genetically modified ingredients or feed.[90]

Fish farming is not without its problems. Early fish farms, such as shrimp farms, were quite destructive, involving the clearing of mangrove forests and the pollution of waterways by chemicals and nutrients.[91] But, over time, their negative environmental impact has significantly declined through better siting of fish and shrimp farms and the cofarming of species such as scallops and mussels with seaweed and microalgae.

The scientist who first raised the concern that genetically engineered fish could threaten wild fish stocks is today one of its most outspoken advocates. "I won't argue that a genetically engineered salmon will never find its way into the ocean," he said. "But there's nothing in this fish that would last more than a single generation because of its low fitness."[92]

As for Trader Joe's and Whole Foods, former AquaBounty CEO Ron Stotish was optimistic that he could change their minds. "We are hopeful that over time, they will embrace our product."[93]

But five years later, neither the environmental groups supposedly worried about the future of wild fish, nor Trader Joe's, Whole Foods, Costco, Kroger, and Target, had changed their minds.[94]

8. Class War

Today, Cesare Marchetti is in his early nineties and lives as a "gentleman farmer near Florence," says his friend and coauthor, Jesse Ausubel, "with olive groves, grapevines, goats, black cats," and his typewriter collection.[95]

Ausubel, who works at Woods Hole Research Institute and Rockefeller University, has been friends with Marchetti since they met in the 1970s at IIASA. Today, the two men are trying to sequence the genome of Leonardo da Vinci out of trace DNA they've been able to gather from the books and other items that belonged to the great Renaissance artist. "If you look at da Vinci's drawings of storms and clouds, he understood the immense indifference of nature," said Ausubel, "and a lot of the human enterprise."

I asked Ausubel why he thought Marchetti's model of energy transitions had been so off in terms of timing, even if it was broadly accurate on the direction. "You can look at the long term and the dynamics win," he said. "But you can look at any phenomenon and find interruptions, hiatus, digressions, and diversions. That's what happened with energy."

Opposition to the new fuel usually comes from the wealthy. In Britain, elites called coal the "devil's excrement," something that many people believed to be literally true, given its sulfuric smell.[96] Coal smoke smelled bad against the sweet smell of wood-burning. The upper-class of Victorian England resisted the transition from wood to coal as long as they could.[97]

It was educated elites who similarly waged the war on fracking. The key antagonists were *The New York Times*, Bill McKibben, and well-financed environmental groups, including the Sierra Club and Natural Resources Defense Council.

Ausubel describes how coal interests fought against greater exploration of natural gas in the 1970s. "People fight tenaciously to hold onto their position. That was true in the U.S. of the coal industry, which developed an alliance between, in the West, [Republican] Senator [Alan] Simpson in Wyoming and in the East, [Democratic] Senator [Robert] Byrd in West Virginia. At the national political level, they were able to do a lot of things that superannuated the coal industry."

Ausubel pointed to the election of President Jimmy Carter in 1976, who, with the support of major environmental groups, pushed for more coal instead of nuclear and natural gas.

Ausubel believes the concern about energy independence in the 1970s was misplaced. He said that the "worries that exporting gas would somehow hurt our national security [were] insane. In fact, having a healthy large industry gives you more national power."

Ausubel notes that scientists knew natural gas was abundant, particularly in the oceans. "Everybody in the American Association of Petroleum Geologists by the early to mid-80s knew there were vast amounts of gas offshore on the continental margins and the methane hydrates. I wrote about them in a 1983 National Academy of Sciences report.

"Geologists allowed the idea to persist that natural gas was precious and that you had to save it because there wasn't that much of it. Today, the [oil and gas] majors see themselves as natural gas more than petroleum companies going forward. But in many places, that could have happened twenty to thirty years ago and didn't." [98]

Happily, the war on fracking failed. When it came to fracking shale for natural gas, the United States interfered less than other countries and benefited enormously as a result. The United States allows property owners the mining and drilling rights to the Earth beneath them. In most other nations, those rights belong to the government, which is a major reason why fracking hasn't taken off in other countries.

Politics even interfered with saving the whales. While environmentalists often blame capitalism for environmental problems, it was communism that made whaling worse than it needed to be. After the fall of communism, historians found records that the Soviet Union was whaling at far higher numbers than they had admitted. It did so even though it was no longer profitable to do so, thanks to Soviet central planning. "Ninety-eight percent of the blue whales killed globally after the ban in 1966 were killed by Soviet whalers," wrote a historian, "as were 92 percent of the 1,201 humpbacks killed commercially between 1967 and 1978." [99]

And had there been freer markets, nations like Japan and Norway might have switched from whale oil to vegetable oil much sooner. "What probably sustains the whaling industry against the inroads of vegetable oil," reported *The New York Times* in 1959, "is the desire of the whaling nations to conserve their foreign exchange. In general, they do not produce enough vegetable oil for their own needs and hence must either catch whales or buy fats and oil abroad." [100]

The moral of the story is that economic growth and the rising demand for food, lighting, and energy drive product and energy transitions, but politics can constrain them. Energy transitions depend on people wanting them. When it comes to protecting the environment by moving to superior alternatives, public attitudes and political action matter.

7

HAVE YOUR STEAK AND EAT IT, TOO

1. Eating Animals

When Jonathan Safran Foer was nine years old, he asked his babysitter why she wasn't eating the chicken that he and his brother Frank were having.

"I don't want to hurt anything," she said.

"*Hurt* anything?" Foer asked.

"You know that chicken is chicken, right?" she said.

"I put down my fork," Foer wrote in his 2009 vegetarian memoir-manifesto, *Eating Animals*.

What about his brother? "Frank finished the meal and is probably eating a chicken as I type these words."[1]

Many of us who eventually became vegetarians have similar stories. When I was four years old, I told my parents I wouldn't eat pigs because I had just met one.

The environmental case for vegetarianism appears to have only grown stronger. In 2019, the Intergovernmental Panel on Climate Change published a special report on food and agriculture. "Scientists say that we must immediately change the way we manage land, produce food and eat less meat in order to halt the climate crisis," reported CNN.[2]

IPCC scientists expect the demand for food to outpace population growth by more than 50 percent by 2050. If that's the case, then Americans and Europeans need to reduce consumption of beef and pork by 40 percent and 22 percent, respectively, they said, in order to feed ten billion people.[3]

"We don't want to tell people what to eat," said the scientist who co-chaired the IPCC's working group on climate impacts and adaptation. "But it would indeed be beneficial, for both climate and human health, if people in many rich countries consumed less meat, and if politics would create appropriate incentives to that effect."[4]

"We need a radical transformation, not incremental shifts, towards a global land-use and food system that serves our climate needs," said the head of an environmental philanthropy. "It's really exciting that the IPCC is getting such a strong message across."[5]

If everyone followed a vegan diet, which excludes not only meat but also eggs and dairy products, land-based emissions could be cut by 70 percent by 2050, said IPCC.[6]

To reduce the consumption of meat, the best strategy is to make it more expensive, say some environmental groups.[7] One estimated that the cost of beef and dairy to consumers would increase by 30 percent, if it were to account for its climate impacts.[8]

Eating less meat would not only benefit the climate but also human health, say many scientists. In 2018, according to the U.S. Department of Agriculture, Americans were expected to eat a record 222 pounds of red meat and poultry, up from 217 pounds per consumer in 2017. In fact, Americans were consuming about ten ounces of meat per day, which is about twice as much as government nutritionists recommend.[9]

"Eating less red meat might be good for the planet, but it could also help your health," reported CNN. "Earlier research has found that eating red meat is tied to increased risks of diabetes, heart diseases and some cancers."[10]

In response to the science linking meat to climate change, some climate activists, including Greta Thunberg, have sworn off meat, and have persuaded their parents to become vegetarian and even vegan.

By reducing meat consumption, ending industrial agriculture, and committing ourselves to free-range, grass-fed, pasture meat, say many scientists and environmentalists, we will return more of Earth to nature.[11]

But would we, really?

2. The Meat-Free Nothingburger

Even though I have been researching and writing on climate and energy policy for nearly two decades, I had failed to recognize that the headline number in the IPCC's 2019 report—70 percent reduction in emissions by 2050—referred only to *agricultural emissions,* which comprise a fraction of total greenhouse emissions. I suspect many others similarly thought the number was referring to all emissions.[12]

One study found that converting to vegetarianism might reduce *diet*-related personal energy use by 16 percent and greenhouse gas emissions by 20 percent but *total* personal energy use by just 2 percent, and total greenhouse gas emissions by 4 percent.[13]

As such, were IPCC's "most extreme" scenario of global veganism to be realized—in which, by 2050, humans completely cease to consume animal products and all livestock land is reforested—total carbon emissions would decline by just 10 percent.[14]

Another study found that if every American reduced her or his meat consumption by one-quarter, greenhouse emissions would be reduced by just 1 percent. If every American became vegetarian, U.S. emissions would drop by just 5 percent.[15]

Study after study comes to the same conclusion. One found that, for individuals in developed nations, going vegetarian would reduce emissions by just 4.3 percent, on average.[16] And yet another found that, if every American went vegan, emissions would decline by just 2.6 percent.[17]

Plant-based diets, researchers find, are cheaper than those that include meat. As a result, people often end up spending their money on things that use energy, like consumer products. This phenomenon is known as the rebound effect. If consumers respent their saved income on consumer goods, which require energy, the net energy savings would only be .07 percent, and the net carbon reduction just 2 percent.[18]

It is for that reason that reducing carbon emissions in *energy,* not food or use of land more broadly, matters most. And energy includes electricity, transportation, cooking, and heating, nearly 90 percent of which globally are fossil fuels.

None of this is to suggest that people in rich nations can't be persuaded to change their diets. For example, since the 1970s, Americans and others in developed nations have been eating more chicken and less beef. The global output of chicken meat has grown nearly fourteen-fold, from eight metric megatonnes to 109 metric megatonnes, between 1961 and 2017.[19]

But the thing that makes chicken production environmentally superior to beef production is the very thing Foer most laments: the higher density of meat production allowed for by factory farming. After visiting a chicken facility, Foer writes, "It's hard to get one's head around the magnitude of thirty-three thousand birds in one room."[20]

3. The Nature of Meat

While meat production is a relatively modest contributor to climate change, it represents humankind's single largest impact on natural landscapes. Today, humans use more than one-quarter of Earth's land surface for meat production. And the spread of pasture for cattle and other domesticated animals continues to threaten many endangered species, including mountain gorillas and yellow-eyed penguins.

During the last 300 years, an area of forests and grasslands almost as large as North America was converted into pasture, resulting in massive habitat loss and driving the significant declines in wild animal populations. Between 1961 and 2016, pastureland expanded by an area almost the size of Alaska.[21]

The good news is that the total amount of land humankind uses to produce meat *peaked* in the year 2000. Since then, the land dedicated to livestock pasture around the world, according to the Food and Agriculture Organization of the U.N., has decreased by more than 540 million square miles, an area 80 percent as large as Alaska.[22]

All of this happened without a vegetarian revolution. Today, just 2 to

4 percent of Americans are vegetarian or vegan. About 80 percent of those who try to become vegetarian or vegan eventually abandon their diet, and more than half do so within the first year.[23]

Developed nations like the United States saw the amount of land they use for meat production peak in the 1960s. Developing nations, including India and Brazil, saw their use of land as pasture similarly peak and decline.[24]

Part of this is due to the shift from beef to chicken. A gram of protein from beef requires two times the energy input in the form of feed as a gram from pork, and eight times a gram from chicken.[25]

But mostly it is due to efficiency. Between 1925, when the United States started producing chicken indoors, and 2017, breeders cut feeding time by more than half while more than doubling the weight.[26]

Meat production roughly doubled in the United States since the early 1960s, and yet greenhouse gas emissions from livestock *declined* by 11 percent during the same period.[27]

Throughout *Eating Animals*, Foer argues that factory farms are far worse for the natural environment than free-range beef. He writes, "[I]f we consumers can limit our desire for pork and poultry to the capacity of the land (a big if), there are no knockdown ecological arguments against [free-range] farming."[28] But would switching to free-range farming really be better for nature in what Foer calls "our overpopulated" world?[29]

Consider that pasture beef requires *fourteen to nineteen times* more land per kilogram than industrial beef, according to a review of fifteen studies.[30] The same is true for other inputs, including water. Highly efficient industrial agriculture in rich nations requires less water per output than small farmer agriculture in poor ones.[31] Pasture beef generates 300 to 400 percent more carbon emissions per kilogram than industrial beef.[32]

This difference in emissions comes down to diet and lifespan. Cows raised at industrial farms are typically sent from pastures to feedlots at about nine months old, and then they are sent to slaughter at fourteen to eighteen months. Grass-fed cattle spend their entire lives at pasture and aren't slaughtered until between eighteen to twenty-four months of age. Since grass-fed cows gain weight more slowly and live longer, they produce more manure and methane.[33]

In addition to their longer lifespans, the roughage-heavy diets typical of organic and pasture farm systems result in cows releasing more methane. These facts combined tell us that the global warming potential of cows fed concentrates is 4 to 28 percent lower for cows fed roughage.[34]

Attempting to move from factory farming to organic, free-range farming would require vastly more land, and thus destroy the habitat needed by mountain gorillas, yellow-eyed penguins, and other endangered species. Foer unwittingly advocates nineteenth-century farming methods that, if adopted, would require turning wildlife-rich protected areas like Virunga Park into gigantic cattle ranches.

Farmers make this point to Foer in *Eating Animals*. "You simply can't feed billions of people free-range eggs. . . . It's cheaper to produce an egg in a massive laying barn with caged hens," says one. "It's more efficient and that means it's more sustainable. . . . Do you think family farms are going to sustain a world of ten billion?"[35]

4. Meat = Life

In 2000, a journalist named Nina Teicholz started writing restaurant reviews for a small newspaper in New York. "It didn't have a budget to pay for meals," she says, "so I usually ate whatever the chef decided to send out to me." Teicholz found herself eating foods she had long avoided, such as beef, cream, and foie gras.[36]

For twenty years she had been eating a mostly vegetarian diet. "When the Mediterranean diet was introduced in the 1990s, I added olive oil and extra servings of fish while cutting back on red meat," she says. "Avoiding the saturated fats found in animal foods, especially, seemed like the most obvious measure a person could take for good health."[37]

Still, she had struggled to lose stubborn extra weight for two years, even while consuming a recommended diet loaded with vegetables, fruits, and grains and exercising daily. Then, after eating the high-fat meals chefs were serving her, something strange happened: she lost ten pounds over two months, despite eating animal products high in saturated fat. And Teicholz

found she loved eating the high-fat dishes. "They were complex and remarkably satisfying," compared to the high-carbohydrate Mediterranean
diet she had been following.[38]

Around the same time, the magazine *Gourmet* asked Teicholz to write an
article about the growing controversy over trans fats, which are made from
vegetable oils. But the more she read about the issue, "the more I became
convinced that the story was far larger and more complex than trans fats."[39]
And so she decided to do more research.

Nine years later, in 2014, Simon & Schuster published Teicholz's
best-selling book, *The Big Fat Surprise: Why Butter, Meat and Cheese Belong
in a Healthy Diet.* She reported on a body of evidence, particularly a series
of clinical trials conducted in the 1950s and 1960s, which challenge the nutrition consensus that diets high in animal fats result in heart disease and
obesity. The evidence suggested there was either no effect, or that diets high
in saturated fats might be beneficial.

"There are now at least seventeen systematic reviews looking at the clinical trials and nearly all conclude that saturated fats have no impact on mortality," she explained.

Teicholz's book built on a significant body of research unearthed by science journalist Gary Taubes, whose 2007 book, *Good Calories, Bad Calories*,
was one of the first to challenge the anti-fat conventional wisdom.

While writing for *Science* and *The New York Times Magazine* in the early
2000s, Taubes unearthed studies finding that a high-fat diet would lead to
weight loss and improvements in heart disease risk factors compared to the
kind of low-fat, plant-rich diets that the American Heart Association and
the U.S. government had been advising us to eat since the 1960s and 1980s,
respectively.

"You had obesity researchers saying that obesity had to be a hormonal
regulatory defect for decades before the revelation in the 1960s that insulin
was regulating fat accumulation," noted Taubes, "which won the researchers the Nobel Prize. One of them said, 'If insulin regulates fat accumulation,
then rather than obesity causing diabetes, wouldn't mild diabetes cause obesity?'"

Yet, for decades, the scientific consensus remained that high-fat diets were dangerous. That consensus led many governments to promote a diet high in carbohydrates, low in animal protein, and very low in animal fats.

Teicholz and Taubes believe that the conventional wisdom that "a calorie is a calorie," which is known as the energy balance theory, is wrong because our bodies process fats radically differently from how they process carbohydrates. When we eat carbohydrates, they believe, the body works to keep the fat locked away in storage. Obesity and diabetes are a result of hormone imbalance caused by eating more carbohydrates than the body can handle, they say.

The people who were challenging the orthodoxy were in the minority, but they were by no means fringe scientists. "You had the best, leading authority on childhood obesity in the mid-twentieth century, and the leading endocrinologist of that era, both making the same argument," notes Taubes.

"You had British researchers saying that obesity, heart disease, and cancer all appeared in other populations when they switched to Western diets and started eating sugar and refined grains, which would have unique effects on insulin secretion," he explained.

"And then you had the discovery that metabolic syndrome—which is a cluster of abnormalities, including weight gain and high blood pressure, which affects around half of middle-aged men and women in the U.S.—is linked to the carb content of the diet, not to the fat content. This syndrome is directly linked to diabetes and obesity."

5. Death for Life

By using fire to cook meat, rather than eating it raw, our prehuman ancestors were able to consume much larger amounts of protein. That, in turn, may have allowed their intestines to grow smaller, since less digestion was required, and their brains to grow bigger, according to the emerging theory of human evolution.

Our brains grew so big that prehumans began delivering their bigger-brained babies prematurely, at nine months instead of twelve months, like other primates. Mothers carried their "premature" babies by strapping them

to their bodies with animal bladders and skins. Those technologies effectively allowed babies to finish what is sometimes called their "fourth trimes-
ter" outside the womb. The final outcome was the human brain. It demands
two to three times as much energy by mass as the brains of other primates.[40]

Around the world, hunter-gatherers as far back as two million years ago
valued animal fat more than protein or carbohydrates. The reason is obvious: animal fats contain two to five times as much energy by mass as protein
and ten to forty times as much as fruits and vegetables. Those higher densities allowed early humans to gain more energy with less work than carbohydrates.[41]

"The consumption of meat has always been associated with masculine
power, carnality, testosterone, and your sexuality," said Teicholz. "It fuels
strength. It gives you the protein and nutrients you need to be strong. So it's
seen as being connected to sexual and masculine desires."

As hunter-gatherers settled down, they domesticated animals to grow
quickly and efficiently. People primarily domesticated those animals that
consumed foods humans could not eat, like ruminants, which have special
protozoa in their intestines to digest grasses mammals cannot.[42]

Even today, meat remains a key source of energy for most people. "My me-
tabolism needs meat and eggs," says animal welfare expert Temple Grandin.
"If I do not eat animal protein, I get lightheaded and have difficulty thinking.
I have tried eating a vegan diet, and I cannot function on it."[43]

I am the same way. During the decade I was vegetarian, I grew tired
most afternoons after eating a carb-heavy lunch, no matter how much sleep
I got the night before. It was only after eating meat again that I could work
through the afternoon without feeling sleepy.

Some studies find that vegans and vegetarians are more prone to fatigue,
headaches, and dizziness because of the deficiency of vitamin B12 and iron
in the absence of red meat.[44]

People are vegetarian for many different reasons and in many different ways.
Some people are vegetarian for ethical reasons, others for health, others for
the environment, and some, like Bernadette, because she can't afford meat.[45]

Like many vegetarians, my motivations shifted over time. On the one hand, I wanted to save the rainforests. On the other hand, when people asked about it, I often identified health as the reason, partly because I wanted to avoid the argument that could ensue if I named ethical reasons.

My experience is not unusual. Liberals and environmentalists are far more likely than conservatives to become vegetarians. And while women are more likely to become vegetarian than men, both men and women are likelier to become vegetarians as adolescents or young adults than at any other age. [46]

"I gave up meat at fourteen or fifteen," said Eric, a professional acquaintance of mine in his mid-forties. "I was in a straight-edge band. You know straight edge? Fugazi?"

I said I did. Fugazi is an influential post-hardcore band from the 1990s.

"No girls. No weed. No beer. Vegetarian," said Eric. "But it only lasted for four weeks because we made friends with a pot dealer and became a ska band. Still, I stuck with vegetarianism because it made my life better."

Eric said he is disgusted by the texture of meat. "A chef once told me that my problem is I don't like to grind things. I find a raw tomato is gross. I once cooked a roast and held the roast in my hand and felt disgusted by how it felt like a newborn baby. Even the [meatless] 'Impossible Burger,' which I ordered, was gross because it reminded me of all the things I hate about meat."

For decades, psychologists have been interested in the relationship between vegetarianism and the emotion of disgust. A study of adolescent British girls found that vegetarians associated meat with cruelty, killing, the ingestion of blood, and disgust.[47]

"Yeah, the disgust thing," said Eric. "I'm a contamination vegetarian. I won't eat pizza if it's half-cheese, half-pepperoni because the cheese side might have been touched by a pepperoni."

An Italian team of psychologists recently found vegetarians view meat as "the representation of death as a contaminating essence."[48] The theme comes up again and again in vegetarian literature. "When we eat factory-farmed meat we live, literally, on tortured flesh," writes Foer. "Increasingly, that tortured flesh is becoming our own."[49]

In 1989, when I arrived at college, animal rights activists were eager to

share horrifying videos of factory farming conditions. "We know if someone offers to show us a film on how our meat is produced," writes Foer, "it
will be a horror film."[50]

It was videos like those, made and distributed by groups like PETA before the Internet, that led people like me, my college girlfriend, and many of
the people around us at our Quaker college in Indiana to stop eating meat in
the late 1980s.

And it is videos like those that continue to motivate young people to go
vegetarian. "I became a vegetarian in fifth grade," my colleague Madison,
who turned twenty-five in 2020, said. "My debate topic was vegetarianism
and my goal was to persuade the class. I watched a bunch of videos on animal cruelty and animal farming and it really disturbed me. It was the main
thing for me. Twelve years went by, and I didn't feel I needed meat."

By the 1990s, PETA had learned the power of going after big brands. It
distributed videos showing McDonald's supplier farms abusing animals.

In 1999, McDonald's hired the animal welfare expert Temple Grandin to
audit the farms of its suppliers. She was appalled by what she found. "It was
terrible," she said, "broken stunning equipment, yelling and screaming and
hitting cattle, poking them repeatedly with electric prods."[51]

Grandin was already a leading authority on humane treatment of animals. In the 1970s she designed equipment for slaughterhouses that reduced
stress for cattle going to slaughter. In 1993, she edited a textbook on how to
best handle livestock.

Grandin said it was easy for her to imagine how cows felt because she
was autistic. "My nervous system was hyper-vigilant. Any little thing out
of place, like a water stain on the ceiling, would set off a panic reaction, and
cattle are scared of the same thing."[52]

Scientists define autism as a developmental disorder, one that involves a
difficulty in communicating and interacting socially, and thinking and behaving in ways that are restricted and repetitive.

But Grandin found her autism also gave her unique insights. It made her
sensitive in the same way animals, including beef cows, are to noises and
visual stimulation. "Animals don't think in language," she said. "They think
in pictures."[53]

In *Eating Animals,* Foer argues that it's "plainly wrong to eat factory-farmed pork . . . poultry or sea animals. . . . With feedlot-raised beef, the industry offends me less (and 100 percent pasture-raised beef, setting aside the issue of slaughter for a moment, is probably the least troubling of all meats . . .)"[54]

But Grandin didn't find that cattle needed to be raised on grass-fed pastures in order to be calm. Rather, she found that what cattle most wanted was cleanliness and predictability. "Keeping the pens dry and keeping cattle clean—that's really important," she said.[55]

Grandin discovered that cows were being made nervous by visual and auditory surprises that had until then been ignored, such as swinging chains and loud, high-pitched banging. Things that felt out of the ordinary signaled danger to cows, and stressed them.[56]

Because "[e]thical reasons alone were not sufficient to convince the manager to change the practices," Grandin explained, she had to find things that resulted in both more humane treatment and lower costs. She and a student soon proved that cattle that remained calm during handling had higher weight gains than stressed cattle. Stress hormones damage meat, which is another reason it is in the farmer's interest to reduce his livestock's fear.[57]

Grandin's audits eventually made more than fifty farms more humane and more efficient.[58] She didn't eliminate all problems. In 2009, ten years after she began her audits, Grandin found that one-quarter of all beef and chicken slaughterhouses should not have passed inspection.[59] Even so, she had made progress. "Compared to the bad old days, it's drastically improved, and I mean drastically," said Grandin.[60]

6. The Nature of Death

One of the questions people frequently ask vegetarians is why it would be unethical for humans to eat animals but not unethical for animals to eat animals.

"Eating meat may be 'natural,' and most humans may find it acceptable—humans certainly have been doing it for a very long time—but these are not moral arguments," says a PETA spokesperson. "In fact, the entirety of human society and moral progress represents an explicit transcendence of what's 'natural.' " [61]

As a college undergraduate attempting to make sense of my vegetarianism, I remember finding this argument persuasive. We prohibit rape and murder not because they are unnatural but because they are immoral.

And yet many seemingly hard, moral arguments for animal rights are, in reality, animal *welfare* arguments.

Take the oft-used comparison of slavery and meat. The consequence of deciding slavery is immoral is making people free. The consequence of deciding meat is immoral is *not* making animals *free*. It's not making *animals*.

Is it more ethical to never create life than to create it and take it away? Foer, to his credit, doesn't claim there is a single right or wrong answer to that question. Instead, he returns to the issue of cruelty.

Foer quotes Grandin's report on factory farms from when she first started doing her work. Grandin documented "deliberate acts of cruelty," notes Foer. "*Deliberate* acts," he emphasizes, "occurring on a *regular* basis . . ." [62]

But one can find many more acts of cruelty in nature than in the slaughterhouse.

"Out on a western ranch, I saw a calf that had its hide ripped completely off on one side by coyotes," writes Grandin. "It was still alive and the rancher had to shoot it to put it out of its misery. If I had a choice, going to a well-run modern slaughter plant would be preferable to being ripped apart alive." [63]

From the perspective of the calf, the deliberate, regulated, and painless modern slaughterhouses may be better than the random, painful, and instinctual cruelty of nature. Either way, the ethics of meat are unavoidably subjective. They aren't something anyone should be dogmatic about.

And yet some vegetarian journalists, activists, and scientists have sought to demand that others follow their personal preferences in the name of environmental protection, particularly as it relates to climate change, and often in stealth fashion.

"Ninety percent of the climate scientists and environmentalists I've met are vegetarian," Foer told *Huffington Post* in 2019. "And the ones that aren't eat very little meat. It's something that seems to go without saying. I wish they would talk about it more, but it's been heartening to see." [64]

But it may be that scientists don't talk about it because people would rightly wonder if their vegetarianism biased their scientific objectivity. In my research I kept coming across cases of vegetarian activists who kept their motives hidden.

"A friend of mine had an experience a few years ago where two young guys came and asked if they could take footage for a documentary about farm life," a farmer told Foer. "Seemed like nice guys. But then they edited it to make it look like the birds were being abused. . . . Things were taken out of context." [65]

"When I started this research in the 1990s, and was looking at dietary salt," said science journalist Gary Taubes, "I interviewed a Harvard nutritionist who told me about going into nutrition science as a vegetarian and student at Berkeley in the late sixties so he could demonstrate to people that his way of eating was correct." [66]

Foer notes that PETA activists used the former head of the IPCC Rajendra Pachauri as a scientific authority on climate change because "he argues that vegetarianism is the diet that everyone in the developed world should consume, purely on environmental grounds." [67]

Sometimes Foer condemns animal farming for reasons that appear to have more to do with anti-capitalist ideology than the environment. The "economics of the market inevitably leads toward instability," he writes. [68]

Such a logic leads Foer to attack farmed salmon as worse for the environment than wild salmon, even though, as we saw, not only are farmed salmon of equal nutritional value as wild salmon, they substitute for wild salmon, and open up the potential of reducing overfishing, one of humankind's largest, and least-discussed, impacts on wild animals. [69]

"I have to say there is part of me that envies the moral clarity of the vegetarian," writes University of California journalism professor Michael Pollan, in a passage from his 2007 book, *The Omnivore's Dilemma*. "Yet part of me pities him, too. Dreams of innocence are just that; they usually depend on a denial of reality that can be its own form of hubris." [70]

The trouble with dogmatic vegetarianism is the same as with dogmatic environmentalism. It ends up alienating the very people needed for improving conditions for animals and reducing the environmental impact of farming.

"In the eighties, the industry tried to communicate with animal groups and we got burned real bad," a farmer told Foer. "So the turkey community decided there would be no more of it. We put up a wall and that was the end. We don't talk, don't let people onto the farms. Standard operating procedure. PETA doesn't want to talk about farming. They want to end farming. They have absolutely no idea how the world actually works."[71]

Taubes and Teicholz seemed partly vindicated in late summer 2019, when the *British Medical Journal* published a review of the nutritional science that upended decades of orthodoxy.

"Diets that replace saturated fat with polyunsaturated fat do not convincingly reduce cardiovascular events or mortality," it found. The authors said we "must consider that the diet-heart hypothesis is invalid or requires modification."

One of the authors had coined the term "the French paradox" to explain why the French could eat so much fatty food without getting fat. The *BMJ* article, and the science collected by Taubes and Teicholz, suggested it was not a paradox, after all.[72]

One month later, just as vegetarian critics were starting to respond to the *BMJ*, the prestigious U.S. scientific journal *Annals of Internal Medicine* published two of the largest and more rigorous studies of meat consumption to date. They found that any negative health impacts of eating red meat, to the extent they exist at all, would be too small to matter.[73]

"It should certainly not be interpreted as a license to eat as much meat as you like," wrote a medical columnist for *The New York Times*, a newspaper that for fifty years has advocated diets low in saturated fats. "But the scope of the work is expansive, and it confirms prior work that the evidence against meat isn't nearly as solid as many seem to believe."[74]

The pro-carb, anti-fat crusade turned out to be as bad for the environment as it was for people. By making pigs less fatty, breeders made them less efficient in converting feed into body mass. More grain and thus more land was required under the low-fat regime than would have been required under a normal-fat one.[75]

Much of the public's concerns about meat have thus been misplaced. Consumers continue to express anxiety over things like the use of growth-promoting hormones in beef, even though the Food and Drug Administration, World Health Organization, and Food and Agriculture Organization have all concluded that meat produced with them is safe for human consumption. The evidence suggests we should have been more concerned by the absence of fat in our meat than by the use of hormones in its production.[76]

7. Don't Eat Wild Meat

The hunting and consumption of wild game remains one of the primary causes of the decline of wild animals in poor and developing nations. Recall that the number of wild animals in the world declined by half in the fifty years between 1960 and 2010. Forests in Africa, Asia, and Latin America that were recently populated with wild animals suffer today from "empty forest syndrome" due to the killing of wild animals.[77] More than 50 percent of all mammal taxa (units for classifying organisms) in the Congo Basin are being hunted unsustainably.[78]

Poor nations like the Congo desperately need to both provide more protein to their people and increase the productivity of meat production to take pressure off the habitats of mountain gorillas, yellow-eyed penguins, and other endangered species.

While people in developing countries increased their per capita meat consumption from 10 kilograms per year to 26 kilograms between 1964 to 1999, people in the Congo and other sub-Saharan African nations experienced no change in per capita meat consumption.[79] When I asked Bernadette how often she and her family ate meat, she sighed wistfully and said, "Maybe once a year at Christmas."

While the people of the Congo do not eat mountain gorillas, they still kill and eat an astonishing 2.2 million tons of wild animals every year because they lack cheap, domesticated meat.[80]

Creating cheap and easily obtainable substitutes in the form of domesticated meat should thus be a high priority for conservationists. Reducing the amount of land required for meat production will allow for more land for people and wildlife.

"In some parts of eastern Congo there have been efforts to introduce alternatives, like fish farms, to reduce bushmeat consumption," the primatologist Annette Lanjouw told me. "Although people were happy to have cabbages and carrots on their plates, the only commodity they had that was valuable enough to transport and that they could sell (so that they could have cash) was meat. They could smoke and dry the meat and it could be transported long distances to urban centers."[81]

The most efficient meat production in North America requires twenty times less land than the most efficient meat production in Africa. Replacing wild animal meat with modern meats like chicken, pork, and beef would require less than 1 percent of the total land used globally for farming.[82]

The technical requirements for creating what experts call "the livestock revolution" are straightforward. Farmers need to improve breeding of animals, their diet, and the productivity of grasses for foraging. Increasing meat production must go hand-in-hand with increasing agricultural yields to improve and increase feed.

In northern Argentina, farmers were able to reduce the amount of land used for cattle ranching by 99.7 percent by replacing grass-fed beef with modern industrial production.[83]

We must change our thinking, too. Just as we overcame our preference for authentic furs, ivory, and tortoiseshell, we must retrain our preferences toward domesticated meats and away from wild meats, including fish, for wild animals once again to flourish.

8. Beyond Food and Evil

Whatever its psychological origins, vegetarianism appears to stem less from a rational consideration of the evidence than an emotional rejection of killing animals, something Foer acknowledges. "Food is never simply a calculation about which diet uses the least water or causes the least suffering."[84]

Indeed, when I returned to meat eating, it was an almost entirely instinctual, not intellectual decision. I hadn't spent any time reading about or discussing the ethical questions. My wife was pregnant and cooking filet mignon. It smelled amazing, and so I ate some.

Others rationalize their carnivorous desire. "When I moved to the Bay, I was aware of its reputation as an incredible place for food and felt that I would be missing out because of the limits I put on myself," said my colleague Madison. "I thought, 'Seafood is clearly different, ethically, and so I'll eat seafood.' But then I went to Paris and accidentally tried pâté and I was like, 'Forget it.' "

"But doesn't the act of killing animals still bother you?" I asked.

"Yes, it does," she said. "I try not to think about it."

"But you must have decided it was okay, ethically?" I asked.

"As I've grown up, things don't seem as black and white as they did when I was a kid," she said. "When I learned that it wasn't having the impact I thought on combating climate change, I decided it wasn't worth it. If you're not actually helping the planet, the calculation definitely changes. Besides, now I more clearly see a separation between humans and animals. Killing a chicken is not the same as murdering a human. There's an important difference there."

Foer recognizes that. And even on the fundamental question of whether it is ethical or unethical to eat animals, Foer accepts that there is not a single moral answer that is true for all of us, concluding, "If it were unhealthy to stop eating animals, that might be a reason not to be vegetarian. . . . Of course there are circumstances I can conjure under which I would eat meat—there are even circumstances under which I would eat a dog—but these are circumstances I'm unlikely to encounter."[85]

As such, we are left simply with personal preference. And, around the world, most people prefer to eat meat.

Even most vegetarians, it turns out, aren't actually vegetarians. A majority of vegetarians in Western countries tell researchers that they eat fish, chicken, and even red meat, on occasion.[86]

Even after Foer and his wife made vows of vegetarianism the same week they got engaged to be married, "We did occasionally eat burgers and chicken soup and smoked salmon and tuna steaks. But only every now and then. Only whenever we felt like it." [87]

Foer is at his best when he embraces a more empathetic position. "The question of eating animals," he writes, "is ultimately driven by our intuitions about what it means to reach an ideal we have named, perhaps incorrectly, 'being human.' " [88]

So, too, is PETA. "As Dr. Grandin has shown us," said PETA's founder, Ingrid Newkirk, "giving a little comfort and relief to animals who will be in those cages their whole lives is worth fighting for, even as some of us are demanding that those cages be emptied." [89]

In the end, Foer wants to be read as an individual telling his story, not as a moralizer. "My decision not to eat animals is necessary for me, but it is also limited, and personal," says Foer. "It is a commitment made within the context of my life, not anyone else's."

8

SAVING NATURE IS BOMB

1. The End of Nuclear Energy

On March 11, 2011, a tsunami triggered by a major earthquake struck the Fukushima Daiichi nuclear plant on the eastern coast of Japan. The forty-nine-foot wave disabled electrical power and inundated backup diesel generators. Without electricity, the plant's pumps could not maintain the steady flow of cooling water over hot uranium fuel inside three reactor cores. Within hours, uranium fuel rods had overheated and melted, triggering the worst nuclear accident since the 1986 disaster at Chernobyl in Ukraine.

But nuclear energy was on the decline well before the Fukushima accident: not a single new nuclear reactor had begun construction in the United States since the 1979 Three Mile Island accident, and the U.S. nuclear fleet was aging even then with no replacements in sight.

Even so, Fukushima accelerated nuclear's descent by putting the brakes on plans to build new plants, slowing the construction of already-approved ones and prompting Germany, Taiwan, and South Korea to phase out their use of nuclear energy entirely. And Fukushima turned public opinion even further against nuclear.

Every effort to make nuclear plants safer makes them more expensive, according to experts, and higher subsidies from governments are required to make them cost-effective. Those soaring subsidies, combined with the financial cost of accidents like Fukushima, estimated to be between 35 trillion yen and 81 trillion yen ($315 billion to $728 billion) by one private Japanese think tank, make nuclear one of the most expensive ways to generate electricity.[1]

Meanwhile, from Finland and France to Britain and the United States, nuclear plants are way behind schedule and far over budget. Two new nuclear reactors at Britain's Hinkley Point C were estimated to cost $26 billion but will now cost as much as $29 billion.[2] Expansion of a nuclear plant near Augusta, Georgia, which was supposed to take four years and cost $14 billion for two new reactors, is now expected to take ten years and cost as much as $27.5 billion.[3] All of this makes nuclear too slow and expensive to address climate change, many experts say.[4]

Nuclear has what energy experts call a "negative learning curve," meaning we get worse at building it the more we do it. Most technologies have a positive learning curve. Take solar panels and wind turbines, for instance. Their costs declined 75 percent and 25 percent, respectively, since 2011.[5] The more we make of them, the better we get at it and the cheaper they become. By contrast, average construction time of nuclear reactors in the United States and France ongoing or completed since 2000 is twelve years—twice as long as it took before the 1979 meltdown at Three Mile Island.[6]

Today, the developed world is abandoning nuclear. Germany is almost done phasing it out. France has reduced nuclear from 80 percent to 71 percent of its electricity and is committed to reduce it to 50 percent. In the United States, nuclear could decline from 20 percent to 10 percent of its electricity by 2030. Belgium, Spain, South Korea, and Taiwan are all phasing out their nuclear plants. While the nuclear industry promotes small new reactors, replacing a project like the four large reactors in one plant being built by Korea for the United Arab Emirates with the leading American design would require about one hundred small reactors, or eight power plants of twelve small reactors each.

Future generations may very well look back to 1996, when nuclear generated 18 percent of global electricity, as the peak of the technology. In 2018 it was at just 10 percent. Within a few years, it could be at 5 percent.

Before anyone realizes it, nuclear energy could be just a distant memory, or a collective bad dream about a time when humankind tried to redeem its invention of atomic weapons when what it should have done was scrap the technology entirely.

At least that's how the story goes. While all of the above is technically accurate, I carefully excluded key facts in order to be misleading in the same ways that antinuclear campaigners have been for fifty years.

2. "That Could Be Quite Nasty"

"There was never any doubt I would go into science," Gerry Thomas told me.[7] At the time, we were sitting in the backyard of my sister's house in Brookline, Massachusetts, a neighborhood of the larger Boston area, in late July 2018.

Gerry—short for Geraldine—is an expert on radiation and health in general, and the nuclear accidents in Fukushima and Chernobyl in particular. She is a professor of molecular pathology at Imperial College, London, and started the Chernobyl Tissue Bank.

"My parents met at the hospital," she said. "My mum worked in histology, the study of tissues, and dad in hematology, the study of blood—or it was the other way around?" She laughed. "I can't remember."

I got to know Gerry after calling her several times with questions about nuclear accidents for writing and speeches I had given. We were in Boston at the same time, and I asked if I could interview her about her work and life.

Gerry suffered an early tragedy when she was eleven years old. "After swim class, a classmate of mine turned to me in the changing room and said to me, 'Your mum has leukemia and is going to die.' I responded, 'No, no, it's not cancer, it's secondary anemia.' She said, 'Yes it is!' and ran off."

A few months later, during a four-hour drive to a summer camp in North Wales, Gerry's father broke the horrible news: her mother did indeed have leukemia and was going to die soon.

"I cried for an hour, and then he dropped me off," she said. "I now think it was because he didn't want to deal with a crying kid."

That September, after Gerry returned to school, she thought her mother was getting better. In truth, she was nearing the end.

"The last time I saw her she didn't know who I was," Gerry recalled. "She was my size but by then was little more than her skeleton. I helped her to the loo. A few days later, she died."

The early tragedy inspired Gerry to want to do something important in her life.

"I think going through all of that as a kid can give you the determination to do something with your life," she said. "You realize your life can be short, and you shouldn't wait. And I didn't. I suspect I was seen as pretty hard by my classmates, but I needed to be brave for my brother, who, though only two years younger, suffered from a severe developmental disability and didn't understand what had happened, and for my father, who had to keep it together so that he could go to work every day."

Gerry decided to study the science of medicine. As a university student she learned the dangers of air pollution. "Part of our course was to go see a post-mortem in the hospital, and we went down as a group—the first dead body I had ever seen, an elderly male. When it came to a resection of the lungs, he took the lungs out, cut into the lungs, and you could see black horrible stuff oozing out of the lungs."

Gerry asked the pathologist if the deceased had been a smoker. They replied, " 'No, that's just the effect of the pollution.' As a group, we were all extremely surprised. We had all thought he was a heavy smoker, and to see that instead it was because he lived in a hollow, a part of a city where the smoke sinks."

In 1984, Gerry developed a technique to evaluate both thyroid and breast cancer. Her supervisor offered her an opportunity to study thyroid cancer.

"We were studying pesticide impacts on animal cells," she said. "We were trying to understand unwanted side effects in animals, and whether they were coming from a single cell, and was it relevant to human health. We saw that you'd need high and prolonged exposure to radiation to create cancers in animals."

Then, when she was twenty-six years old in 1986, she saw news coverage of the Chernobyl disaster on TV.

"I remember thinking, 'That could be quite nasty.' I didn't give it too much more thought. Then, in 1989, my boss, a preeminent endocrine pathologist, and an Italian endocrine clinician, were asked to travel to Belarus. When he returned he was visibly shaken by how many childhood thyroid cancers there were."

The 1986 Chernobyl nuclear accident in modern-day Ukraine (then part of the Soviet Union) was the worst nuclear energy accident in history. Plant operators lost control of an unauthorized experiment that caused a reactor to catch fire. There was no containment dome, and radioactive particulate matter escaped.

Gerry went to Belarus and Ukraine and kept going back regularly to study patients who developed thyroid cancer. She eventually created the Chernobyl Tissue Bank to preserve removed thyroid glands and make them widely available to researchers seeking to understand radiation's impact.

According to the United Nations, twenty-eight firefighters died after putting out the Chernobyl fire, and nineteen first responders died in the next twenty-five years because of "various reasons" including tuberculosis, cirrhosis of the liver, heart attacks, and trauma.[8] The U.N. concluded that "the assignment of radiation as the cause of death has become less clear."

While the death of any firefighter is tragic, it's worth putting that number in perspective. Eighty-four firefighters died in the United States in 2018, and 343 died during the September 11, 2001, terrorist attacks.[9]

Gerry points out that the only public health impact from Chernobyl beyond the deaths of the first responders were twenty thousand documented cases of thyroid cancer in those aged under eighteen at the time of the accident. In 2017, the U.N. concluded that only 25 percent, five thousand, can be attributed to Chernobyl radiation.[10] In earlier studies, the U.N. estimated there could be up to sixteen thousand cases attributable to Chernobyl radiation by 2065, while to date there have been five thousand.

Since thyroid cancer has a mortality rate of only 1 percent, that means the expected deaths from thyroid cancers caused by Chernobyl will be just 50 to 160 over an eighty-year lifespan.[11]

"Thyroid cancer is not what most people think of as a cancer," said Gerry, "because it has such a low mortality rate when treated properly. Suddenly it becomes something you shouldn't be so scared of. It's not a death sentence. It shouldn't reduce a patient's life. The key is replacement hormones, and that wasn't an issue because thyroxine is dirt cheap."

What about non-thyroid cancers? The 2019 HBO miniseries *Chernobyl* claimed there was "a dramatic spike in cancer rates across Ukraine and Belarus."[12] That assertion is false: residents of those two countries were "exposed to doses slightly above natural background radiation levels," according to the World Health Organization (WHO). If there are additional cancer deaths they will be "about 0.6 percent of the cancer deaths expected in this population due to other causes." [13]

The WHO claims on its website that Chernobyl could result in the premature deaths of four thousand liquidators, but, says Gerry, that number is based on a disproven methodology. "That WHO number is based on LNT," she explained, using the acronym for the *linear no-threshold* method of extrapolating deaths from radiation.

LNT assumes that there is no threshold below which radiation is safe, but people who live in places with higher background radiation, like my home state of Colorado, do not suffer elevated rates of cancer. In fact, residents of Colorado, where radiation is higher due to its altitude and its elevated soil concentration of uranium, enjoy some of the lowest cancer rates in the United States. [14]

In Fukushima, Thomas says, nobody will die from radiation they were exposed to because of the nuclear accident. The Japanese government awarded a financial settlement to a Fukushima worker's family, after he claimed the accident caused his cancer. But the worker's cancer was highly unlikely to have come from Fukushima, Gerry says, because the level of radiation that workers were exposed to was simply too low.

Similar to Fukushima, a meltdown occurred in 1979 at Unit Two of Pennsylvania's Three Mile Island nuclear plant. The incident created a national panic that contributed to the halting of nuclear energy's expansion, despite neither killing anyone nor elevating anyone's risk of cancer.

It is difficult to find other major industrial accidents that kill nobody. In 2010, the Deepwater Horizon oil drilling rig caught fire, killed eleven people, and emptied more than 130 million gallons of oil into the Gulf of Mexico, keeping the Gulf contaminated for months. [15] Four months later, a Pacific Gas and Electric (PG&E) natural gas pipeline exploded just south of San Francisco and killed eight people. [16]

The worst energy accident of all time was the 1975 collapse of the Banqiao hydroelectric dam in China. It collapsed and killed between 170,000 and 230,000 people. [17]

It's not that nuclear energy never kills. It's that its death toll is vanishingly small. Here are some *annual* death totals: walking (270,000), driving (1.35 million), working (2.3 million), air pollution (4.2 million).[18] By contrast, nuclear's known *total* death toll is just over one hundred.[19]

Nuclear's worst accidents show that the technology has always been safe for the same inherent reason that it has always had such a small environmental impact: the high energy density of its fuel. Splitting atoms to create heat, rather than splitting chemical bonds through fire, requires tiny amounts of fuel. A single Coke can of uranium can provide enough energy for an entire high-energy life.[20]

As a result, when the worst occurs with nuclear—and the fuel melts—the amount of particulate matter that escapes from the plant is insignificant in comparison to the particulate matter from fossil- and biomass-burning homes, cars, and power plants, which killed eight million people in 2016.[21]

Nuclear is thus the safest way to make reliable electricity.[22] In fact, nuclear has saved more than two million lives to date by preventing the deadly air pollution that shortens the lives of seven million people per year.[23]

For that reason, replacing nuclear energy with fossil fuels costs lives. A study published in late 2019 found that Germany's nuclear phase-out is costing its citizens $12 billion per year, with more than 70 percent of the cost resulting from 1,100 excess deaths from "local air pollution emitted by the coal-fired power plants operating in place of the shutdown nuclear plants."[24]

3. France Beats Germany

With France and Germany, we can compare two major (sixth and fourth largest) economies, which are highly proximate geographically and at similarly high levels of economic development, on a decades-long time scale.[25]

France spends a little more than half as much for electricity that produces one-tenth of the carbon emissions of German electricity.[26] The difference is that Germany is phasing out nuclear and phasing in renewables, while France is keeping most of its nuclear plants online.

Had Germany invested $580 billion into new nuclear power plants instead of renewables like solar and wind farms, it would be generating 100 percent of its electricity from zero-emission sources and have sufficient zero-carbon electricity to power all of its cars and light trucks, as well.[27]

Nuclear has long been one of the cheapest ways to make electricity in the world. In most of the world, including Europe and Asia, nuclear electricity is usually cheaper than electricity from natural gas and coal.[28]

At a global level, there has been a natural experiment since 1965. Between 1965 and 2018, the world spent about $2 trillion for nuclear, and $2.3 trillion for solar and wind. At the end of the experiment, the world received about twice as much electricity from nuclear as it did from solar and wind.[29]

It's true that new nuclear plants are behind schedule and above costs, but this is typical for large construction projects, and has often been the case for nuclear plants, including many highly profitable ones operating today. Because nuclear plants are relatively inexpensive to run, the importance of cost overruns declines over time. This is particularly true as the lives of nuclear plants are extended from forty to eighty years.

As for nuclear waste, it is the best and safest kind of waste produced from electricity production. It has never hurt anyone and there is no reason to think it ever will.

When most people refer to nuclear waste, they are referring to the used nuclear fuel rods. After they cool for two to three years in spent fuel pools in nuclear plants, they are put in steel and concrete canisters and stored on land in a manner known as dry cask storage. This makes nuclear the only form of electricity that internalizes its waste product. All other forms externalize their waste onto the natural environment.

One of the best features of nuclear waste is that there is so little of it. All the used nuclear fuel ever generated in the United States can fit on a single football field stacked less than seventy feet high.[30]

If an airplane crashed into the canisters of used fuel, the plane would explode and the cement-sealed steel canisters would likely remain intact. Even were some used fuel to escape, it would not be the end of the world. Emergency workers could easily recover it.

There is no realistic way used nuclear fuel rods could contaminate a river or some other body of water. They are closely monitored and protected on land inside heavily guarded nuclear plants. It's hard to imagine how one would ever fall into a river. Even if one somehow did, there is little reason to believe the fuel would be exposed to the water. Even if the used fuel were exposed to water, the impact would be immeasurably small. Nuclear plant workers sometimes put on dive suits and enter the pools where used fuel is cooling. They are safe because the water shields them from dangerous levels of radiation.

When I talk to people who fear the waste, they often can't articulate why they believe it is dangerous, but it appears to emanate from a conscious or unconscious fear of nuclear weapons. However, to turn used fuel rods into a bomb would require transporting the giant casks to massive and complicated facilities that only exist in a few countries in the world, or building such a facility, to turn into weapons material.

And it is impossible to imagine a realistic scenario in which terrorists could break into a nuclear plant, use a crane to raise the 100-ton canister of used fuel rods onto an 18-wheel truck, drive it out of the plant along the highway to a coastal port, send it by boat to somewhere with a reprocessing plant, unload it, and then reprocess it. In the real world, the terrorists would be gunned down before getting through the nuclear plant's entrance.

Between 1995 and 2018, a period of large and unprecedented subsidies for solar and wind, the share of energy globally coming from zero-emission energy sources grew just two percentage points, from 13 percent to 15 percent. The reason is that the increase in energy from solar and wind barely made up for the decline in nuclear.[31]

And electricity is just one-third of total energy use, globally. The remaining two-thirds of primary energy consumption is dominated by fossil fuels, which are used for things like heating, cooking, and transportation.

Only nuclear, not solar and wind, can provide abundant, reliable, and inexpensive heat. Thus, only nuclear can affordably create the hydrogen gas and electricity that will provide services such as heating, cooking, and transportation, which are currently provided by fossil fuels.

And only nuclear can accommodate the rising energy consumption that will be driven by the need for things like fertilizer production, fish farming, and factory farming—all of which are highly beneficial to both people and the natural environment.

And yet the people who say they care and worry the most about climate change tell us we don't need nuclear.

Consider the case of climate activist Bill McKibben. Along with Vermont senator and 2020 Democratic presidential candidate Bernie Sanders, he urged Vermont legislators in 2005 to commit to reducing emissions 25 percent below 1990 levels by 2012, and 50 percent below 1990 levels by 2028, through the use of renewables and energy efficiency.[32] Vermont's main electric utility helped customers go "off-grid" with solar panels and batteries,[33] and the state's aggressive energy efficiency programs ranked fifth best in the nation for five years in a row.[34] But instead of falling 25 percent, Vermont's emissions actually rose 16 percent between 1990 and 2015.[35]

Part of the reason emissions rose in Vermont is that the state closed its nuclear power plant, something McKibben advocated. "I believe Vermont is completely capable of replacing (and far more) its power output with renewables, which is why my roof is covered with solar panels," he wrote.[36]

I emailed McKibben in early 2019 to ask if he regretted advocating for Vermont Yankee Nuclear Power Station's closure. He told me its demise "didn't, I think, lead to big increases in emissions from Vermont electricity." He pointed to a *New York Times* data tool, which he said showed "the state replaced the [nuclear] power by buying lots and lots more hydro from Quebec."[37]

But that's not what the data show. In reality, Vermont's utilities couldn't replace the lost electricity from Vermont Yankee with in-state generation, and turned to electricity imports from the New England power pool, which is primarily from natural gas.[38]

McKibben's opposition to nuclear is the rule not the exception among environmentalists. While referring to the Green New Deal, the office of New York Rep. Alexandria Ocasio-Cortez said in early 2019 that "the plan is to transition off of nuclear . . . as soon as possible."[39]

A few weeks later, environmental activist Greta Thunberg wrote on Facebook that nuclear is "extremely dangerous, expensive, and time-consuming,"[40] even though, as we have seen, the best-available science shows the opposite.

"They can't have it both ways," said MIT climate scientist Kerry Emanuel. "If they say this [climate change] is apocalyptic or it's an unacceptable risk, and then they turn around and rule out one of the most obvious ways of avoiding it [nuclear power], they're not only inconsistent, they're insincere."[41]

All of which raises a question: if nuclear power is so good for the environment and necessary for replacing fossil fuels, why are so many of the people who say they most fear climate change so against it?

4. Naturally Nuclear

In the early 1960s, Kathleen Jackson, an artist living on the central coast of California, launched an effort to protect the nearby Nipomo Dunes. Her strategy was to bring the state's most powerful people to see their beauty for themselves. One of them was the president of the Sierra Club, Will Siri, a biophysicist from the University of California, Berkeley. "I didn't know it looked like this," Siri told her when he visited in 1965. "It is magnificent."[42]

Siri was famous in conservation circles as a world-class mountaineer. In 1954, Siri led the first American climbing expedition to the Himalayas. They climbed Makalu, the fifth highest mountain in the world. En route, Siri came upon a member of the climbing team of Sir Edmund Hillary, who the year before had made history by becoming the first man, along with his Nepalese Sherpa guide, to reach the summit of Everest. One of their men had fallen into a crevasse, and Siri rescued him.[43]

During his eighteen years on the Sierra Club board, Siri played a key role in transforming the group from a gentlemanly San Francisco–based hiking club to one of the country's most powerful environmental organizations. "We couldn't play the role of country gentlemen," Siri recalled. "We were activists and had a lot of battles to win; and we couldn't always pull our punches to spare acquaintances in government bureaus."[44]

Under Siri's leadership, the Sierra Club won major victories to protect

ancient redwood forests, the Grand Canyon, and a Northern California valley that Walt Disney wanted to turn into a ski resort.[45]

Most of the Nipomo Dunes were undeveloped, but the economically depressed county of San Luis Obispo had zoned them industrial, and was actively seeking their development. "We toured the dunes, and it was clear that
they had to be preserved," recalled Siri later. "Some of the flora and fauna
was rare; it could not be found in many other places."[46] But now PG&E was
considering building a nuclear power plant at the location.

Siri and Jackson met with officials from PG&E and proposed a compro-
mise: PG&E, they said, could build the plant about a mile from the water.
That turned out to not make sense economically since the plant needed close
access to the water. Siri was unmoved. "We want the dunes preserved," Siri
told them, "go find another place."[47]

PG&E officials left and came back with a new proposal: it would build
six reactors at a single plant near Avila Beach on the coast. Doing so would
avoid the need for building more power plants along the coast. "If you were
going to wreck a piece of coast, one unit will do it as well as two," Siri explained later. "The object was to find a site where they could put multiple
units."[48]

Siri and Jackson brought the swap for Nipomo Dunes to the Sierra Club
board of directors for consideration. They debated the matter for a day and a
half. Siri argued that a high-energy society was a prerequisite to saving and
enjoying nature.

"Nuclear power is one of the chief long-term hopes for conservation, perhaps next to population control in importance," he wrote. "Cheap energy
in unlimited quantities is one of the chief factors in allowing a large rapidly
growing population to preserve wildlands, open space, and lands of high
scenic value. . . . Even our capacity and leisure to enjoy this luxury is linked
to the availability of cheap energy."[49]

It may have helped that Siri was a veteran of the Manhattan Project, the
research and development team that produced the first nuclear bombs during
World War II, and a biophysicist who understood the relative health risks of
coal versus nuclear plants.

But Siri was not alone in his thinking. In the 1960s, most conservationists favored nuclear plants as a clean energy alternative to coal plants and hydro-electric dams. So, too, did most Democrats and liberals. Indeed, there was widespread popular support for nuclear energy among Americans, Europeans, and others around the world, who viewed it as a clean, energy-dense form of effectively limitless energy.

"Ansel Adams [the great nature photographer and Sierra Club board member] was certainly strong in his support," said Siri. Adams was "absolutely adamant in feeling that [nuclear] was a reasonable solution to a difficult problem."[50]

Siri's arguments proved persuasive, and in 1966 the Sierra Club's board voted nine-to-one to not oppose PG&E's plans. The nuclear power plant would go forward with the tacit blessing of one of the most powerful organizations in California. And it would be given the name granted to the place by Spanish explorers: Diablo Canyon.[51]

5. Atoms for Peace

In early 1953, Robert Oppenheimer, the creator of the first atomic bomb, gave a speech to the Council on Foreign Relations. Though America's newly elected president, General Dwight D. Eisenhower, wasn't in the room, he was the person at whom Oppenheimer aimed his remarks.

In his speech, Oppenheimer explained that nuclear weapons had created a revolution in foreign policy. No defense against them was possible, only deterrence, or frightening away adversaries through the threat of assured destruction.

"We may anticipate a state of affairs in which two great powers will each be in a position to put an end to the civilization and life of the other, though not without risking its own," said Oppenheimer. "We may be likened to two scorpions in a bottle, each capable of killing the other, but only at the risk of his own life."[52] President Eisenhower needed to level with the American people about the terrifying new reality, he said.[53]

Eisenhower respected Oppenheimer's views and acted upon them. The president asked his speechwriter to prepare a major speech to the American people.[54] But the early drafts were too morbid. They were filled with graphic descriptions of atomic war. They left "everybody dead on both sides with no hope anywhere," Eisenhower complained. The president's advisors jokingly called it the "Bang! Bang!" speech.

Eisenhower felt that, in its current form, the speech would undermine his efforts to reduce military spending. It might even do what Oppenheimer feared, which was frighten the American people into demanding a preventative war.[55]

The more Eisenhower thought about the revolutionary power of nuclear energy, the more his desire grew for something bigger than mere "candor." What he really wanted to do was to make the Soviets a "fair offer" for disarmament, one that would also "contain a tremendous lift for the world—for the hopes of men everywhere."[56]

Seeking to solve the problem, Eisenhower hosted an after-dinner conversation with his advisors. They discussed the need to inform the public that the potential of nuclear holocaust was real.[57] Eisenhower's science advisor "immediately took up the case for scaring the people into a big tax program to build bomb defenses," noted one of the participants later. This made Eisenhower despair. "Is this all we can do for our children?" he asked the group.[58]

At that point, "Ike became greatly spirited," wrote one of the meeting's participants later, using the president's nickname, "and said that our great advantage was our spiritual strength—this was our greatest offensive and defensive weapon."[59] But the president's enthusiasm wasn't shared by the group. Eisenhower had to look elsewhere for inspiration.

As a boy raised by pacifist Mennonites, a Protestant sect that rejects war, it is possible that Eisenhower called upon his faith. He may have thought back to a famous passage from the Book of Isaiah, "He shall judge among the nations, and shall rebuke many people: and they shall beat their swords into plowshares, and their spears into pruning hooks: nation shall not lift up sword against nation, neither shall they learn war any more."[60]

As it became clear to Eisenhower and his advisors that the nuclear arms race was of universal and not just national importance, the president requested the opportunity to address the United Nations General Assembly. But before giving a speech to the world, Eisenhower wanted to win over European allies and so he met with British Prime Minister Winston Churchill. "Men needed power everywhere," Eisenhower told him. "If we could give hope, it would give these nations a stronger feeling of participation in the struggle of East and West, and such a feeling of participation would be on our side, and hope might be engendered from a fairly insignificant start."[61] Churchill was enthusiastic.[62]

The next day, December 8, 1953, Eisenhower stood before the United Nations General Assembly with a message of hope. With its seventy-five-foot ceiling and pew-like seating, the U.N.'s Assembly Hall created the experience of being in a cathedral. But behind the American president hung not a cross but rather the seal of the United Nations, a wreath of olive branches, a symbol of peace, embracing a map of the whole Earth.

As a hush fell over the congregants, the former military man began what would become one of the most consequential speeches of the twentieth century.

Eisenhower began by stating that he wasn't there to recite "pious platitudes," but rather to address issues of universal importance. "If a danger exists in the world," he explained, "it is a danger shared by all." The former Army general opened his address by acknowledging the "awful arithmetic" that humankind had enough atomic firepower to destroy itself. The rules of the game had irrevocably changed, Eisenhower explained.

The United States was no longer the sole owner of the terrible power. The "secret is possessed by our friends and allies," too, he said, and "the knowledge now possessed by several nations will eventually be shared by others, possibly all others." Conventional military superiority no longer guaranteed a nation's safety. "Let no one think that the expenditure of vast sums for weapons and systems of defense can guarantee absolute safety for the cities and citizens of any nation."[63]

Having reached the dark midpoint of his speech, Eisenhower declared that the possibility for annihilation wasn't the end of the story. "To stop there would be to accept helplessly the probability of civilization destroyed," he said. "My country's purpose is to help us move out of the dark chamber of horrors into the light." But how? Arms reduction wouldn't be sufficient, Eisenhower said. What was the point of peace if billions remained in poverty?[64]

Humankind could only redeem itself from the scourge of nuclear weapons by realizing the dream of universal prosperity—and that required cheap and abundant energy. "Experts would be mobilized to apply atomic energy to the needs of agriculture, medicine, and other peaceful activities," Eisenhower said. "A special purpose would be to provide abundant electrical energy in the power-starved areas of the world. Thus the contributing Powers would be dedicating some of their strength to serve the needs rather than the fears of mankind." [65]

Eisenhower closed by extending an olive branch to the other scorpion under glass. "Of those 'principally involved,' " he said, "the Soviet Union must, of course, be one."

Eisenhower's vision was at once material and spiritual, patriotic and internationalist, altruistic and self-interested. "The United States," he said, "pledges . . . to devote its entire heart and mind to find the way by which the miraculous inventiveness of man shall not be dedicated to his death, but consecrated to his life." [66]

After Eisenhower finished, there was a brief silence in the Assembly Hall, and then something extraordinary happened: representatives from every nation—communist and capitalist, Muslim and Christian, black and white, rich and poor—rose to their feet and applauded as a single chorus for ten minutes.

"Atoms for peace," as the speech, and the big, humanistic idea at the heart of it, was born.

The American people grew optimistic that the bright vision for nuclear energy that Eisenhower offered would redeem the creation of such a horrible weapon, and the positive reception of Atoms for Peace by people around the world made Eisenhower as happy as he ever was as president.

But the atomic hope wouldn't last. Within ten years, the war on nuclear power would begin.

6. The War on Nuclear

In 1962, a young Sierra Club staffer named David Pesonen visited Bodega Head, a site in northern California where PG&E wanted to build a nuclear power plant. Just a few years earlier, the California legislature had voted to turn the same stretch of coastline into a public park, and the University of California had announced plans to build a marine laboratory nearby. But PG&E had managed to make all those plans go away, and some of the local people were upset.

Pesonen told the Sierra Club's board of directors that he could stop the plant's construction but that typical conservationist arguments about the place's beauty wouldn't be enough. To win, Pesonen argued, they would need to convince the local people the nuclear plant would contaminate the countryside with radiation.[67]

Pesonen was inspired in part by publicity generated from a 1961 study published in the journal *Science*, which found that levels of strontium-90, a cancer-causing radioactive isotope, were fifty-five times higher in children's teeth born during nuclear weapons testing than before.[68] The amount was about 200 times less than the levels known to cause cancer, but enough to generate headlines. Parents demanded that U.S. President John F. Kennedy negotiate an end to weapons testing with the Soviet Union, which he did in 1963.[69]

One of the men who drew attention to radioactive fallout from weapons testing was Barry Commoner, a World War II veteran, socialist, and botanist at Washington University in St. Louis. Commoner had come to fame in the early 1950s when he helped the Nobel Prize–winning chemist and peace activist Linus Pauling to circulate a petition calling for a moratorium on weapons testing. Their argument was that the testing risked contaminating the public.[70]

Commoner viewed nuclear power plants as a "non-warlike excuse for continuing the development of nuclear energy . . . a kind of political Potëmkin Village." Nuclear energy, Commoner argued, was constructed in order to justify President Eisenhower's nuclear arms testing. "It's the most expensive charade in history," Commoner argued.[71]

The Sierra Club board of directors was taken aback by Pesonen's proposal. "Don't you dare mention public safety," one of them warned Pesonen. "The Sierra Club can talk about scenic beauty, and maybe the loss of scenic beauty, but not about public safety. That's not our job."[72]

Few on the board had any problem with nuclear power per se, and several others, including Will Siri, were advocates of it. One director called Pesonen an "extremist."[73]

He quit working with Sierra Club and started a new organization. Pesonen produced and distributed a report claiming the proposed nuclear plant would create "death dust," like nuclear fallout, that would contaminate local milk.[74]

Pesonen and his allies then attached notes to hundreds of helium balloons and released them from Bodega Head. The notes read, "This balloon could represent a radioactive molecule of Strontium-90 or Iodine-131—tell your local newspaper where you found this balloon."[75] The dairy farmers, alarmed by the apparent danger, started donating money to his cause.[76]

Pesonen and Commoner tapped into significant anxieties over nuclear weapons among baby boomers who had been subjected to duck-and-cover drills, where teachers ordered them to prepare for the apocalypse by hiding under their desks as schoolchildren, not to mention both government and Hollywood propaganda films.

Consequently, some activists who were originally focused on nuclear weapons disarmament began displacing their anxieties on nuclear reactors instead.[77] Displacement is a psychological concept very similar to scapegoating. The idea is that we take our negative emotions out on weaker objects because we fear the more powerful object. If the boss yells at us, we kick the dog because talking back to the boss is too dangerous. In this case, the nuclear weapons were the boss and nuclear power plants were the dog.

In the 1970s, groups like the Union of Concerned Scientists went from seeking nuclear disarmament to blocking the construction of nuclear power plants, eventually joining forces with other anti-nuclear groups, Friends of the Earth (FOE), Natural Resources Defense Council (NRDC), Sierra Club, and Greenpeace. All of those groups were as focused and perhaps more focused on stopping the construction of nuclear power plants as they were on any other cause in the 1970s, including and especially stopping coal power plants, which were the main alternative to nuclear at the time.[78]

In 2019, a friend of mine, who worked at Global Exchange, reached out to me to express his desire to advocate for nuclear energy in order to address climate change, something he had opposed in the 1970s. My friend, who is about a decade older than me, told me about participating in anti-nuclear weapons advocacy in the 1970s, and it seamlessly transitioned to campaigning against nuclear energy.

"What did you guys think?" I asked him. "That if you got rid of nuclear power plants then for some reason we would get rid of the bomb?"

He paused and looked into the distance for a few seconds before chuckling.

"I don't think we really thought about it that much," he said.

Opposition to nuclear power started rising in the mid-1960s. From 1962 to 1966, only 12 percent of applications by electric utilities to build nuclear plants were challenged. By the beginning of the 1970s, 73 percent of applications to build nuclear plants would be challenged.[79]

Despite these rumblings, nuclear power in the early 1970s still seemed like a promising way to deal with air pollution in places like Ohio, a major industrial state. Air pollution was so bad that people had to turn on their car headlights some days to see through the smoke.[80] On particularly bad days, people had to brush soot off their cars and rewash clothes they had hung out to dry.

Everybody agreed something had to be done, and so Ohio's electric utilities sought to build eight reactors across four different nuclear power plants.[81]

In 1970, the people of Ohio knew they needed nuclear power if they wanted cleaner air. That year, during a public hearing for a new nuclear plant, it appeared that the initial public wariness toward nuclear had given way to support. "People just aren't afraid of atomic energy anymore," a gas station attendant told the *Pittsburgh Press*.[82]

By 1971, the antinuclear faction had taken over the Sierra Club, which threw its full weight behind an effort to kill nuclear plants in Ohio. It hired lobbyists, filed lawsuits, and frightened local parents about the shipment of used fuel rods. [83]

Club lawyers were secretive about their work. "We're going to maintain our lawsuit and certain other plans that cannot be disclosed right now," one told Ohio's *Evening Review* in 1971.[84]

The Sierra Club was joined by a charismatic and aggressive young attorney named Ralph Nader, who had won the public's trust in the mid-1960s, and become a household name, after criticizing the safety of American cars. It is hard to overstate his influence in turning the public against nuclear energy. "A nuclear accident could wipe out Cleveland," Nader told an Ohio newspaper in 1974, "and the survivors would envy the dead." [85]

Antinuclear groups publicized a report written by a professor at the University of Pittsburgh, which claimed 400,000 infants had died from radioactive fallout from weapons testing.[86]

Nader and other antinuclear activists claimed that nuclear was far worse for the environment than fossil fuels. While the water that comes out of nuclear plants is clean, it is also warm, which could change the environment and impact animals. While nuclear plants produce no air pollution, they could overheat and melt, killing as many people as a bomb, they said. And while nuclear waste is the only waste from energy safely stored at the site of production, Nader and other campaigners constantly implied that it could somehow poison waterways or be used as a bomb.

Against the picture of antinuclear campaigners as ineffectual and inconsequential hippies, their ranks included Ivy League lawyers, well-paid lobbyists, and powerful Hollywood celebrities. The anti-nuclear movement raised significant amounts of money, which it used to finance protests, lobbying, and lawsuits and sow fear.

Few individuals did more to frighten the public about nuclear energy than the actress Jane Fonda. She starred in the anti-nuclear disaster film, *The China Syndrome*, and provided the leadership to get it made. In the film, a scientist famously claims that an accident at a nuclear plant "could render an area the size of the state of Pennsylvania permanently uninhabitable." The Three Mile Island nuclear power plant in Pennsylvania had an accident twelve days after the film's premier.[87]

It would be difficult to exaggerate Hollywood's role in turning the public against nuclear energy. Nuclear is the go-to scary technology for makers of films and television, and not just the bombs, nor even just the power plants, but even the largely harmless used fuel rods. By the early 1990s, the television show *The Simpsons* depicted nuclear waste as a green sludge leaking from drums. In a 1990 episode, Lisa and Bart catch a three-eyed fish from the river near the nuclear plant.[88] At the time, I thought the episode was funny in part because I thought it was, at least partly, true.

The nuclear industry in the West, especially its most powerful members, the electric utilities that own and operate nuclear plants, were taken aback by the cultural power of the anti-nuclear movement, and could barely muster a response. What the technology needed most were humanistic and environmentalist defenders like Siri. Instead, it was defended by nuclear engineers and utility executives, who came across as patronizing and uncaring. The industry retreated from public engagement and spent the next forty years focused on maintaining support from the communities around nuclear plants. Nuclear energy's scientific and technical associations retreated into university nuclear engineering departments and government laboratories.

Little wonder then that, after two decades of widespread misinformation that went largely unanswered by anyone, people who were already nervous about nuclear weapons came to imagine that nuclear plants generally and nuclear waste specifically constituted a significant threat to public safety.

Nader, the Sierra Club, and others insisted that nuclear wasn't needed because electricity consumption could be reduced, at a profit, through energy efficiency and conservation. "[Conservation] is all we need to do," claimed the Sierra Club's Brower in 1974, "plus doing a little bit better with the alternative technologies we have. If you then take what solar could do, solar and wind, by the end of the century you're in pretty good shape."[89]

But energy efficiency did not obviate the need for power. And per capita U.S. electricity consumption in the 1970s ended up rising almost as much as it had in the 1960s, and the overall population grew 14 percent between 1970 and 1980. As a result, when an electric utility didn't build a nuclear plant, it usually built one that burned coal, instead.[90]

Antinuclear environmentalists openly favored coal and other fossil fuels over nuclear. "We do not need nuclear power," said Nader. "We have a far greater amount of fossil fuels in this country than we're owning up to . . . the tar sands . . . oil out of shale . . . methane in coal beds . . ."[91] The Sierra Club's energy consultant, Amory Lovins, wrote, "Coal can fill the real gaps in our fuel economy with only a temporary and modest (less than twofold at peak) expansion of mining."[92]

The Sierra Club deliberately sought to make nuclear plants expensive. "We should try to tighten up regulation of the [nuclear] industry," wrote the organization's executive director, in a 1976 memo to the board of directors, "with the expectation that this will add to the cost of the industry and render its economics less attractive."[93]

Shortly after, the anti-nuclear advocacy by Sierra Club and Nader helped motivate President Jimmy Carter to advocate new coal plants rather than new nuclear ones.[94]

It's not that nobody knew of coal's dangers. In 1979, *The New York Times* published a front-page article noting that coal's death toll would rise to fixty-six thousand if coal instead of nuclear plants were built.[95]

Antinuclear groups sought additional regulations and sued to halt and slow construction. The antinuclear strategy to drive up costs by adding new regulations, or simply demanding new regulations be considered in order to create uncertainty and delay, worked.

"The economics are still there," a utility chief said in 1979, upon announcing they wouldn't build two new reactors at Davis–Besse Nuclear Power Station. "But when you burden it with all the regulatory requirements and delays, then it becomes pretty iffy."[96]

All told, antinuclear groups killed six nuclear reactors in Ohio, including Zimmer, which was 97 percent complete before being converted into a coal plant. Environmental groups Sierra Club, NRDC, and Environmental Defense Fund (EDF) accepted the conversion of Zimmer from nuclear to coal without complaint.[97]

And their work wasn't limited to Ohio. In Haven, Wisconsin, Sierra Club advocacy forced utilities to convert a nuclear plant under construction into a coal plant. All in all, the antinuclear movement managed to help kill in planning or cancel during or after construction half of all nuclear reactors that utilities in the United States had planned to build, even when it was known and acknowledged by everyone, including the environmental groups, that coal plants would be built instead.[98]

Were the antinuclear activists themselves really so afraid of nuclear? There are reasons to doubt it. A Sierra Club member who led the campaign to kill Diablo Canyon confessed, "I really didn't care [about nuclear plant safety] because there are too many people in the world anyway. . . . I think that playing dirty, if you have a noble end, is fine."[99]

Pesonen adopted the Machiavellian view that the ends justify the means. He scolded an ally for *not* lying. "If you had been as unscrupulous as [the opposition] just this once," said Pesonen, "it would have strengthened our position immeasurably."[100]

"If you're trying to get people aroused about what is going on," said one of Pesonen's antinuclear colleagues, "you use the most emotional issue you can find."[101]

The experience left Sierra Club board member and landscape photographer Ansel Adams bitter. "It shows how people can be really fundamentally dishonest at times," he said.[102]

7. The Power in Danger

As soon as the antinuclear groups won, they insisted they had nothing to do with the lawsuits, delays, protests, and regulations that had made nuclear plants so expensive to build. Nuclear had died from "an incurable attack of market forces," said Lovins.[103] They credited energy efficiency, even though electricity demand in Ohio rose in the 1970s nearly as much as it had in the 1960s.[104]

Today, antinuclear groups continue to deceive and frighten the public about nuclear energy in their efforts to shut down nuclear plants in the United States, Europe, and around the world. They do so with an eye to triggering fears of nuclear apocalypse. They claim nuclear is not necessary because of renewables. In reality, whenever nuclear plants aren't in use, fossil fuels must be used and emissions rise. They claim that used nuclear fuel rods and the plants themselves attract terrorists, when in reality the only ones who have attacked nuclear plants have been antinuclear activists. And they claim that radiation is cartoonishly potent.[105]

The public's fear of nuclear technology remains the main obstacle to its expansion. Surveys of people around the world find that nuclear is slightly less popular than coal, less popular than natural gas, and far less popular than solar and wind.[106]

People and nature have paid a high price for the war on nuclear and for our continuing fears of the technology. Air pollution from coal power shortened millions of lives that could have been saved with nuclear energy.

Fear of nuclear led to panic and negative mental health consequences in the former Soviet Union and Japan. The notion that people exposed to radiation are contagious was first used to stigmatize people in Hiroshima and Nagasaki.[107] History repeated itself in Chernobyl. Women as far away from the accident as Western Europe were misled to believe that Chernobyl radiation had contaminated them, which led them to terminate 100,000 to 200,000 pregnancies in a panic.[108] Adults who were near Chernobyl at the time of the accident, as well as the liquidators, were two times more likely to report "post-traumatic stress and other mood and anxiety disorders." [109]

In response to Fukushima, the Japanese government shut down its nuclear plants and replaced them with fossil fuels. As a result, the cost of electricity went up, resulting in the deaths of a minimum of 1,280 people from the cold between 2011 and 2014.[110] In addition, scientists estimate that there were about 1,600 (unnecessary) evacuation deaths and more than four thousand (avoidable) air pollution deaths per year.[111]

The problem started with the over-evacuation of Fukushima prefecture. About 150,000 people were evacuated but more than 20,000 have yet to be allowed to return home. While some amount of temporary evacuation might have been justified, there was simply never any reason for such a large and long-term evacuation. More than one thousand people died from the evacuation, while others who were displaced suffered from alcoholism, depression, post-traumatic stress, and anxiety.[112]

"With hindsight, we can say the evacuation was a mistake," said Philip Thomas (unrelated to Gerry Thomas), a professor of risk management at the University of Bristol, who in 2018 led a major research project on nuclear accidents. "We would have recommended that nobody be evacuated."[113]

The Colorado plateau is more naturally radioactive than most of Fukushima was after the accident.[114] "There are areas of the world that are more radioactive than Colorado and the inhabitants there do not show increased rates of cancer," said Gerry. And whereas radiation levels at Fukushima declined rapidly, "those [other] areas stay high over a lifetime, since the radiation is not the result of contamination but of natural background radiation."

Even residents living in areas of Fukushima with the highest levels of soil contamination were unaffected by the radiation, according to a major study of nearly eight thousand residents in the three years since the accident.[115]

In summer 2017, a team of nuclear plant managers gave me a tour of South Korea's two most recently built nuclear reactors, Shin Kori 3 and 4, as well as of two new reactors, Shin Kori 5 and 6, which are under construction. While there, I interviewed, through a professional translator, three senior construction managers. These managers had been building a very similar kind of nuclear reactor since the 1980s. They built eight reactors together during a thirty-five-year period. Each man was in his early- to mid-sixties.

In 2015, two French economists, Michel Berthélemy and Lina Escobar Rangel, identified the causes of nuclear plant cost escalation in both the United States and France using comprehensive data sets and econometric methods to separate causation from correlation. They found that only by sticking with the same design and the same team were builders able to shorten construction times and reduce costs over time.[116]

I asked the construction managers what had changed between South Korea's earlier and more recent reactor designs. They named incremental changes: the containment domes were thicker, the steel in the reactor vessels stronger, the doors were waterproof, they added portable generators, and they had improved the intake of cooling water so as to reduce the number of fish occasionally sucked into the plant.

I asked the men how they were building the two new reactors differently. Had there been any breakthroughs in construction methods? They insisted that they were building the plant the same way.

"Surely you must have done something differently!" I protested.

One of the men paused for a moment and then said, "We are using more of the smaller cranes."

It's not the case that nuclear only has a "negative learning curve." Standardization gives construction managers like the Koreans I met the opportunity to "learn by doing" and build each consecutive nuclear reactor a little faster and a little cheaper. The best available, peer-reviewed data set of nuclear construction costs shows that the cheapest plants are the ones that people have the most experience building and operating.[117]

Building multiple reactors at a single site, as PG&E wanted to do with Diablo Canyon, can also significantly reduce costs, economists Berthélemy and Rangel found, in both construction and operation.

Some experts have argued that manufacturing big chunks of reactors or even entire nuclear plants, called "modules," in factories might lower costs, and so I asked the Koreans if they had considered it. They told me that they already manufacture key plant components in factories on assembly lines, including reactor vessels, steam generators, and coolant pipes, and didn't think doing more would make a big difference.

The South Koreans had managed to increase the size of their reactors by 40 percent, from 1,000 megawatts to 1,400 megawatts, which constituted a massive leap forward in terms of efficiency and thus economics all while avoiding the terrible delays seen in France as reactors jumped up in size. In the United States and France, increasing the size of reactors had slowed construction time, Berthélemy and Rangel had found. While "larger nuclear reactors take longer to be built," they wrote, "they are also cheaper" when measured according to the electricity they produce. That's because a 40 percent increase in power did not require a 40 percent increase in the size of the labor force.[118]

Nuclear power plants have dramatically improved their productivity in other ways. Decades' worth of knowledge passed down means nuclear plants today can stay open and run at full capacity for much longer between refueling and maintenance, as compared to the nuclear plants of the 1960s.

Operating experience has also changed expectations of nuclear plant life. While regulators in the 1960s thought nuclear plants might run for only forty years, it is today clear that they can run for at least eighty.[119]

8. The Peace Bomb

As we have seen, fears of nuclear weapons have long contributed to fears of nuclear energy. In the climactic scene of HBO's 2019 *Chernobyl* series, the lead character claims, "Chernobyl reactor number four is now a nuclear bomb."[120] The claim was egregiously false, but many viewers undoubtedly believed it was true. I myself believed nuclear plants could explode like a nuclear bomb until I was an adult.

But was the invention of nuclear bombs the apocalyptic event so many people feared?

I asked the Pulitzer Prize–winning author of *The Making of the Atomic Bomb*, Richard Rhodes, if the invention of nuclear weapons had traumatized humankind. Rhodes and I became friends after being featured together in Robert Stone's 2013 documentary film, *Pandora's Promise*, which was about environmentalists who had changed their minds about nuclear energy.

"I remember once talking with Victor Weisskopf, a wonderful theoretical physicist from Austria, a Jew who had escaped Nazi Germany," said Rhodes. "He said, 'We were there at Los Alamos in the darkest part of physics.' I assume that's a reference to the potential for killing other human beings, the mass killing aspect of working on a nuclear weapon. He then said, 'And then [Danish physicist Niels] Bohr arrived and he gave us the possibility that there was hope at the end of all this.'

"How did Bohr do that? He did that by saying that nuclear was a fundamental change in our relationship with the natural world. Inevitably, it's going to change the way nation-states relate to each other. They will no longer be able to dominate one another. Now it would be possible for even a small state to deter a large state that wanted to dominate it. Of course, there was a dark side. But the fact that there's been no [nuclear] war since 1945 shows how correct Bohr was."

The closest the world came to nuclear war occurred in 1962, a tender thirteen years after the Soviet Union developed the bomb. It was then that the U.S. government discovered the Soviets had transferred missiles to Cuba.

President John F. Kennedy demanded that Soviet Premier Nikita Khrushchev remove the missiles and imposed a U.S. naval blockade. During the crisis, Air Force colonel Curtis LeMay, who would be lampooned a little more than a year later in Stanley Kubrick's *Dr. Strangelove*, pressured Kennedy to bomb Cuba. "We don't have any choice but direct military action," said LeMay. "I see no other solution." Kennedy rejected LeMay's advice.

The president and Khrushchev instead agreed that the United States would remove its missiles from Turkey, at a later date, in exchange for the Soviets removing their missiles from Cuba. New research shows that Kennedy would have publicly committed to removing American missiles from Turkey had Khrushchev insisted upon it.[121] That fact, among others, suggests that the two sides may not have been as close to war as historians had previously believed.[122]

Whatever the case, the stand-off resulted in an orchestrated relaxing of tensions between the United States and the Soviet Union, and a major effort to improve communications, including with China.

One of America's leading historians of the Cold War, John Lewis Gaddis, credits nuclear weapons with keeping the peace between the United States and the Soviet Union for so many decades. "It seems inescapable that what has really made the difference in inducing this unaccustomed caution," he said in 1986 speech, "has been the workings of the nuclear deterrent."[123]

The intensity and scale of major wars had risen in fits and starts for 500 years from the wide-scale introduction of firearms and artillery in the 1400s, until the death toll from battles and wars peaked in World War II at tens of millions of military and civilian deaths. And then from a post-war peak of more than 500,000 deaths in 1950, battle deaths in 2016 were 84 percent lower despite a tripling in the world population.[124]

Even if one gives no credit to nuclear weapons for the "Long Peace," it must be acknowledged that the apocalyptic fears about nuclear have been unrealized, and that we are further from global nuclear war now than at any other point in the last seventy-five years since the invention and use of the bomb.

After the Cold War, many experts in the West feared nuclear war between India and Pakistan. In 2002 the risk seemed high. The two nations mobilized one million troops along their shared border as part of a long-running dispute over territorial claims on the region of Kashmir. "Many of the political, technical, and situational roots of stable nuclear deterrence between the United States and the Soviet Union," worried one U.S. expert, "may be absent in South Asia, the Middle East, or other regions to which nuclear weapons are spreading."[125]

But then political leaders in India and Pakistan considered the likely impacts of nuclear war and frightened each other into peace, just as the United States and Soviet Union had done before them. "In South Asia [the bomb] has, for all practical purposes, done away with the prospect of full scale war," said an India-Pakistan military expert recently. "It's just not going to happen. The risks are so great as a consequence of the nuclearization of the subcontinent that neither side can seriously contemplate starting a war."[126]

Americans and Europeans today worry about North Korea, which has nuclear weapons, and Iran, which most experts believe wants them. But even the most hawkish experts believe they will act like other nuclear-armed nations.

In 2019, a former director of the U.S. nuclear weapons laboratory in Los Alamos concluded that North Korea is "less dangerous today than it was at the end of 2017."[127]

Yes, North Korean missiles can still reach Japan and South Korea, and experts believe it won't ever give up its nuclear arsenal. But relations between the United States and North Korea have stabilized, just as U.S. relations did between the Soviet Union and China.

Iran is aware that Israel has been nuclear-armed since the 1960s. Just because a regime is sometimes violent and cruel doesn't make it suicidal. "Nuclear weapons and terrorist groups have both existed for nearly seventy years," wrote Matthew Kroenig of Georgetown University, "and no state has ever provided nuclear capabilities to a terrorist organization. . . . It is likely that Iran would show similar restraint . . ."[128]

Since 1945, leading experts have echoed the argument of International Relations founder Kenneth Waltz, that the idea of humans ever doing away with nuclear weapons is "fanciful." If two nations dismantled their atomic bombs and then went to war with each other, they would simply reenter the "mad scramble to rearm."[129]

"There is no permanent method of exorcising atomic energy from our affairs, now that men know how it can be released," concluded a 1952 report for President Eisenhower that Robert Oppenheimer oversaw. Said Oppenheimer, "It is hard to see how there could be any major war in which one side or another would not eventually make and use atomic bombs."[130]

Even advocates of disarmament agreed. "Whatever agreements not to use H-bombs had been reached in time of peace, they would no longer be considered binding in time of war," acknowledged Albert Einstein and British philosopher Bertrand Russell in 1955, "for, if one side manufactured the bombs and the other did not, the side that manufactured them would inevitably be victorious."[131]

Today, just 25 percent of Americans say they believe nuclear weapons can be eliminated.[132]

When a *New York Times* reporter asked Oppenheimer how he felt after the bomb was tested on July 16, 1945, the father of the atomic bomb said, "Lots of boys not grown up yet will owe their life to it."[133]

After the bombings of Hiroshima and Nagasaki, Oppenheimer put out word that "the atomic bomb is so terrible a weapon that war is now impossible."[134]

9

DESTROYING THE ENVIRONMENT
TO SAVE IT

1. "The Only Path"

In spring 2015, Elon Musk walked on stage to loud applause from an audience of hundreds of supporters and invited guests. "What I'm going to talk about tonight," he said, "is a fundamental transformation of how the world works, about how energy is delivered across Earth.

"This is how it is today—it's pretty bad." He showed a graph of rising carbon dioxide concentrations in the atmosphere. "I think we collectively should do something about this and not try to win the Darwin Award."

The crowd laughed and Elon smiled, before continuing. "We have this handy fusion reactor in the sky called the Sun. You don't have to do anything, it just works. It shows up every day and produces ridiculous amounts of power."

Musk said there was no need to worry about land use requirements. "Very little land is needed to get rid of all fossil fuel electricity generation in the United States," Musk told the crowd. "It's really not much and most of that area is going to be on rooftops. You won't need to disturb land, you won't need to find new areas, it's mostly just going to be on the roofs of existing homes and buildings.

"Now, the obvious problem with solar power is that the Sun does not shine at night," Musk continued. "I think most people are aware of this. So, this problem needs to be solved. We need to store the energy that is generated during the day so that you can use it at night, and also even during the day the energy variation varies. There's a lot more energy generated in the middle of the day than at dawn or dusk. So it's very important to smooth out that energy generation and retain enough so that you can use it at night."

Hence the need for Tesla's new product: Powerwall, a battery that hangs on the wall of a garage. Musk said his batteries and panels would provide cheap and reliable electricity for those living in remote parts of the world, where energy is intermittent and expensive.

"In fact, I think what we'll see is something similar to what happened with cell phones versus landlines where the cell phones actually leapfrogged the landlines and there wasn't a need to put landlines in a lot of countries or in remote locations," Musk predicted. "People in a remote village or an island somewhere can take solar panels, combine it with a Tesla Powerwall, and never have to worry about having electricity lines."

With only 160 million Powerpacks, Musk said, solar could power the United States. With 900 million, they could power the entire world.[1]

"You can basically make all electricity generation in the world renewable and primarily solar," Musk said. "The path I've talked about, solar panels and the batteries, is the only path that I know of that can do this. And I think it's something that we must do and that we can do and that we will do."

Musk ended his talk to roaring applause and walked off stage.[2]

2. Unreliables

The dream of a world powered by solar panels and batteries inspired me to advocate a New Apollo Project, the precursor to the Green New Deal, as a renewable energy advocate beginning in 2002. In the same way we had gained access to revolutionary power with our smartphones, I thought, we were about to gain access to similarly revolutionary power with our solar panels and batteries.

Why, then, didn't we?

While there has been modest growth in demand for Tesla Powerwalls, notes one analyst, "It is unclear if Tesla's storage business is driven by availability of batteries or if there is a demand surge for residential energy storage." In truth, there is little evidence that demand among homeowners has risen to buy Powerwalls.[3]

The cost of buying and installing the latest Tesla Powerwall is more than $10,000. The cost of installing solar panels on top of that ranges from $10,000 to $30,000.[4] Helen and I pay about $100 per month for electricity. It would thus take at least 200 months, or more than seventeen years, for us to recoup our investment.

It's possible the solar panels and batteries system might pay for itself after that for a few years. But consider that the amount of electricity coming from panels declines every year, which is why most people say the lifetime of a system should be considered twenty or twenty-five years. Plus, if Helen and I decide to move to a different house before then we are unlikely to recoup our investment. Why would we put money into a speculative investment on solar panels and batteries rather than our retirement savings?

And if we can't afford solar panels and batteries, how could Suparti, much less Bernadette?

Even if they could, it's not clear that they would provide sufficient electricity. In Uganda, Helen and I stayed at an eco-lodge equipped with solar panels and batteries. But after a single day of cloudy weather, we quickly drained the lodge's batteries charging our laptops, cameras, cell phones, and other devices. When we told the lodge manager that we needed more electricity, he did what small businesses across sub-Saharan Africa do, and fired up a diesel generator.

Even so, energy analysts are bullish on renewables. The U.S. government estimates renewables will be a larger source of electricity than natural gas in the United States by 2050. Globally, it estimates renewables will rise from being 28 percent of the world's electricity in 2018 to nearly 50 percent in 2050.[5]

But those numbers are misleading. While renewables in 2018 globally generated 11 percent of total primary energy, 64 percent of it (7 percent of total primary energy) came from hydroelectric dams.[6] And dams are largely maxed out in developed nations, while their construction is opposed by environmentalists in poor and developing ones.

Despite the hype, the shares of global primary energy from solar and wind in 2018 was just 3 percent, the share coming from geothermal was 0.1 percent, and tidal was too small to measure.[7]

But won't solar and wind take off now that the costs of batteries are rapidly declining?

The costs of batteries are declining, but the progress has been gradual, not radical. The switch from nickel-cadmium to lithium-ion batteries during the last several decades has been wonderful. It has allowed for the proliferation of cordless phones, cell phones, laptops, wireless electric appliances, and an array of electric vehicles, small and large. But it has not allowed for the cheap storage of the grid's electricity.

Consider Tesla's most famous battery project, a 129 megawatt-hour lithium battery storage center in Australia. It provides enough backup power for 7,500 homes for four hours.[8] But, there are nine *million* homes in Australia, and 8,760 hours in a year.

One of the largest lithium battery storage centers in the world is in Escondido, California. But it can only store enough power for about twenty-four thousand American homes for four hours.[9] There are about 134 *million* households in the United States.

To back up all the homes, businesses, and factories on the U.S. electrical grid for four hours, we would need 15,900 storage centers the size of the one in Escondido at a cost of $894 billion.[10]

Various studies have shown that the cost of integrating unreliable wind energy is high and rises as more wind is added to the system. For example, in Germany, when wind is 20 percent of electricity, its cost to the grid rises 60 percent. And when wind is 40 percent, its cost rises 100 percent.[11]

This is because of all the power plants, often natural gas, that must be standing by and ready to fire up the moment wind dies down, the extra power lines that have to be built to remote renewable energy locations, and all of the other extra equipment and personnel required to support fundamentally unreliable and often unpredictable forms of energy.

Another study by a group of climate and energy scientists found that when taking into account continent-wide weather and seasonal variation, for the United States to be powered by solar and wind, while using batteries to ensure reliable power, the battery storage required would raise the cost to more than $23 trillion.[12] That number is $1 trillion higher than U.S. gross domestic product was in 2019.

Does that seem likely? Consider that in 2018, the Associated Press and University of Chicago conducted a survey and found that 57 percent of Americans were willing to pay $1 per month to combat climate change, 23 percent were willing to pay $40 per month, and just 16 percent were willing to pay $100 each month to combat climate change. The survey found 43 percent were unwilling to pay anything.[13]

Even the leading advocates of renewable energy recognize that batteries will not solve the problems created by the daily and seasonal cycles of solar and wind, and they look elsewhere for storage solutions.

The most influential proposal for 100 percent renewable energy was created by Stanford professor Mark Jacobson, who points out that most renewable energy proposals only try to replace the third of U.S. energy coming from the electricity grid.

Jacobson's proposal for all energy, not just electricity, relied on the conversion of existing hydroelectric dams into the equivalent of giant batteries. The idea is that when the sun shines or the wind blows, vast quantities of surplus power would be used to pump water uphill in some cases or entire river systems simply stopped almost entirely for large amounts of time. The water would be stored as long as necessary, and then released to flow downhill back through turbines to produce electricity when needed.[14]

Jacobson's studies and proposals became the basis of the energy plans of many American states, as well as that of Democratic presidential candidate Senator Bernie Sanders.[15]

But in 2017, a group of scientists pointed out that Jacobson's proposal rested upon the assumption that we can increase the amount of instantaneous power from U.S. hydroelectric dams more than ten-fold when, according to the Department of Energy and other major studies, the real potential is just a tiny fraction of that. Without all that additional hydropower, Jacobson's 100 percent renewables proposal falls apart.[16]

Even though California is a world leader when it comes to renewables, the state hasn't converted its extensive network of dams into batteries. You need the right kind of dams and reservoirs, and even then it's an expensive retrofit. There are also many other uses for the water that accumulates behind dams, namely irrigation and cities. And because the water in California's rivers and reservoirs is scarce and unreliable, the water from dams for those other purposes is becoming ever-more precious.

Without large-scale ways to back up solar energy, California has had to block electricity coming from solar farms when it's extremely sunny, or even pay neighboring states to take it in order to avoid blowing out the Californian grid.[17]

Germany is investing billions to develop a way to use its solar and wind electricity to make hydrogen, which would be stored and then burned or used to generate energy through fuel cells at a later date or time.[18]

But it's turning out to be too expensive. "From a business perspective, it isn't worth it," reported Der Spiegel in 2019 on what had seemed to be a promising hydrogen storage project. "Much of the energy is lost in the process of turning wind into electricity, electricity into hydrogen, and then hydrogen into methane—efficiency is below 40 percent. It isn't enough for a sustainable business model."[19]

Even if much more storage were available, it would still make electricity more expensive. The low cost of electricity comes from the fact that it is mostly made in large, efficient plants and distributed with very low losses across a shared grid connecting producers and consumers. While current electrical systems allow for a modest amount of storage, it is on the order of a few minutes, not the days and weeks that would be required for entirely renewable electricity. Every new energy conversion, such as from electricity to dam, battery, or hydrogen gas, and back again, imposes a massive physical and economic cost.

The big oil and gas companies know perfectly well that batteries can't back up the grid. The places integrating large amounts of solar and wind onto electricity grids are relying more and more on natural gas plants, which can be ramped up and down quickly to cope with the vagaries of the weather.

France is a perfect example. After investing $33 billion during the last decade to add more solar and wind to the grid,[20] France now uses less nuclear and more natural gas than before, leading to higher electricity prices and more carbon-intensive electricity.[21]

Between 2016 and 2019, the five largest publicly traded oil and gas companies—ExxonMobil, Royal Dutch Shell, Chevron Corporation, BP, and Total—invested a whopping $1 billion into advertising and lobbying for renewables and other climate-related ventures.[22]

Their ad blitz has targeted the global elite in airports and on Twitter. "Natural gas is the perfect partner for renewables," say airport ads run by Norwegian oil and gas giant Statoil.[23] "See why #natgas is a natural partner for renewable power sources," tweets Shell.[24]

In 2017, I attended the U.N. climate change talks in Munich, Germany, on the invitation of the climate scientist James Hansen. When I got off the plane, I was confronted by airport ads paid for by Total, the French oil and gas company, reading, "Committed to Solar" and "Committed to Natural Gas."[25]

3. Renewables Predator, Wildlife Prey

"I grew up in the sixties during a time when the planet really was dirty," says Lisa Linowes. "Our rivers were filthy. People would just throw their trash out on the streets. I became a very strong environmentalist at a time when it meant you really did care about literally picking up the garbage."[26]

Linowes is a lifelong environmental activist and a leader in a growing grassroots movement fighting the expansion of industrial wind turbines in North America and Europe. When I interviewed her in late 2019, she was finishing a review of the science documenting wind energy's impact on birds and bats.

In 2002, Linowes and her husband purchased property in New Hampshire. They soon learned there was a wind farm being built near town. "And I was like everyone else and was like, 'Wind? What's the problem with wind?'"[27]

"We were all indoctrinated into the idea that renewables are better than fossil fuel and the only reason renewables haven't taken off is because the oil and gas industry squeezed them out of the market," she said. "When we came to understand the enormity of the project on the landscape and the impacts on the environment, we knew we had to mobilize."[28]

Linowes and others learned that a wind farm requires roughly 450 times more land than a natural gas power plant.[29]

It didn't take Linowes and her husband very long to learn that nearly everyone else in her small New England town of about 500 residents felt the same way about the wind farm.

"But it took everything to win: working with the community, understanding the laws, beating them on the laws, and making the more convincing arguments," she said.[30]

Linowes learned early on that, in many countries, wind turbines pose the single greatest threat to bats after habitat loss and white-nose syndrome. "The wind industry is well aware of the problem yet vigorously resists even modest mitigations known to reduce bat mortality at operating wind facilities," she said. "The result is that many of our bat species are on a path to extinction. Even five years ago they were abundant in large numbers, particularly the hoary bat, but they've declined heavily."[31]

In some places such as Texas, where white-nose syndrome, a deadly fungus, has only recently arrived, wind turbines are the single greatest threat to bats. "There are no other well-documented threats to populations of migratory tree bats that cause mortality of similar magnitude to that observed at wind turbines," one scientist wrote.[32]

By occupying large areas of migratory habitat, wind turbines have also emerged as one of the greatest threats to large, threatened, and high-conservation value birds.[33]

"Look at the whooping crane," said Linowes. "The wind industry wants to expand to its habitat," she explained.

"With just 235 whooping cranes in the wild, their gene pool is very limited. A rule of thumb is that you need at least a thousand individuals to make sure the gene pool will grow and so you don't get inbreeding and lose diversity."

Wind energy also threatens bird species, including golden and bald eagles, burrowing owls, red-tailed hawks, Swainson's hawks, American kestrels, white-tailed kites, peregrine falcons, and prairie falcons, among many others, Linowes said. The expansion of wind turbines has particularly harmed the golden eagle in the western United States, where its population is at a very low level.[34]

The wind industry claims house cats kill more birds than wind turbines, but whereas cats mainly kill small, common birds, like sparrows, robins, and jays, wind turbines kill big, threatened, and slow-to-reproduce species like hawks, eagles, owls, and condors.[35]

In fact, wind turbines are the most serious new threat to several important bird species to emerge in decades. The rapidly spinning blades act like an apex predator that big birds never evolved to deal with. "Birds have evolved over hundreds of years to fly certain paths to migrate," Lisa explained. "You can't throw a turbine up in the way and expect them to adapt. It's not happening."

And because big birds have much lower reproductive rates than small birds, their deaths have a far greater impact on the overall population of the species. For example, golden eagles will have just one or two chicks in a brood, and usually less than once a year, whereas a songbird like a robin could have up to two broods of three to seven chicks a year.

Wind turbines may have a significantly larger impact in poor and developing nations rich in wildlife.

Scientists calculate that a single new wind farm in Kenya, inspired and financed by Germany, will kill hundreds of endangered eagles because it will be located on a major flight path of migratory birds. "It's one of the three worst sites for a wind farm that I've seen in Africa in terms of its potential to kill threatened birds," one biologist said.[36]

In response, the wind farm's developers have done what industries have long done, which is to pay the organizations that ostensibly represent the doomed animals to collaborate rather than fight.[37]

No nation has done more to support renewables than Germany. For the last twenty years it has been going through what it calls an *Energiewende*, or energy transition, from nuclear and fossil fuels to renewable energy sources. It will have spent $580 billion on renewables and related infrastructure by 2025, according to energy analysts at Bloomberg.[38]

And yet, despite having invested nearly a half-trillion dollars, Germany generated just 42 percent of its electricity from wind, solar, and biomass, as compared to the 71 percent France generated from nuclear in 2019. Wind and solar were just 34 percent of German electricity, and relied upon natural gas as back-up.[39]

Germany spent about thirty-two billion euros on these renewables every year between 2014 and 2019, or 1 percent of its GDP a year, the economic equivalent of the United States spending $200 billion annually, to increase its share of electricity from solar and wind from 18 percent of electricity to 34.[40]

And yet, in the fall of 2019, the consulting giant McKinsey announced that Germany's *Energiewende* posed a significant threat to the nation's economy and energy supply. "Problems are manifesting in all three Dimensions of the energy industry triangle—climate protection, security of supply, and economic efficiency," wrote McKinsey.[41]

Germany's electricity grid came close to having blackouts for three days in July 2019. Germany had to import emergency power from neighboring nations to stabilize its grid. "The supply situation will become even more challenging in the future," McKinsey wrote.

The cost to consumers of renewables has been staggeringly high. Renewables contributed to electricity prices rising 50 percent in Germany since 2007.[42] In 2019, German electricity prices were 45 percent higher than the European average.

It's been a similar story in the United States. "Cumulatively," wrote the authors of a University of Chicago report on renewables, "consumers in the twenty-nine states studied paid $125.2 billion more for electricity than they would have in the absence of the policy."[43] Electricity prices in renewables-heavy California have risen six times faster than in the rest of the United States since 2011.[44]

In the end, there is no amount of technological innovation that can solve the fundamental problem with renewables. Solar and wind make electricity more expensive for two reasons: they are unreliable, thus requiring 100 percent backup, and energy-dilute, thus requiring extensive land, transmission lines, and mining.

In other words, the trouble with renewables isn't fundamentally technical—it's natural.

The physical demands of renewables thus spark local environmental opposition around the world. Of the 7,700 new kilometers of transmission lines Germany needed for the energy transition, only 8 percent have been built; in 2019, the deployment of renewables and related transmission lines slowed rapidly.[45]

As goes Germany so may go the world. Globally, 2018 was the first year since 2001 that growth in renewables failed to increase.[46] Many are pessimistic that the renewables expansion can continue, for physical, environmental, and economic reasons. "The wind power boom is over," concluded German newsmagazine *Der Spiegel* in 2019.[47]

4. Powering Utopia

The idea that a prosperous society could be powered by renewables was first proposed in 1833 by a man named John Etzler. That year he published his utopian manifesto: *The Paradise within the Reach of all Men, without Labor, by Powers of Nature and Machinery.*

With a precision and passion similar to today's renewable energy enthusiasts, Etzler laid out a plan for scaling up concentrated solar power plants, gigantic wind farms, and dams to store the power when neither wind nor sun was available. "I promise to show the means for creating a paradise within ten years, where everything desirable for human life may be had for every man in superabundance," he wrote.[48]

Etzler anticipated the objection that sunlight and wind are unreliable. "[I]t will now be objected, that there is not always sunshine, that the nights and cloudy or foggy weather interrupt the effect." He claimed that by storing power for use later through pumping water uphill or winding it up as in a clock, "the interruption of sunshine . . . is therefore immaterial."[49]

The utopian zeal and specificity of Etzler's renewables vision are eerily similar to the style and manner of both Lovins and Jacobson. He proposed a land-based wind farm of sails exactly 200 feet high and a mile long, for example. When the sails were put at a perfectly right angle, he explained, its operators would generate one horsepower for every hundred square feet.

Just as today's advocates claim that renewables are light on the land and allow humans to break free from "extractive" industries, Etzler said that his wind, water, and solar machines work "without consuming any material."

But conservationist Henry David Thoreau was horrified by the amount of landscape Etzler's vision would require. "Could he not heighten the tints of flowers and the melody of birds?" asked Thoreau, sarcastically. "Should he not be a god to them?"[50]

Thoreau need not have worried. Etzler's supposedly superior wind- and water-powered plow—what he called the "Satellite" because it used wind to spin around an axis—proved grossly impractical and broke down. Soon after, so did his utopian commune.[51]

From time to time, people find niche applications for solar energy. In 1911, an inventor used parabolic troughs to concentrate sunlight and drive an engine, albeit at a prohibitively high cost.[52] In 1912, a plant was built in Egypt to use solar power to pump water for agriculture, but oil proved cheaper and easier.[53] Before 1941, half of all homes in Miami used solar water heaters. But they broke down often and, by the 1970s, had been replaced by a more reliable energy source, natural gas.[54]

Solar homes became a fad in the 1940s. Popular enthusiasm for renewables inspired President Harry Truman to appoint a blue-ribbon commission, headed by the CEO of CBS, which concluded that thirteen million solar homes could be built by 1975. But it would be just another unrealized utopian vision.[55]

After World War II, many intellectuals conjured visions of a world powered by renewables. The key to ending humankind's alienation from nature, the influential German philosopher Martin Heidegger argued in 1954, was for societies to use *unreliable*, not reliable, renewables. He condemned hydroelectric dams, which created large reservoirs of water that allowed for energy to be created whenever humans needed it. By contrast, he praised windmills.[56]

In 1962, American socialist writer Murray Bookchin denounced cities for spreading over the countryside like a rampant "cancer" and praised renewables as an opportunity for bringing land and city into a "synthesis of man and nature." Bookchin recognized that his proposal "conjures up an image of cultural isolation and social stagnation, of a journey backward in history to the agrarian societies of the medieval and ancient worlds." Still, he insisted his vision was neither reactionary nor religious.[57]

The antinuclear activist Barry Commoner similarly saw renewables as the key to bringing modern civilization, or the "technosphere," into harmony with the "ecosphere." Commoner invented the basic outline of the Green New Deal that was introduced first by European Greens and then by Rep. Alexandria Ocasio-Cortez in 2019. Commoner viewed the transition to a low-energy, renewable-powered economy as key to "massively redesigning the major industrial, agricultural, energy, and transportation systems . . ."[58]

Commoner's vision will sound familiar: farmers should go organic; we should use biofuels and other bioenergies; our cars should be smaller; homes and buildings should be made more energy efficient; and we should reduce our use of plastic.[59] He endorsed public subsidies for alternative energy and other technologies, along with military-style procurement, like the kind that resulted in microchips, for solar energy.[60]

Advocates claimed renewables could replace fossil fuels and nuclear. In 1976, Amory Lovins wrote in *Foreign Affairs* that the obstacles to a renewables economy are "not mainly technical, but rather social and ethical." Like Etzler, Lovins dismissed concerns over reliability. "Directly storing sunlight or wind," he explained, "is easy if done on a scale and in an energy quality matched to most end-use needs." He pointed to low-tech solutions like "water tanks, rock beds, or perhaps fusible salts" and "wind-generated compressed air."[61]

Lovins's policy framework became the policy agenda of nearly all of the country's environmental organizations, from the Sierra Club to Environmental Defense Fund (EDF), the country's largest environmental philanthropies, U.S. presidents Bill Clinton and Barack Obama, and all of the major 2020 Democratic presidential candidates.

5. What a Waste

In spring 2013, a solar-powered airplane called Solar Impulse flew across the United States, "proving that unfueled, clean flight is possible," according to news reports.[62] The pilot flew from San Francisco to Phoenix to Dallas to St. Louis to Washington, DC. "Who Needs Fuel When the Sun Can Keep You Afloat?" asked a headline.[63]

But Solar Impulse underscored the inherent limits of energy-dilute fuels. Solar Impulse's wingspan was the same length as a Boeing 747, which carries 500 people at close to 1,000 kilometers an hour.[64] Solar Impulse could only carry one person, the pilot, and fly less than 100 kilometers an hour, which is why it took two months to complete the trip.

The dilute nature of sunlight means that solar farms require large amounts of land and thus come with significant environmental impacts. This is true even for the world's sunniest places. California's most famous solar farm, Ivanpah, requires 450 times more land than its last operating nuclear plant, Diablo Canyon.[65]

Solar panels can become more efficient and wind turbines can become larger, but solar and wind have hard physical limits. The maximum efficiency of wind turbines is 59.3 percent, something scientists have known for more than one hundred years.[66] The achievable power density of a solar farm is up to 50 watts of electricity per square meter. By contrast, the power density of natural gas and nuclear plants ranges from 2,000 to 6,000 watts per square meter.[67]

And building a solar farm is a lot like building any other kind of industrial facility. You have to clear the whole area of wildlife. In order to build Ivanpah, the developers hired biologists to pull threatened desert tortoises from their burrows, put them on the back of pickup trucks, transport them, and cage them in pens where many ended up dying.[68]

Solar panels and wind turbines also require far more in the way of materials and produce more in the way of waste. Solar panels require sixteen times more materials[69] in the form of cement, glass, concrete, and steel than do nuclear plants, and create three hundred times more waste.[70]

Solar panels often contain lead and other toxic chemicals that cannot be removed without breaking apart the entire panel. "I've been working in solar since 1976 . . . and that's part of my guilt," a veteran solar developer told Solar Power World in 2017. "I've been involved with millions of solar panels going into the field, and now they're getting old."[71]

California is in the process of determining how to divert discarded solar panels from landfills, which is where they currently go, because solar panel disposal in landfills is "not recommended," concluded a group of experts, "in case modules break and toxic materials leach into the soil."[72]

It is far cheaper for solar manufacturers to buy raw materials than recycle old panels. "The absence of valuable metals/materials produces economic losses," wrote a team of scientists in 2017.[73] "If a recycling plant carries out every step by the book," a Chinese expert concluded, "their products can end up being more expensive than new raw materials."[74]

Since 2016, many solar companies have gone bankrupt.[75] When that happens, the public inherits the burden of managing, recycling, and disposing of photovoltaic waste.[76]

Governments of poor and developing nations are often not equipped to deal with an influx of toxic solar waste, experts say. Poor and developing nations are at higher risk of suffering the consequences because richer nations historically have sent their older panels to them.[77]

The attitude of some Chinese solar recyclers appears to feed this concern. "A sales manager of a solar power recycling company," the South China Morning Post reported, "believes there could be a way to dispose of China's solar junk, nonetheless. 'We can sell them to Middle East,' said the manager. 'Our customers there make it very clear that they don't want perfect or brand-new panels. They just want them cheap.' "[78]

According to the United Nations Environment Programme, somewhere between 60 percent to 90 percent of electronic waste is illegally traded and dumped in poor nations. They found that "thousands of tonnes of e-waste are falsely declared as second-hand goods and exported from developed to developing countries, including waste batteries falsely described as plastic or mixed metal scrap, and cathode ray tubes and computer monitors declared as metal scrap."[79]

In 2019, the *New York Times* reported: "As solar energy booms in the region, so do expired lead-acid batteries for rooftop solar panels and lithium batteries for solar lamps. E-waste can damage the environment by leaking dangerous chemicals into groundwater and harm people who scavenge recyclable materials by hand."[80]

Cities require concentrated energies. Today, humankind relies upon fuels that are up to one thousand times more power-dense than the buildings, factories, and cities they power. The low power densities of renewables are thus a problem not only for protecting the natural environment but also for maintaining human civilization.

Human civilization would have to occupy one hundred to one thousand times more space if it were to rely solely on renewables. "This power density gap between fossil and renewable energies," writes energy analyst Vaclav Smil, "leaves nuclear electricity generation as the only commercially proven non-fossil high-power-density alternative."[81]

What about Elon Musk's claim that an apparently tiny square of solar panels could power the United States? It was deeply misleading.

If the only requirement was producing the same total electricity the United States currently does, regardless of time of day or season, Musk underestimated the required land area by 40 percent. Even if the solar panels were placed in the sunniest area of his sunniest option, the ecologically sensitive Sonoran Desert of Arizona, his solar farm would require an area larger than the state of Maryland.[82]

Musk misrepresented the amount of energy that would need to be stored.

His square of solar desert would generate only two-fifths of its annual electricity in the autumn and winter months, but the United States consumes almost 50 percent of its total annual electricity during the colder portion of the year.[83]

What that means is that roughly 10 percent of yearly demand in the United States, around 400 terawatt-hours, would need to be stored from one half of the year for use in the other in batteries (which would only charge and discharge once per year). At current lithium battery prices, that adds up to $188 trillion.[84]

That's an enormous cost, but we can solve that issue by overbuilding his solar farm by 30 percent, so that it takes up an area of eighteen thousand square miles. This would be 80 percent larger than Musk's original calculation, equivalent to the area of Maryland and Connecticut combined. Doing this, we can get much closer to Musk's claim that we'd "only" need sixteen terawatt-hours of storage[85] at a cost of $7.5 trillion.

Musk has also claimed the batteries would require just one square mile of land,[86] but if we use the brand-new facility in Escondido, California, as the model, with its 120 megawatt-hours of batteries requiring 1.2 acres, his recommended sixteen terawatt-hours would actually take up *250 square miles*.

These calculations only consider electricity. If we move beyond electricity to include all energy, space requirements quickly get out of hand. For example, if the United States were to try to generate all of the energy it uses with renewables, 25 percent to 50 percent of all land in the United States would be required.[87] By contrast, today's energy system requires just 0.5 percent of land in the United States.[88]

Solar panels and wind turbines just don't return enough energy for the energy invested to create them, especially when the need to store energy is considered.

One pioneering study found that in the case of Germany, where nuclear and hydroelectric dams produce seventy-five and thirty-five times more in energy, respectively, than is required to make them, solar, wind, and biomass produce just 1.6, 3.9, and 3.5 times more.[89] Coal, gas, and oil return about thirty times more energy than they require.[90]

Just as the far higher power densities of coal made the industrial revolution possible, the far lower power densities of solar and wind would make today's high-energy, urbanized, and industrial civilization impossible. And, as we have seen, for some advocates of renewables, that has always been the goal.

In its 2019 exposé, *Der Spiegel* concludes that Germany's renewable energy transition was just done incorrectly,[91] but that's misleading. The transition to renewables was doomed because modern industrial people, no matter how romantic they are, do not want to return to pre-modern life.

6. Why Dilute Energy Destroys

Since the 1970s, when the renewable energy agenda was proposed as an alternative to nuclear, most scenarios for 100 percent renewables depended heavily on burning biomass when the sun wasn't shining and the wind wasn't blowing. Biomass became a key component of European renewable energy, with giant coal plants like Drax in Britain converted to burn wood pellets, often shipped from American forests, and agricultural lands in Germany taken out of food production to grow energy crops.

But conservationists have been turning against the use of biomass and biofuels since 2008, when they started to understand their full environmental impact.

To supply a one-thousand-megawatt wood-burning biomass power plant operating for 70 percent of the year requires 3,364 square kilometers of working forest land per year.[92] If just 10 percent of the electricity in the United States were to come from wood-burning biomass power plants, the fuel to power them would require an area of forest land the size of Texas.

Previous calculations of the emissions from bioenergy had not taken into account the emissions created from converting forests into farmlands in different parts of the world to make up for the lost agricultural land in the countries switching to biomass and biofuels. Direct emissions from burning combined with these adverse land use changes mean that the amount of carbon dioxide released from producing and burning biomass and biofuels is higher than from burning fossil fuels.[93]

Scientists now know that corn making and using ethanol emits twice as much greenhouse gas as gasoline. Even switchgrass, long touted as more sustainable, produces 50 percent more emissions.[94]

The main problem with biofuels—the land required—stems from their low power density. If the United States were to replace all of its gasoline with corn ethanol, it would need an area 50 percent larger than all of the current U.S. cropland.[95]

Even the most efficient biofuels, like those made from soybeans, require 450 to 750 times more land than petroleum. The best performing biofuel, sugarcane ethanol, widely used in Brazil, requires 400 times more land to produce the same amount of energy as petroleum.[96]

When I cofounded the New Apollo Project in 2002, we thought "advanced biofuels" from cellulosic ethanol would be a major improvement. They weren't. The power density of cellulosic ethanol turned out to be no better than Brazilian sugarcane ethanol.[97] American taxpayers poured an astonishing $24 billion into failed biofuels experiments from 2009 to 2015.[98]

Governments rarely stop wind projects or require changes in wind turbine locations or operations. Nor do governments require that wind developers disclose when they kill birds and bats, or count the dead. Wind developers have even sued to prevent the public from accessing data about bird kills.[99]

Scientists say bird deaths are being undercounted because scavengers like coyotes quickly eat them, and because body parts are often outside the search radius.[100] "I recently found two golden eagles mortally injured by modern wind turbines just after I had been watching them," wrote a scientist in 2018. Both eagles "ended up outside the maximum search radius, and both left no evidence of their collisions within the search radius."[101]

As such, the mainstream practice of limiting death counts to search radii "is analogous to excluding highway fatalities," the scientist wrote, "when fatalities are found beyond the road verge."[102]

Wind developers are allowed to self-report violations of the Migratory

Bird Treaty Act, the Endangered Species Act, and the Bald and Golden
Eagle Protection Act. Only Hawaii requires bird and bat mortality data to
be gathered by an independent third party and to be made available to the
public on request.[103]

The "U.S. Fish and Wildlife Service has encouraged wind developers
to avoid prosecution for killing eagles," reported *The New York Times*, "by
applying for licenses to cover the number of birds who might be struck by
wind turbines."[104]

In the rare circumstances when governments require the wind industry
to mitigate its impact, such as by setting aside land elsewhere, there is often
little to no enforcement, scientists say. In other circumstances, wind developers do not follow through on their promises and in some cases lie.

Virginia-based Apex Clean Energy claimed on its 2017 application to the
New York Electric Generation Siting Board that there were no known bald
eagle nests where it planned to build.[105] But, later, Apex flew a helicopter
over an eagle's nest, even though doing so posed a direct threat to the birds.
"They destroyed an active eagle's nest," said Lisa Linowes, a local conservationist.[106]

The wind industry has actively discouraged investigations into the killing of bats.

"Bat mortality at wind farms was not understood until around 2003
when researchers visited a wind facility in West Virginia," said Lisa.
"Walking around the turbines, they discovered bat carcasses strewn
throughout the site. When the project owner recognized what they were
looking at, the researchers were escorted off the site and the gate was
locked behind them."[107]

Curtailment, the intentional halting of turbine blades, can reduce the killing of birds, bats, and insects, but few wind farm developers are willing to
curtail because it means losing money. A U.S. National Renewable Energy
Laboratory study found that curtailment levels are lower than 5 percent of
the total wind energy generation.[108]

And curtailment often isn't enough to stop the killings. "In fact, red-
tailed hawk fatalities peaked at the 50 percent of turbines that never operated
during the three years of monitoring," reported a scientist.[109] He calls the

most-studied wind farm in California, Altamont Pass, a "population sink for golden eagles as well as burrowing owls."[110]

In 2018, a scientist with Germany's leading technology assessment research institute announced that industrial wind turbines appeared to be contributing significantly to that country's insect die-off. "Wind-rich migration trails used by insects for millions of years are increasingly seamed by wind farms," wrote Dr. Franz Trieb of the Institute of Engineering Thermodynamics, in a major report.[111]

Scientists have reported the significant build-up of dead insects on wind turbine blades for three decades, and in different regions around the world. In 2001, researchers found that the build-up of dead insects on wind turbine blades can reduce the electricity they generate by 50 percent.[112]

Trieb concluded that a "rough but conservative estimate of the impact of wind farms on flying insects in Germany" is a "loss of about 1.2 trillion insects of different species per year," which "could be relevant for population stability."

The German wind insect death toll is an astonishing one-third of the total annual insect migration in southern England, a comparison that scientists say "shows that losses of a trillion per year certainly have a relevant order of magnitude."

Because insects migrate, the impact of German wind farms is "not limited to local populations, but includes species like the ladybird beetle (*C. septempunctata*) and the painted lady butterfly (*V. cardui*) that travel hundreds and even thousands of kilometers through Europe and Africa."[113]

Insects cluster at the same altitudes used by wind turbines. In Oklahoma, a major wind energy state, scientists found that the highest density of insects is between 150 to 250 meters.[114] Large new turbine blades stretch from 60 to 220 meters above the ground.

And wind turbines may be killing insects during a "critical, vulnerable period" because "a strong lever action is applied on total insect population by killing a mature insect during migration just before breeding, as hundreds of potential successors from the next generation will be destroyed."[115]

While much of the media coverage has blamed industrial agriculture, it is notable that the biggest insect population declines are being reported in Europe and the United States, where the land area dedicated to agriculture has declined, over the last two decades. What have spread are wind turbines.[116]

When I emailed Dr. Trieb to request an interview, he replied, "Unfortunately, I cannot give any interviews on that topic." When I asked an institute spokesperson why Dr. Trieb was unavailable to speak with me, she referred me to a couple of articles and said, "Please accept that neither DLR nor Dr. Trieb are available for further comment on this topic."[117]

7. Defenders of Wind Wildlife

In the United States, Linowes has often found herself defending wildlife from the biggest environmental groups. Sierra Club claims, falsely, "the toll from turbines is far from a major cause of bird mortality."[118] NRDC endorses a massive expansion of wind turbines on the Great Lakes against opposition from local wildlife experts, birders, and conservationists who noted the lakes are one of the world's most important sanctuaries to many migratory bird species.[119]

And Environmental Defense Fund repeats wind industry misinformation by claiming that "wind turbines kill far fewer songbirds than building collisions or cats," and that "technological solutions are in the works."[120]

All three environmental organizations are advocating the rapid expansion of wind farms in New York State, even though they pose a direct threat to bald eagles.[121]

Lisa has stuck to her conservationist principles. "For whatever reason," Linowes said, "I wasn't able to give it up. It just seemed so wrong. The wind industry had the same arguments and talking points and misrepresentations for everyone."

During the last few years, conservationists and biologists have joined Linowes in speaking out against renewables. In 2012, a Science Advisory Board to the Environmental Protection Agency concluded bioenergy was not "carbon neutral," which was supported by more than ninety leading scientists in an open letter to the European Union's Environment Agency.[122]

In 2013, U.S. wildlife officials outraged conservationists and bird enthusiasts when they took the unprecedented step of informing industrial wind energy developers that they would not prosecute them "for inadvertently harassing or even killing endangered California condors."[123]

Said a spokesperson for Audubon, "I can't believe the federal government is putting so much money into a historic and costly effort to establish a stable population of condors, and at the same time is issuing permits to kill them. Ludicrous."[124]

Some scientists understand the implications of energy-dilute energy sources for land use. "[R]enewable energy sources like wind and solar face real-world problems of scalability, cost, material and land use," wrote a group of seventy-five conservation biologists in a 2014 open letter.[125]

Also that year, a conservation biologist who consulted for Ivanpah, the $2.2 billion solar farm in California, told *High Country News*, "Everybody knows that translocation of desert tortoises doesn't work. When you're walking in front of a bulldozer, crying, and moving animals and cacti out of the way, it's hard to think that the project is a good idea."[126]

Biologists fought another solar farm in the Mojave a year later, arguing that it "would likely add another nail in the coffin of [bighorn] sheep by precluding the reestablishment of a critical migration corridor across Interstate 15."[127]

In 2015, the novelist and birder Jonathan Franzen questioned whether the emphasis on climate change was sacrificing nature. "To prevent extinctions in the future," argued Franzen in *The New Yorker*, "it's not enough to curb our carbon emissions. We also have to keep a whole lot of wild birds alive right now."[128]

"Wind turbines are among the fastest-growing threats to our nation's birds," a scientist with American Bird Conservancy said a few weeks later, and "industry players have worked behind the scenes to try to minimize state and federal regulations and to attack important environmental legislation, such as the Migratory Bird Treaty Act. . . . Attempts to manage the wind industry with voluntary as opposed to mandatory permitting guidelines are clearly not working." [129]

Bat scientists have been raising the alarm for over fifteen years. In 2005, leading bat scientists warned federal regulators that wind turbines threatened migratory bat species.[130] In 2017, a team of scientists warned that the hoary bat, a migratory species, could go extinct if the expansion of wind farms continues.[131]

Meanwhile, local communities and environmentalists successfully blocked the construction of transmission lines from the windy north to the industrial south of Germany. "By the first quarter of 2019, only 1,087 kilometers of the planned 3,600 kilometers of power lines were completed," wrote McKinsey. At that rate, McKinsey notes, "the 2020 target will not be reached until 2037." [132]

"The politicians fear citizen resistance," *Der Spiegel* reported in 2019. "There is hardly a wind energy project that is not fought."

Germany is just the most high-profile place struggling to reduce its emissions with renewables. As we saw in the last chapter, Vermont not only failed to reduce emissions by 25 percent, its emissions rose 16 percent between 1990 and 2015, in part due to the closure of the state's nuclear plant, and in part due to the inadequacy of renewables.[133]

The only wind farm to be built in the entire state since the closure of Vermont Yankee was the Deerfield Wind Project. None are planned.[134] And it took from 2009 to 2017 for the project to be completed because of litigation stemming from the fact that the giant wind farm near Readsboro, Vermont, "is sited in the middle of critical black bear habitat." [135]

The number of wind farms the size of Deerfield that would be needed to replace the lost annual electricity from Vermont Yankee, one of the smallest nuclear plants remaining in the United States when it was closed, would have been fifty-six. At that rate, Vermont will make up for the clean energy lost from Vermont Yankee sometime around the year 2104.[136]

8. The Starbucks Rule

The fact that the energy density of fuels, and the power density of their extraction, determine their environmental impact should be taught in every environmental studies class. Unfortunately, it is not. There is a psychological and ideological reason: the romantic appeal-to-nature fallacy, where people imagine renewables are more natural than fossil fuels and uranium, and that what's natural is better for the environment.

Just as people imagined "natural" products from tortoiseshell and ivory to wild salmon and pasture beef are better than "artificial" alternatives, people imagine that "natural" energy from renewables like solar, wood, and wind is better than fossil fuels and nuclear.

At the same time, it is notable that the advocacy for industrial wind energy comes from people who don't live near the turbines, which are almost invariably loud and disturb the peace and quiet.

Those communities that have proven most able to resist the introduction of a wind farm tend to be more affluent. In 2017, the upper-class residents of Cape Cod, for example, defeated an effort by a wind developer to build a 130-turbine farm, despite the developer having spent $100 million on the project.[137]

"Turns out there's something called the Starbucks Rule when it comes to siting wind farms," reported *BusinessWeek* in 2009. Wind developers "plot where Starbucks are in the general area and then make sure their project is at least thirty miles away. Any closer and there'd be too many NIMBYs who'd object to having their views spoiled by a cluster of 265-foot-tall wind towers."[138]

10

ALL ABOUT THE GREEN

1. Fossil-Funded Denial

In summer 2019, a think tank hosted a fundraising gala typical of Washington, DC–based research and advocacy organizations. The theme was *Game of Thrones*, after the hugely popular HBO series. The organization, Competitive Enterprise Institute (CEI), is viewed by many people as the most influential climate denial organization in Washington.

After he was elected, President Donald Trump picked CEI director Myron Ebell to oversee the transition of staff at the U.S. Environmental Protection Agency. President Trump had called climate change a hoax in 2015, and the next year he told *The Washington Post*, "I think there's a change in weather. I am not a great believer in man-made climate change." [1]

In 1998, Ebell helped start the "Cooler Heads Coalition," funded by fossil fuel companies, which described its mission as "dispelling the myths of global warming." Ebell predicted victory once the public recognized "uncertainties in climate science." [2]

The New York Times obtained a copy of the invite list to the *Game of Thrones* dinner, and was thus able to, for the first time, figure out who was financing CEI.

"It's difficult to figure out who's funding climate denial because many of the think tanks that continue to question established climate science are nonprofit groups that aren't required to disclose their donors," the *Times* wrote. "So, the program for a recent gala organized by the institute, which included a list of corporate donors, offered a rare glimpse into the money that makes the work of these think tanks possible." [3]

It was little surprise that many of the institute's donors were companies in the business of selling fossil fuels. Some had an interest in killing regulations that might hurt their business. "The fuel and petrochemical group, which lobbies for gasoline producers, pushed to weaken car fuel economy standards, one of the Barack Obama administration's landmark climate policies," noted the *Times*.[4]

No fossil fuel company appears to have had a greater impact on deceiving the public and squashing climate legislation than ExxonMobil. Internal Exxon documents show that the company has known since the 1970s that fossil fuels were warming the planet. But instead of warning the public, Exxon donated tens of millions to climate skeptic organizations to confuse the public by emphasizing scientific uncertainty.[5]

"We're not in this mess because we heat our homes or drive our children to school," explained a leader of the #ExxonKnew campaign, "which is what the fossil fuel companies want us to think. We're victims of a small group of gargantuan companies that recklessly and deliberately ignored the implications of their own science and unabashedly worked to deceive the public."[6]

But if funding from fossil fuel interests is corrupting politics and killing the planet, why do climate activists take so much of it?

2. The Power of Hypocrisy

In mid-January 2020, climate activists held up signs reading, "Pete Takes Money from Fossil Fuel Billionaires," at a campaign rally for Democratic presidential candidate Pete Buttigieg.

"We are really concerned about candidates who have taken money from fossil-fuel executives," said protest organizer Griffin Sinclair-Wingate. "As a young person who's really concerned about climate change and knows that our lives are threatened by the climate crisis, we cannot have a president who is taking money from fossil-fuel executives."

Buttigieg responded defensively, saying, "I took the fossil fuel pledge," but Sinclair-Wingate pointed out that the candidate "hosted a fundraiser in a wine cellar or wine cave with Craig Hall, who runs a firm that funds fossil fuel infrastructure projects."[7]

Sinclair-Wingate identified himself to reporters as a spokesperson for the New Hampshire Youth Movement. But he was also a paid staff member of 350.org, a group started and led by climate activist Bill McKibben.[8] And 350.org, it turns out, is funded by "Fossil Fuel Billionaire" Tom Steyer, who also happened to be running for president.[9]

Much and perhaps most of Steyer's wealth derives from investments in all three main fossil fuels—coal, oil, and natural gas. Steyer's firm, Farallon Capital Management, reported *The New York Times* in 2014, was "like an anchor in the Indonesian coal industry," an industry colleague of Steyer's said. "By drawing money to an overlooked sector, they helped expand the coal industry there."[10]

As we have seen, substituting coal for wood in poor and developing nations like Indonesia can be positive for human and environmental progress. What is inappropriate is accepting fossil fuel funding while attacking others for doing the same. Even less appropriate is lying about it.

When Steyer announced he was running for president in July 2019, 350.org founder Bill McKibben and Sierra Club head Mike Brune were effusive in their praise. Steyer was a "climate champ," tweeted McKibben, who added that "his just-released climate policy is damned good!"[11] Brune tweeted, "@TomSteyer has been a climate leader for yrs & I'm glad to see yet another climate champ join the primary."[12]

But many Democrats and climate activists did not share their enthusiasm. "Please urge this guy to step down. Huge waste of money," one said.[13] "Bill," wrote another person, "if you're seriously in cahoots with Steyer, you lose any and all credibility in this fight."[14]

Forms filed to the Internal Revenue Service by Steyer's philanthropic organization, the TomKat Charitable Trust, show that it gave $250,000 to 350.org in 2012, 2014, and 2015. Steyer may have given money to 350.org in 2013, 2016, 2017, 2018, 2019, and 2020, as well. The reason to think so is that 350.org thanked either Steyer's philanthropy, TomKat Foundation, or his organization, NextGen America, in each of its annual reports since 2013.[15] In 2018, 350.org reported revenues of nearly $20 million.[16]

Steyer spent $250 million running for president and at least $240 million to influence elections at the federal level since 2013. Steyer has also contributed to Sierra Club, Natural Resources Defense Council (NRDC), Center for American Progress (CAP), and Environmental Defense Fund (EDF) since 2012.[17]

McKibben told *The Washington Post* in 2014 that "he was not bothered" that Steyer had made much of his fortune in fossil fuels because Steyer had promised to divest himself from those investments.[18]

The Washington Post article did not mention that Steyer funded McKibben's organization. "The environmentalist said that Steyer deserves praise for his willingness to give up a lucrative career," reported the *Post*, "and that even many of the most ardent anti-fossil-fuels activists have benefited from the industry at some point."[19]

In July 2014, Steyer told *The Washington Post* that he would divest from fossil fuels by the end of that month, but in August 2014 his spokesperson admitted to *The New York Times* that he had not done so. "Farallon is still invested in carbon-generating industries and [Steyer's] aides declined to say whether Mr. Steyer had asked it to sell those holdings, a request that would presumably hold significant sway given his role as a founder."[20]

But in the same article, Steyer's aides added, "He remains a passive investor, though they declined to describe the size of his investment."

Steyer misrepresented his investments for five more years. In July 2019, Steyer told ABC News, "Look, in our business we invested in every part of the economy, including fossil fuels. When I realized what a threat this was to our environment and to the people of the United States and people around the world, I changed. I divested from all that stuff. I left my business."[21]

But just a few weeks later, when pressed on the issue at a campaign rally, Steyer said, "There's probably some dregs left" of fossil fuel investments. In fact, Bloomberg News discovered that Steyer had retained millions in investments in coal mining, oil pipelines, and fracking for petroleum and natural gas.[22]

Does it matter? After all, where climate skeptic groups like CEI work to kill climate policies, 350.org and other environmental groups use Steyer's money to *support* clean energy, not kill it, right?

Wrong. Not only are 350.org, Sierra Club, NRDC, and EDF all funded by fossil fuel billionaires, they are also all trying to kill America's largest source of carbon-free electricity, nuclear power.[23]

3. Green on the Inside

McKibben is one of America's most influential environmental activists and, as we have seen, successfully advocated closing Vermont's nuclear plant, which contributed to the state's emissions rising 16 percent rather than declining 25 percent, as planned.

But McKibben isn't the only climate activist who has successfully forced a nuclear plant to be closed and replaced by fossil fuels. Every major climate activist group in America, including NRDC, EDF, and Sierra Club, has been seeking to close nuclear plants around the United States while taking money from or investing in natural gas companies, renewable energy companies, and their investors who stand to make billions if nuclear plants are closed and replaced by natural gas.

Killing nuclear plants turns out to be a lucrative business for competitor fossil fuel and renewable energy companies. That's because nuclear plants generate large amounts of electricity. During a ten-year period, Indian Point's owner could bring in $8 billion in revenue. During forty years, revenues could easily be $32 billion. If the plant closes, those billions will flow to natural gas and renewables companies.[24]

Sierra Club, NRDC, and EDF have worked to shut down nuclear plants and replace them with fossil fuels and a smattering of renewables, since the 1970s. They have created detailed reports for policymakers, journalists, and the public purporting to show that neither nuclear plants nor fossil fuels are needed to meet electricity demand, thanks to energy efficiency and renewables. And yet, as we have seen, almost everywhere nuclear plants are closed, or not built, fossil fuels are burned instead.[25]

The Sierra Club Foundation has taken money directly from solar energy companies. Barclays's renewable energy investment banking chief, a director and assistant general counsel for SolarCity, the founder and CEO of Sun Run, the CEO of Solaria, and others have all served on Sierra Club Foundation's board of directors.[26]

EDF's board of trustees and advisory trustees have also included investors and executives from oil, gas, and renewable energy companies, including Halliburton, Sunrun, Northwest Energy, and many others.[27]

NRDC helped create and put $66 million in a Black Rock "Ex-Fossil Fuels Index Fund" stock fund that—in fact—invests heavily in natural gas companies. And in a 2014 financial report, NRDC disclosed that it had nearly $8 million invested in four separate renewable energy private equity funds.[28]

"If an environmentalist were to take a gander at [NRDC's] holdings," a reporter for an environmental website wrote in 2015, "she might raise a quizzical eyebrow: 1,200 shares of Halliburton, 500 of Transocean, 700 of Valero. Marathon, Phillips 66, Diamond Offshore Drilling—they're in there, too."[29]

The founding donor of Friends of the Earth was oilman Robert Anderson, the owner of Atlantic Richfield. He gave Friends of the Earth the equivalent of $500,000 in 2019 dollars. "What was David Brower doing accepting money from an oilman?" his biographer wondered.[30] The answer is that he was pioneering the environmental movement's strategy of taking money from oil and gas investors and promoting renewables as a way to greenwash the closure of nuclear plants.

Natural Resources Defense Council even helped Enron to distribute hundreds of thousands of dollars to environmental groups. "On environmental stewardship, our experience is that you can trust Enron," said NRDC's Ralph Cavanagh in 1997.[31] Enron executives at the time were defrauding investors of billions of dollars in an epic criminal conspiracy, which in 2001 bankrupted the company.[32]

From 2009 to 2011, lawyers and lobbyists with EDF and NRDC advocated for and helped write complex cap-and-trade climate legislation that would have created and allowed some of their donors to take advantage of a carbon-trading market worth upwards of $1 trillion.[33]

Climate activists have long suggested that fossil fuel–funded climate denial organizations are massively outspending them, but are they really? It is an easy thing to check, given that the government requires nonprofits to disclose their revenues publicly.

Climate activists massively outspend climate skeptics. The two largest U.S. environmental organizations, EDF and NRDC, have a combined annual budget of about $384 million compared to the mere $13 million of the two largest climate skeptic groups, Competitive Enterprise Institute and Heartland Institute. That amount of money, $384 million, is significantly more than all of the money Exxon gave to climate-skeptical organizations for two decades.[34]

It might be objected that there are other organizations that criticize and oppose climate *policy*, including the Heritage Foundation ($86 million in revenue; actual amount is $86,808),[35] the American Enterprise Institute ($59 million in revenue),[36] and the Cato Institute ($31 million in revenue).[37]

But those three organizations all accept that humans are changing the climate, even if they oppose many of the proposed ways of dealing with it. The American Enterprise Institute has endorsed both a carbon tax and government research and development for clean energy innovation.[38]

And there are many other organizations, including the Nature Conservancy ($1 billion in revenue, 2018) and Center for American Progress ($44 million in revenue, 2018), that advocate for renewables and against nuclear energy.[39]

4. Fracking Nukes

In spring 2016, Elizabeth Warren, senator from Massachusetts and 2020 Democratic presidential candidate, traveled to Chicago to speak at a gala fundraising dinner hosted by Illinois's leading environmental group, the Environmental Law and Policy Center (ELPC). Its founder, Howard Learner, was well-connected in Democratic Party circles. Learner had served as senior advisor for energy and environmental issues for President Obama from

2007 to 2008.[40] In addition to Warren, U.S. Senator Dick Durbin (D-IL) also spoke at the event.

At the time, ELPC was seeking to halt legislation in the Illinois legislature that would have extended to nuclear plants a small fraction of the subsidy the state already gave to wind and solar developers. Learner was a longtime antinuclear crusader, having helped to kill new nuclear plants and shut down existing ones for at least thirty years, going back to the 1980s.

"Everybody looks with excitement when a new natural gas plant is built," he said, referring to the alleged increase in jobs.[41] But even that was misleading. Where an average-size nuclear plant tends to employ one thousand people, a similarly sized natural gas plant tends to employ fewer than fifty.

Anything that shuts down nuclear plants is good business for natural gas and renewables companies and investors. And so it is little wonder that when ELPC asked those business interests that would benefit directly from killing nuclear plants to cosponsor its gala dinner, many of them were more than happy to make donations.

The most important name on the list was Invenergy, a combined natural gas and industrial wind developer. For Invenergy, funding ELPC may have been part of its effort to close Illinois's nuclear plants. During the same period, Invenergy was aggressively lobbying in the state legislature, making campaign contributions, and promoting renewables.

In February 2012, the new executive director of the Sierra Club went to *Time* with a confession: his organization had accepted more than $25 million from natural gas investor and fracking pioneer Aubrey McClendon.[42] The former executive director of the Sierra Club had regularly traveled across the United States with McClendon promoting the environmental benefits of natural gas.[43] The new executive director of Sierra Club, Michael Brune, denounced the acceptance of McClendon's funding and decided Sierra Club would no longer accept money from natural gas interests.

Brune told *Time* that, in 2010, "when he discovered the natural gas do-
nation, he knew it risked tainting the organization," and demanded the club
refuse any future funding. Brune described his decision as a difficult one but
also the right one. "The [additional] money would have been a quarter of
our budget for an entire year. It wasn't just a throwaway check. But there
were clear reasons why we needed to do that," Brune said.[44]

Why, the *Time* reporter asked, did Brune only reveal the secret natural
gas donations "more than a year and a half after the decision to cut financial
ties with the gas industry was made?"

> He says that he's concerned by the prominence that natural gas and
> oil drilling received in President Obama's recent State of the Union
> speech, when Obama said that gas drilling would "create jobs and
> power trucks and factories that are cleaner and cheaper, proving that
> we don't have to choose between our environment and our econ-
> omy."[45]

That could be, but another possible reason Brune went to *Time* with
the story is that Sierra Club members opposed to fracking had complained to a left-wing journalist who was about to break the story. The
reporter, Russell Mokhiber of Washington, D.C.–based *Corporate Crime
Reporter*, emailed the Sierra Club to ask if it took money from fracking
companies.

> . . . I get an email from Maggie Kao, the spokesperson for the Sierra
> Club.
>   On Tuesday, Kao writes to me: "We do not and we will not take
> any money from any natural gas company."
>   I write back: I understand you do not and will not. But have you
> taken money from Chesapeake?
>   That was Tuesday. All day Wednesday goes by. All day Thursday
> goes by. And I can't get an answer. Then Thursday night, Kao writes,
> says: Okay, Brune can talk to you at 7:30 pm EST.

The Netherlands became a wealthy nation while farming below sea level. Dutch experts are today helping Bangladesh adapt to rising sea levels.

(frans lemmens/Alamy Stock Photo)

Scientists say the accumulation of wood fuel in forests and the building of homes in fire-prone areas are greater factors behind fires in Australia and the United States, such as in Malibu, California.

(ZUMA Press, Inc./ Alamy Stock Photo)

Mamy Bernadette Semutaga near Virunga National Park in the Democratic Republic of the Congo the day after baboons ate her sweet potatoes.

(Michael Shellenberger)

What most determines how much plastic waste goes into the ocean is whether a nation has a waste collection and management system. Developed nations like Japan (*top*) have such a system, while poor nations like Congo (*bottom*) do not.

(Japan: ton koene/Alamy Stock Photo; Congo: mauritius images GmbH/Alamy Stock Photo)

Plastics made from fossil fuels replaced natural plastics, including hawksbill tortoiseshell used in eyeglasses.

(Glasses: shinypix/Alamy Stock Photo; turtle: Andrey Armyagov/ Alamy Stock Photo)

Suparti left her home in the countryside at the age of seventeen to work in factories in the city. Even when life was hard, she didn't want to go back to farm life.

(Michael Shellenberger)

In poor nations like Congo, women and girls complain more of the time they have to spend chopping, hauling, and burning wood than they do of the smoke it emits.

(Michael Shellenberger)

Temple Grandin says her autism made her more empathetic to stresses of farm animals. McDonald's Corporation and others have adopted her methods for raising cattle more humanely.

(Alison Bert/Elsevier)

California's Ivanpah solar farm would require 450 times more land than Diablo Canyon nuclear power plant to generate the same amount of electricity, demonstrating the inherently high environmental impacts of energy-dilute renewables.

(aerial-photos.com / Alamy Stock Photo)

Whale populations have recovered since they were saved by kerosene in the nineteenth century and palm oil in the twentieth century. This whale swims in front of the Diablo Canyon nuclear power plant.

(John Lindsey)

Wildlife biologists in 2017 warned that the continued expansion of industrial wind turbines threatens the extinction of the hoary bat, which are killed directly and indirectly from pressure on their lungs.

(Rick & Nora Bowers/Alamy Stock Photo)

California governor Jerry Brown protected his family's oil monopoly through state air pollution regulations, while also working to block and close nuclear power plants.

(Associated Press)

Extinction Rebellion activists (also called the Red Rebel Brigade), claiming climate change threatened human extinction, halted traffic in London in the spring and fall of 2019.

(Guy Corbishley/Alamy Live News)

Caleb *(left)* with Daniel at the Virunga dam, which was built to support economic development and reduce the threat to mountain gorillas from wood-fuel use.

(Michael Shellenberger)

In 2014, White House science advisor John Holdren (*left*) claimed Roger Pielke, Jr. (*below*), had misled Congress. One year later, Representative Raúl Grijalva from Arizona announced an investigation of Pielke.

(Holdren: NASA Image Collection/Alamy Stock Photo: Pielke: CSPAN)

Today

ROGER PIELKE
University of Colorado
Environmental Studies Professor

C-SPAN 2

Helen with the gorillas. People protect endangered species like mountain gorillas not because human civilization depends on them but rather because of their spiritual and aesthetic value.

(Michael Shellenberger)

And by the way, Kao says: Check out this story just posted in *Time Magazine*. . . . Turns out, Sierra Club didn't want the story to break in *Corporate Crime Reporter*.[46]

The Sierra Club under Brune never stopped taking money from natural gas interests. In fact, he radically increased how much it accepted from them.

Under Brune, Sierra Club took an additional $110 million from Michael Bloomberg, the former mayor of New York, owner of Bloomberg Media Group, a 2020 Democratic presidential candidate, and major investor in natural gas.[47]

When I published these facts, some people on Twitter objected that Bloomberg was rich from other things, not just natural gas, and besides, he cared about climate change and was seeking to replace coal with natural gas, something I support.

But one cannot have it both ways. Bloomberg has no fewer conflicts of interest than Aubrey McClendon, Tom Steyer, and Exxon. For groups like 350.org and Sierra Club to accept money from two of them while denouncing their opponents for accepting money from the other two is audacious in its hypocrisy.

It also raises a question: how long, exactly, have oil and gas interests been funding environmental groups to shut down nuclear plants?

5. Brown's Dirty War

On July 1, 1979, about thirty thousand people assembled on an airstrip in a remote part of the central California coast to hear Jackson Browne, Bonnie Raitt, Graham Nash, and other pop stars sing at a "No Nukes" concert. The concert was inspired by a growing backlash to nuclear energy sparked first by the blockbuster antinuclear film, *The China Syndrome*, and the nuclear reactor meltdown at Three Mile Island, twelve days later.

California's governor, the forty-one-year-old Jerry Brown, showed up at the event, and requested an opportunity to address the audience. The event's organizers were suspicious of the governor's sincerity and questioned him for an hour. After they decided that he was genuinely antinuclear, they decided to let him address the audience.

Once onstage, Brown flattered the young concert-goers. They represented "triumph of the people over power," Brown said, and "a growing power to protect the earth." Brown made a promise to the crowd: "I personally intend to pursue every avenue of appeal if the Nuclear Regulatory Commission ignores the will of this community." [48]

The audience gave Brown a minute-long standing ovation. Brown ended his speech by leading the crowd in a chant, "No on Diablo! No on Diablo! No on Diablo!" The next day's headline in the local newspaper, the *San Luis Obispo County Telegram-Tribune*, read "Rally Spurs Brown to Oppose Diablo." [49]

The headline, and the initially chilly reception by concert organizers, made it seem to a local journalist reporting on the event that the governor was responding to the antinuclear movement. In reality, Brown had been quietly leading antinuclear efforts for years.

Three years earlier, antinuclear groups had introduced a ballot initiative that would have effectively banned nuclear, while a Brown ally introduced legislation to block new plants from being built until a waste repository was created. When the state's electric utilities came out against the bill, Governor Brown threatened to campaign in favor of the more radical ballot initiative unless they backed down. The utilities caved and allowed the legislation to pass, and Brown signed it into law.

When San Diego Gas & Electric sought to build a nuclear plant in a project called Sundesert, Brown attacked them through the agencies he controlled. His allies at the California Energy Commission argued that future demand should instead be met by burning oil and coal. Brown's California Air Resources Board backed up California Energy Commission and concluded that "new fossil fuel power plants can be built in many parts of California without causing environmental damage." [50]

A departing California Energy Commission commissioner objected to what he viewed as the agency's deliberate and unjustified underestimation of future electricity demand, based on assumptions of massive energy consumption reductions from more energy-efficient buildings and appliances. The man accused Brown and his allies of "purposely attempting to stop nuclear power in the state." [51]

Brown's behavior outraged his fellow Democrats, who believed nuclear plants would lock in cheap, pollution-free electricity for decades. The effort to kill Sundesert was "orchestrated by the Governor for his own reasons," a member of the state assembly charged. "It obviously had nothing to do with the merits of the case." [52]

Brown was proud of what he had achieved. After learning that antinuclear activists were taking credit for blocking the plant from being built, Brown boasted to a reporter, "*I* blocked the Sundesert plant." [53] Between 1976 and 1979, Brown and his allies killed so many nuclear power plants that, had they been built, California would today be generating almost all of its electricity from zero-pollution power plants. [54]

Most people believed Jerry Brown's crusade against nuclear was strictly ideological. He is an environmentalist, after all, active in both climate change and antinuclear weapons causes. [55]

The truth is a tad more complicated.

In the late 1960s, the Indonesian government asked Jerry's father, Edmund "Pat" Brown, who was California's governor from 1959 to 1967, to help recapitalize its state-owned oil company, Pertamina, after the country's bloody civil war. Pat Brown was well-connected on Wall Street. The former governor raised $13 billion, which is more than $100 billion in 2020 dollars.

In exchange for Brown's services, Pertamina gave him exclusive rights to sell Indonesian oil in California. At the time, California burned significant quantities of oil for electricity production, not just for transportation.

What was special about Indonesian oil was that it contained less sulfur than most U.S. petroleum, and thus created less sulfur dioxide, a pollutant that contributes to smog and respiratory problems, including asthma. New air pollution regulations meant that Indonesia oil would enjoy a monopoly in California, which the Brown family would work to guarantee.

Dan Walters, a former *Sacramento Bee* columnist, who discovered

Brown's Indonesian oil connections, told me that he confirmed that Kathleen Brown inherited a share of her father's oil wealth but had not confirmed the same for Jerry.

Shortly after he first became governor, Jerry Brown took actions that protected his family's oil monopoly in California. Brown appointed his former campaign manager, Tom Quinn, as director of the California Air Resources Board, which changed an air pollution regulation, leading Chevron to cancel an oil refinery, which would have introduced Alaskan oil into the California market, and competed directly with the Brown family's oil business. At the very same time, another top Jerry Brown political aide-turned-appointee, Richard Maullin, chairman of the California Energy Commission, began pressuring the state's utilities to burn more oil rather than shift to nuclear energy.[56]

Next, Jerry Brown made the top investment manager for the Getty Oil fortune a state superior court judge. The Getty man had raised political contributions for Brown's father when the latter was governor, and then for Jerry Brown. Then, as superior court judge, the Getty man lobbied for and passed legislation that protected Getty Oil family money from taxation. The Getty Oil man's name was Bill Newsom, and he was the father of California's current governor, Gavin Newsom.[57]

In 1976, the military general who headed the state-owned Pertamina oil company in Indonesia was ousted after widespread corruption under his management was discovered. After the oil boom collapsed in the early 1970s, Pertamina's loans went into default and threatened American banks.

Shortly after, Jerry Brown and his father lobbied aggressively to build a liquefied natural gas terminal in Southern California to import natural gas from Indonesia. The terminal, noted former *Sacramento Bee* reporter Walters, "would have relieved Pertamina's serious money problems, and indirectly bailed out the big U.S. banks that had loaned Pertamina billions of dollars."[58]

Brown's oil and gas ties extended into Mexico. He did business deals with Carlos Bustamante, head of a powerful Mexican oil and gas family, *The New York Times* reported in a front-page investigative article in 1979. The *Times* reported that Bustamante contributed to Brown's campaign.[59]

Brown acknowledged that he urged Mexico's president to approve a power plant project in Baja, California, to provide power to the San Diego Gas & Electric Company. Bustamante was working as a lobbyist for the utility company, and he was listed as the principal investor in a proposed financing plan for the plant; he was also the owner of the land upon which it would sit.

In 1979, the Federal Bureau of Investigation investigated allegations that Jerry Brown's 1974 gubernatorial campaign failed to report Bustamante's contributions. The FBI "received several allegations from Democratic politicians and businessmen of unreported Bustamante contributions . . . ," *The New York Times* reported. "One of the allegations, which reportedly include details about principals in the transactions, is said to tie unreported contributions to gas and oil deals benefitting the Bustamantes."[60]

In 1978, the San Diego Gas & Electric Company canceled the project because it could have been used as a "vehicle for improper payments," in the words of the utility's lawyer. But even after the project was killed for fear of corruption, Jerry Brown's administration sought yet another oil and gas project with the Bustamante family. There were few apparent checks on Brown's power.[61]

Jerry Brown's advocacy for natural gas was part and parcel of his anti-nuclear work, which didn't end when he left office in 1983. Seven years later, two close allies, Bob Mulholland and Bettina Redway, passed a ballot initiative to shut down the Rancho Seco nuclear plant near Sacramento, California, which Brown had also tried to shut down toward the end of his second term as governor.[62]

Shortly after, Brown, as chairman of the California Democratic Party, rewarded Mulholland with a job as political director. Redway's husband, Michael Picker, became one of Brown's closest advisors, later playing a central role in overseeing the closure of California last two nuclear plants.[63]

6. Grazing the Tall Grass

Pat Brown wasn't the first Democratic politician to make money in fossil fuels. After Al Gore, Sr., the senator from Tennessee, lost reelection in 1972, he went to work for a coal power plant owned by Occidental Petroleum. "Since I had been turned out to pasture, I decided to go graze the tall grass," Gore Sr. quipped, years later.[64]

As U.S. senator and vice president, Al Gore, Jr., helped advance the same company's interests. Gore raised $50,000 from the company in phone calls he made from his office, triggering a minor scandal.[65]

The Center for Public Integrity, a nonprofit organization that won the 2014 Pulitzer Prize for uncovering efforts by the coal industry to defeat legal claims that miners had been sickened with black lung disease, investigated Gore's connection to Occidental.

The Center for Public Integrity reported in January 2000 that, "since Gore became part of the Democratic ticket in the summer of 1992, Occidental has given more than $470,000 in soft money to various Democratic committees and causes."[66]

According to the report, two days after the Occidental chairman slept in the Lincoln Bedroom, he donated $100,000 to the Democratic National Committee. The chairman was also a guest at a 1994 White House party for Boris Yeltsin. Occidental had an interest in Russian oil. Earlier that year, the chairman had traveled with President Bill Clinton's commerce secretary on a trade mission to Russia.[67]

Gore personally accepted fossil fuel money in 2013. He and a co-owner sold Current TV to Al Jazeera, which is state-funded by Qatar, the oil-exporting nation whose citizens have the largest per capita carbon footprint in the world. One year earlier, Gore had said the goal of "reducing our dependence on expensive dirty oil" was "to save the future of civilization."[68]

As part of the agreement, Gore reportedly received $100 million.[69] Environmental activists weren't particularly bothered by it. "I don't think the community is too upset," a politically active environmentalist told *The Washington Post* about Gore's deal with Qatar. "My personal sense is he got a good deal."[70] The deal appears to have worked out better for Gore than Al Jazeera, which operated Al Jazeera America television from 2013 to 2016 before closing the operation down, apparently due to low ratings.[71]

Between 2011 and 2019, Jerry Brown served as California governor for a third and fourth term. During that time his sister, Kathleen Brown, was on the board of directors of Sempra Energy, one of the country's largest natural gas companies, and owner of San Diego Gas & Electric.

Brown actively sought to advance oil and gas interests. In 2011, he fired two state regulators because they were enforcing federal fracking regulations to protect California's water quality.[72] In 2013, a lobbyist for Pacific Gas & Electric Company told his boss in an email that Governor Brown told a California Public Utilities Commission (CPUC) commissioner to approve a natural-gas-fired power plant for PG&E.[73]

The next year, Brown ordered California's Division of Oil, Gas and Geothermal Resources to explore his personal land for oil and gas rights. The agency produced a fifty-one-page report complete with satellite images of oil and gas deposits for the area around Brown's ranch. It was a brazen use of government resources for personal gain, but California's newspaper reporters and editorial writers made little of the incident.[74]

Brown appointees' actions kept open Aliso Canyon, a natural gas storage facility owned by Sempra that suffered a catastrophic leak and caused mass evacuation in 2015. As of 2016, Kathleen Brown owned one thousand acres of oil and gas interests in California, and $749,000 worth of stock in real estate and oil company Forestar Group, which owns 700 acres adjacent to Porter Ranch, where the Aliso Canyon blowout occurred. Kathleen Brown was at the time on the board of renewable energy investment company Renew Financial, which directly benefited from California's significant subsidization of renewable energy.[75]

After the Aliso Canyon blowout, Governor Brown took steps to keep the accident's cause a secret. "Months into the efforts to stop the leak," notes Consumer Watchdog, a liberal antinuclear organization, "Brown issued an executive order keeping any investigation of the causes and whether it could or should be shut down secret."[76]

In office as governor for a third and fourth term, starting in 2011, Brown and his allies resumed the effort they began in the 1970s to shut down the state's nuclear plants. It started with a plant called San Onofre Nuclear Generating Station (SONGS) in San Diego County.

In February 2013, Brown appointee and CPUC president Michael Peevey sought out a senior executive of Southern California Edison while the two men were on a junket in Poland.[77] During that meeting, according to the *Los Angeles Times*, Peevey laid out the terms of a deal to close SONGS. The excuse was that that the utility had botched the replacement of the plant's steam generators.

New steam generators could have been purchased and installed for well under $1 billion. The prior steam generators had cost less than $800 million.

Peevey proposed that, instead of replacing the steam generators, Southern California Edison shut down the plant entirely. In exchange, Peevey said, he would make sure CPUC allowed the utility to increase electricity rates.

Peevey proposed that ratepayers pay $3.3 billion to close the plant early, and investors pay an additional $1.4 billion.[78]

The scheme went forward. SONGS closed permanently, natural gas replaced the plant's electrical output, and California's carbon emissions spiked, as did electricity prices. [79]

In November 2014, state and federal agents raided the CPUC's offices in a joint investigation of potential criminal activities related to the permanent closure and settlement proceedings of SONGS. Kamala Harris, California's attorney general at the time, either killed or stalled the investigation. The CPUC refused to turn over sixty or more emails from Governor Brown's office.[80]

In 2014, CPUC attorneys acknowledged that their colleagues might have been destroying evidence related to a criminal investigation into a Pacific Gas & Electric natural gas explosion, which killed eight people. The California legislature passed legislation to reform the CPUC in August 2016, but it was halted at the last minute at Picker's urging, according to reports by the *Los Angeles Times* and the *San Diego Union Tribune*.[81]

State Superior Court Judge Ernest Goldsmith made a strongly worded request for the CPUC to disclose Picker's correspondence related to SONGS. "This is a big deal," Goldsmith said. "This is not a trivial issue to the taxpayers of California. And just like the San Bruno events [natural gas explosion that killed eight people] were not a trivial deal, and when something is big enough, it's just got to come out. It's going to come out, and it's either going to be horribly painful, or you can just do the right thing." [82]

While the dark cloud of criminal investigation hung over CPUC, it has moved forward on closing the state's last surviving nuclear plant, Diablo Canyon. These circumstances include many of the very same actors and groups involved in negotiating the closure of SONGS. One of the key anti-nuclear advocates, Americans for Nuclear Responsibility, was represented by John Geesman, a longtime Brown advisor, the former chairman of the California Energy Commission, and a renewable energy industry advocate. When CPUC chairman Peevey proposed his scheme to shut down SONGS, he specifically asked that Geesman be part of the effort. [83]

7. Bigger than the Internet

By 2006, when *An Inconvenient Truth* won Al Gore an Oscar and a Nobel Prize, renewables were becoming big business. That same year, the venture capitalist John Doerr, an early Google and Amazon investor, cried while giving a TED Talk about global warming. "I'm really scared," Doerr said. "I don't think we're going to make it."

But the upside of crisis was opportunity. "Green technology—going green—is bigger than the Internet," Doerr said. "It could be the biggest economic opportunity of the twenty-first century." [84]

As I mentioned, I cofounded a progressive Democratic, labor-environment push for a New Apollo Project, the predecessor to Rep. Ocasio-Cortez's Green New Deal. We sought $300 billion for efficiency, renewables, electric cars, and other technologies. [85]

In 2007 our efforts paid off when then-U.S. presidential candidate Obama picked up our proposal and ran with it. Between 2009 and 2015, the U.S. government spent about $150 billion on our Green New Deal, $90 billion of it in stimulus money. [86]

Stimulus money wasn't evenly distributed but rather clustered around donors to President Obama and the Democratic Party. At least ten members of Obama's finance committee and more than twelve of his fundraising bundlers, who raised a minimum of $100,000 for Obama, benefited from $16.4 billion of the $20.5 billion in stimulus loans.

Fisker, which produced some of the world's first luxury hybrid vehicles, received $529 million in federal loans; Doerr was one of Fisker's major investors. It eventually went bankrupt, costing taxpayers $132 million.[87]

The people who directed the loan program had been fundraisers for Obama. In March 2011, the U.S. Government Accountability Office wrote a report, calling the loan program's process "arbitrary" and noted that not a single one of the program's first eighteen loans had been documented.

The loans for electric car companies like Tesla and Fisker, each of which received nearly a half-billion dollars, had no performance measures. The Government Accountability Office found that the U.S. Department of Energy (DOE) "treated applicants inconsistently in the application review process, favoring some applicants and disadvantaging others."[88]

But the loans were just one program among many others that funneled money to well-connected Obama donors without creating many jobs. The most famous of the green investments was when DOE gave $573 million to a solar company called Solyndra, 35 percent of which was owned by a billionaire donor and fundraising bundler for Obama, George Kaiser.

Nobody wanted to invest in Solyndra because its panels were too expensive, which independently minded DOE staffers pointed out. They were overruled, however, and the loan was approved.

The people who benefited the most from the green stimulus were billionaires, including Musk, Doerr, Kaiser, Khosla, Ted Turner, Pat Stryker, and Paul Tudor Jones. Vinod Khosla led Obama's "India Policy Team" during the 2008 election and was a major financial contributor to Democrats. His companies received more than $300 million.[89]

However, few Democratic Party donors outperformed Doerr when it came to receiving federal stimulus loans. More than half of the companies in his Greentech portfolio, sixteen of twenty-seven, received loans or outright grants from the government. "Considering that the acceptance rate in most of the Department of Energy programs was often 10 percent or less, this is a stunning record," wrote an investigative reporter. "That $2 million Doerr had invested in politics may have provided the best return on investment he had ever seen."[90]

In 2017, Tesla joined NRDC, EDF, and the Sierra Club in urging California government regulators to close Diablo Canyon Nuclear Power Plant, the state's single largest source of clean, zero-emissions energy. Tesla's statements were transparently self-serving. The nuclear plant, Tesla's lobbyist claimed, could be replaced by (Tesla's) solar panels and (Tesla's) batteries. The government of California, in return, was as of late 2019 subsidizing nearly half the cost of Tesla's $6,700 Powerwall battery.[91]

As we saw with Steyer and Bloomberg, sometimes seemingly altruistic environmental philanthropies have a financial interest in their advocacy.

Consider the case of Sea Change. It gave more $173 million in grants to groups advocating cap and trade, including the Center for American Progress, Sierra Club, NRDC, EDF, World Wildlife Fund, and the Union of Concerned Scientists, from 2007 through 2012.[92]

The man behind Sea Change was venture capitalist Nathaniel Simons, who had investments in seven companies that had received federal loans, grants, or contracts since 2009.[93] Simons "invested substantially in pursuing a cap-and-trade policy," according to a Rockefeller Family Fund report.[94] He and his wife hired a lobbyist to push for the bill.[95]

During the last decade, Stanford's Mark Z. Jacobson eclipsed Amory Lovins, EDF, NRDC, and all other groups in supposedly proving that renewables alone can power the planet and that nuclear energy isn't needed to replace fossil fuels. When I debated him at the University of California, Los Angeles, in 2016, Jacobson claimed it would be cheaper to shut down existing nuclear plants, including Diablo Canyon, and replace them with renewables.[96] He did so as a senior fellow at the Precourt Institute for Energy, which was named after an oil and gas magnate and board member of Halliburton, the oil and gas services firm. Precourt's board is comprised of leading oil, gas, and renewables investors. It is hard to imagine a more direct conflict of interest.[97]

8. Leaving a Legacy

In 2018, Arizona voters considered whether or not to support a ballot initiative that ostensibly promoted renewables but in reality would have resulted in the premature closure of the state's sole nuclear plant, Palo Verde. The plant also happens to be America's largest single source of zero-emissions clean energy.

Had the initiative passed, Palo Verde's operator would have been forced to replace it not just with renewables but also significant amounts of natural gas, and emissions would have risen. In the end, voters rejected the initiative in a landslide: just 30 percent voted in favor while 70 percent voted against it.

The ballot initiative's sponsor was Tom Steyer, who spent $23 million on it and who might have benefited personally. And yet few in the mainstream news media explored Steyer's potential conflict of interest.[98]

During the very same years they were denouncing fossil fuel interests, particularly Exxon and the Koch brothers, for funding their political opponents, and demanding universities stop investing in fossil fuels, 350.org, the Sierra Club, NRDC, and EDF were all accepting money from fossil fuel billionaires Steyer and Bloomberg.[99]

Where the news media have for decades demonized Exxon, the Koch brothers, and climate skeptics, they have largely given a pass to fossil fuel billionaires like Steyer and Bloomberg and the environmentalists they fund.

Steyer and Bloomberg may be motivated to do good in the world, but so may be the Koch brothers. Financial conflicts of interest are no less conflicts of interest just because a person is ideologically committed.

McKibben and Sierra Club chief Brune bent over backward to praise Steyer after he announced he would spend $100 million running for office, and effectively buy his way onto the Democratic presidential election debate stage through TV and Facebook advertising.[100] Steyer and Bloomberg ended up spending $750 million total on their 2020 presidential campaigns.

It is hard to imagine a more "pay-to-play" relationship than the one between Steyer and his grantees. It epitomizes the cynicism of Washington, D.C. And it exposes the news media's double standard.

If Steyer and other fossil fuel and renewable energy investors get their way and kill some or all of the remaining ninety-nine U.S. nuclear reactors, which provide nearly 20 percent of America's electricity, they will not only make a fortune, they will spike emissions and eliminate the only real hope for phasing out fossil fuels before 2050.

If that happens, a major part of Steyer's legacy will be as the fossil fuel billionaire who has killed more clean energy, and increased emissions, more than any other in recent history.

11

THE DENIAL OF POWER

1. Power Tripping

In 2019, some of the world's wealthiest and most powerful people started responding to demands that we act on climate change.

At the end of July, Google brought together celebrities with climate activists in Italy to discuss what they could do to support the cause. Leonardo DiCaprio, Stella McCartney, Katy Perry, Harry Styles, Orlando Bloom, Bradley Cooper, Priyanka Chopra, Nick Jonas, and Diane von Furstenberg reportedly brainstormed ways they could leverage their fame to change behavior.

Some were already taking action for the climate. Perry had made videos for UNICEF, while DiCaprio had narrated a documentary. Prince Harry, who described the climate emergency to the assembled guests, bare-footed, had recently written on Instagram, "With nearly 7.7 billion people inhabiting this Earth, every choice, every footprint, every action makes a difference." [1]

The gathering in Sicily took place at the Verdura, a five-star resort larger than the entire nation of Monaco, and home to six tennis courts, three golf courses, four pools, and a football field. While the resort is world-class, many of the participants chose to stay instead on superyachts floating off the coast, from which they were ferried to the event on the island in Maseratis. By the first night of the event, forty private jets had landed at the resort, with another seventy or so expected to arrive before the weekend was over.

"I've never seen such a pampered bunch," someone present told the *Sun*, a British tabloid. "Everything is laid out for them. It's so extravagant." [2]

"They just don't seem to be aware that they're the ones burning huge amounts of fossil fuels. They turn up with unnecessary entourages in helicopters or fast cars and then preach about saving the world."[3]

Two weeks later, just as the controversy was starting to die down, media reported that Prince Harry and his wife, Meghan Markle, Duchess of Sussex, and their new baby had taken two additional trips on private planes, first to Ibiza, Spain, and then, a few weeks later, to the town of Nice on the French Riviera.[4] While an economy class ticket from London to Nice costs £232 (U.S. $306), a private jet flight costs £20,000 (U.S. $29,000).[5]

"Frankly it's hypocritical. Harry can't be preaching about the catastrophic effects of climate change while jetting around the world on a private plane," said one of the prince's former bodyguards.[6]

BBC calculated that the flights to Ibiza and Nice had produced six times more emissions than the average Briton does each year and more than one hundred times more than the average resident of the African nation of Lesotho, where Prince Harry had lived during his gap year.[7]

Friends of Harry and Meghan rushed to the couple's defense. "I'm calling on the press to cease these relentless and untrue assassinations," Elton John said. "Imagine being attacked," said Ellen DeGeneres, "when all you're trying to do is make the world a better place."[8]

But the problem wasn't that the celebrities were flaunting their high-energy lifestyles. The problem was that they were moralizing for low-energy lives. "Stop lecturing us on how we live our lives," said one Briton, "and live by example."[9]

Al Gore wouldn't have been similarly embarrassed by Associated Press for living in a twenty-room home that used twelve times more energy than the average home in Nashville, Tennessee, had he not claimed, "We are going to have to change the way we live our lives" to solve climate change.[10]   ·

2. Not as We Do

Such hypocrisy understandably upsets many climate activists. "Name a single celebrity who's standing up for the climate!" Greta Thunberg demanded of her mother in 2016. "Name a single celebrity who is prepared to sacrifice the luxury of flying around the world!" [11]

But sacrificing the luxury of flying around the world doesn't necessarily mean one can avoid producing significant amounts of carbon emissions, as Thunberg soon discovered.

In August 2019, Thunberg sailed from Europe to New York to set an example of how to live without emitting carbon. But Greta's renewable-powered sailboat trip across the Atlantic produced four times more emissions than flying. The reason was that sailing required a sailboat crew, who flew back home afterward. [12]

The reason even the most sincere greens consume large quantities of energy is simple: living in wealthy nations and doing things that people in wealthy nations do, from driving and flying to eating and living in a home, requires significant quantities of energy.

As we have seen, there is no energy leapfrogging. Per capita income remains tightly coupled with per capita energy consumption. There is no rich low-energy nation just as there is no poor high-energy one. While Europeans consume less energy than Americans, on average, this is due less to environmental virtue and more to the fact that they rely more on trains and less on cars, due to higher population densities. [13]

And overall, energy consumption has steadily risen in developed nations. Europe's primary energy consumption, which is electricity plus the energy used for heating, cooking, and transportation, rose from about 12,500 to 23,500 terawatt-hours between 1966 and 2018. North America's rose from around 17,000 to 33,000 terawatt-hours during the same period. [14]

True, *per capita* energy consumption in wealthy nations has modestly declined during the last decade, but that was largely due to the loss of energy-intensive industries like manufacturing to countries such as China, not due to energy efficiency or conservation. [15]

When you count the emissions "embedded" in product imports from China, the increase in U.S. emissions between 1990 and 2014 rose from 9 percent to 17 percent, and the 27 percent reduction in British emissions during the same period declines to an 11 percent reduction. [16]

While environmentalists have not had the political power to restrict energy consumption and thus economic growth in rich nations, they have, for fifty years, had enough to restrict it in poorer and weaker ones. Today, the World Bank is diverting funding from cheap and reliable energy sources like hydroelectricity, fossil fuels, and nuclear, to expensive and unreliable ones like solar and wind. And in October 2019, the European Investment Bank announced it would halt all financing of fossil fuels in poor nations by 2021.[17]

While rich-world environmentalists are not the underlying cause of poverty in places like the Congo, they are, at a minimum, making things more difficult for nations already disadvantaged from being the last nations on Earth to industrialize and develop.

In 1976, a twenty-eight-year-old white South African named John Briscoe went to Bangladesh. Raised under apartheid, Briscoe had become radicalized against the country's system of racial segregation. Briscoe had earned a PhD in environmental engineering at Harvard and went to Bangladesh seeking to use his skills to help lift people out of poverty. Briscoe ended up in a village that was flooded by several meters of water for one-third of the year. Locals suffered from disease and malnutrition. Life expectancy was less than fifty.[18]

But when Briscoe heard of plans to build an embankment around the village to protect it from the annual flooding and provide it with irrigation, he opposed it. Briscoe, a Marxist at the time, thought the embankment would simply concentrate wealth in the hands of the rich.

Twenty-two years later, Briscoe returned to the village in Bangladesh and was surprised by what he found. The villagers were healthy, the kids were in school, and instead of rags they wore clothes. Life expectancy had increased to nearly seventy. Women were more independent, and there were vibrant markets for food.

Briscoe asked people what had changed. "The embankment!" they said. It had been built in the 1980s, after Briscoe left. It prevented flooding and allowed for the controlled use of water for irrigation. He asked people what else had made a difference. The bridges, they said, had reduced how much time it took to get to the marketplace. The villagers credited the big infrastructure projects for their prosperity.

Briscoe couldn't help but change his thinking. "Of course, infrastructure is not a sufficient condition for poverty reduction," he told an interviewer in 2011, "but it most certainly is a necessary condition!" [19] Added Briscoe, "Every presently rich country has developed more than 70 percent of its economically viable hydroelectric potential. Africa has developed 3 percent of its potential." [20]

For two decades after World War II, the World Bank, which is financed by developed nations, loaned money to developing nations to build the basic infrastructure of modern societies: dams, roads, and electricity grids. Investments like dams are low-risk investments since they will generate revenue through electricity sales, allowing nations to repay the loan. Much of Brazil's electricity grid is the result of the World Bank, which financed twelve hydroelectric projects there. [21]

But then, in the late 1980s, under the sway of green NGOs like World Wildlife Fund (WWF) and Greenpeace, the United Nations started to promote a radically different development model: sustainable development. Under this new model, poor and developing nations would continue to use small-scale renewable energy, rather than large-scale electrical power plants like dams. The World Bank followed the UN's direction.

By the 1990s, only 5 percent of World Bank financing went to infrastructure. "Water infrastructure, as an instrument for growth and as a precondition for economic growth, had essentially been thrust aside through a process led by the rich countries which already have their infrastructure," explained Briscoe. [22]

3. The Power of Electricity

The United Nations pioneered the notion that poor nations could grow rich without using much energy, in sharp contrast to every other rich nation in the world. They would do this through leapfrogging, the concept that former Shell economist Arthur van Benthem debunked.

In 1987, the United Nations published a book called *Our Common Future*, paying lip service to the problems of overuse of wood fuel in poor nations. [23]

"Both the routine practice of efficient energy use and the development of renewables will help take pressure off traditional fuels," the report claimed, "which are most needed to enable developing countries to realize their growth potential worldwide."[24]

And yet there was never any evidence to support this claim, and a large amount of evidence against it. As we saw, the industrial revolution could not have happened with renewables. Preindustrial societies are low-energy societies. Coal allowed preindustrial humans to escape the organic solar energy economy. There was no example in 1987 of any nation escaping poverty with renewables and energy efficiency.

The fact that developed nations required fossil fuels to grow wealthy could not possibly have been a mystery to the lead author of *Our Common Future*, Gro Brundtland. After all, she was the former prime minister of Norway, a nation that just a decade earlier had become one of the richest in the world thanks to its abundant oil and gas reserves.[25]

In 1998, a Brazilian energy expert argued for leapfrogging in the form of continued wood fuel use in more efficient cook stoves and in ways that avoided both fossil fuels and nuclear.[26]

Two years later, two U.N. agencies noted that while people in the past escaped poverty by moving away from "simple biomass fuels (dung, crop residues, firewood) to . . . liquid or gaseous fuels for cooking and heating and electricity," people in poor nations today could now "leapfrog directly from fuelwood to . . . new renewables" like biomass and solar.[27]

The idea captivated environmental philanthropies, NGOs, and U.N. agencies, which diverted money that in the past had gone into financing infrastructure to leapfrogging experiments in poor nations.

The United Nations and environmental NGOs described their work as helping poor nations "avoid the mistakes made in the industrialised world," in the words of the UN Development Program.[28]

Some of the shift away from infrastructure and manufacturing was justified as a way to fight corruption, a view that former World Bank economist Hinh Dinh said was unsupported by the best available research.

"If some European politicians or technocrats think that Africa could and should develop by eliminating corruption first, that then becomes a national development strategy for many African countries," Hinh Dinh, the former World Bank economist who advocates factories as a way for poor nations to develop, told me. "Never mind that not a single country in the world has become developed through that route."[29]

As climate change emerged as an elite concern in the 1990s, efforts within developed nations to cut off financing for cheap energy, industrial agriculture, and modern infrastructure to poor and developed nations grew stronger.

By 2014, Senator Patrick Leahy of Vermont, the ranking Democrat on the House Committee on Appropriations, sought to cut off U.S. development funding to poor nations seeking to build hydroelectric dams, on the basis that such dams have a "negative impact" on river ecosystems.

Briscoe was outraged. "If Senator Leahy is so adamantly against hydropower," said Briscoe, "let him show his commitment by first turning out the lights of Vermont."[30]

The same year, *The New York Times* claimed, "Many poor countries, once intent on building coal-fired power plants to bring electricity to their people, are discussing whether they might leapfrog the fossil age and build clean grids from the outset," due to Germany's influence.[31]

European governments actively promote bioenergy in poor nations. My first night in Kigali, Rwanda, during my December 2014 visit, I went to a party hosted by the Dutch Embassy to celebrate its promotion of a project, cofinanced with Germany, to harness the biogas from human feces for cooking.[32]

In 2017, Eva Müller, the director of forestry at the U.N. Food and Agriculture Organization, claimed, "Woodfuel is kinder to the environment than fossil fuels, and including charcoal, accounts for roughly 40 percent of current global renewable energy supplies—as much as solar, hydroelectric, and wind power combined."[33]

Not all environmentalists oppose cheap energy, including hydroelectric dams and fossil fuels for poor nations. In my experience, many and perhaps most environmentalists in developed nations believe it is unethical for rich nations to deprive poor ones of the technologies responsible for our prosperity.

But the leadership of Western NGOs and U.N. agencies, as well as many IPCC authors, have participated in efforts during the last two decades to redirect public and private money away from cheap energy and toward unreliable and expensive renewables.

Poor nations, claimed the IPCC in 2018, can leapfrog centralized energy sources like dams, natural gas plants, and nuclear plants to decentralized energy sources such as solar panels and batteries. It did not cite Van Benthem or other economists who have debunked leapfrogging.[34]

In 2019, many NGOs, including the German Urgewald, campaigned for diverting World Bank funding away from large hydroelectric dams and fossil fuels to small-scale renewables like solar and wind.[35]

Toward the end of his life, Briscoe was deeply upset by the success of green NGOs in pressuring Western nations to divert funding away from basic infrastructure and agricultural modernization to various "sustainable development" experiments. "Time and time again I have seen NGOs and politicians in rich countries advocate that the poor follow a path that they, the rich, never have followed," he wrote, "nor are willing to follow."[36]

Why is that?

4. "A Stain on the Race"

In 1793, British philosopher William Godwin published *An Enquiry Concerning Political Justice and Its Influence on General Virtue and Happiness*. In it, Godwin argued that human reason, not political revolution, was the key to progress. We are unique in our ability to control ourselves, including our passions, and improve our societies. Over time, Godwin argued, such rationalism would result in massive reductions in human suffering.[37]

One year later, the Marquis de Condorcet, a French nobleman and mathematician, published a book that forecast infinite human progress. In *Outlines of an Historical View of the Progress of the Human Mind*, Condorcet pioneered what he called "social science." He championed humankind using technology to grow more food on less land in order to support a larger human population. He supported trade between nations as a way to reduce food scarcity. At the heart of his book was the idea that science and reason can be used to advance human progress.[38]

Godwin and Condorcet's combined ideas were a vision of what we now call the Enlightenment, and both thinkers were "humanists" because they believed humans were special through our unique capacity to reason. They had effectively secularized the Judeo-Christian concept that humans were chosen by God to have dominion over Earth.

As feudal dictatorships gave way to capitalist democracies, Enlightenment humanism became the dominant political ideology. "Taken together," Condorcet's vision held, "technological and human advances would allow forever-smaller portions of ground to support ever-larger populations."[39]

Thomas Robert Malthus, an economist, grew so annoyed with Enlightenment optimism that, in his early thirties, he sought to refute Godwin and Condorcet in a 1798 book called *An Essay on the Principle of Population.*

Malthus argued that human progress was unsustainable. While humans might grow more food incrementally (e.g., 1, 2, 3, 4), we reproduced "geometrically" (e.g. 2, 4, 8, 16). The result of progress would thus inevitably be overpopulation and famine. "The poor consequently must live much worse, and many of them be reduced to severe distress," Malthus wrote. "The power of population is so superior to the power of the earth to produce subsistence for man, that premature death must in some shape or other visit the human race."[40]

Lest readers be confused by what he was arguing, Malthus added this remarkable passage to the second edition of his book:

A man who is born into a world already possessed, if he cannot get subsistence from his parents on whom he has a just demand, and if the society do not want his labour, has no claim of right to the smallest portion of food, and, in fact, has no business to be where he is.[41]

Godwin responded that "there are various methods by the practice of which population may be checked," including birth control, which would allow human societies to avert the famines Malthus viewed as inevitable,[42] and using technology to grow more food on less land.[43]

Malthus responded not by arguing that humans *wouldn't* use birth control but rather that they *shouldn't*. Why? Because doing so would be "unnatural."[44] The only way famines could be avoided, Malthus argued, was through a long period of celibacy, followed by a late marriage and bearing few children.

In other words, the only way Malthus's prediction of population outstripping resources could be correct is if everybody in the future subscribed to Malthus's opposition to birth control.

Malthus professed concern for the poor while advocating policies that would keep them poor. He advocated that policymakers maintain the aristocratic system by favoring agriculture over manufacturing, and pointed to the superiority of country life, or rather, the country life that he, as an aristocrat who avoided manual labor, enjoyed.[45]

Some defend Malthus by claiming that he wrote his famous book when it was still too early to know that the industrial revolution would radically increase food production. Malthus came of age in what historians call the "advanced organic economy," which, due to its reliance on renewables, namely wood fuel and waterwheels, "condemned the majority of the population to poverty" for inherently physical reasons.[46]

But the bleakness of Britain's renewable-powered economy hadn't prevented Godwin and Condorcet and other humanists from imagining not only the end of hunger but also universal prosperity. Indeed, there was evidence of success all around them. Had it not been for the continuous improvements to agriculture yields during Malthus's lifetime, along with an expansion of farming from 11 to 14.6 million acres between 1700 and 1850, hunger in the British countryside would have been far worse.[47]

In 1845, a virulent fungus destroyed much of Ireland's potato crop, triggering what would become known as the Great Famine. Between 1845 and 1849, one million people in Ireland starved to death, and another million fled the island.[48]

To this day, when people think of the Great Famine, they tend to focus on the fungus and overlook the fact that, for the next four years, Ireland exported food, including beef, to England. Irish families had to sell their pigs in order to pay the rent, even as their children were starving.

British elites justified letting the Irish starve by blaming them for their fate. The real reason the Irish were starving, held good opinion in Britain during the famine, was the Irish people's lack of moral restraint. Increasing the wages of Irish workers, the The Economist warned, "would stimulate every man to marry and populate as fast as he could, like rabbits in a warren."[49]

The Economist and other British elites simply repeated the thinking pioneered a half-century earlier by Malthus. Back then, he had condemned the overpopulating tendencies of the Irish as a consequence of inexpensive food. "The cheapness of this nourishing root [potatoes]," he wrote, "joined to the ignorance and barbarism of the [Irish] people, have encouraged marriage to such a degree that the population has pushed much beyond the industry and present resources of the country."[50]

Ultimately, the problem in Ireland was one of overpopulation, Malthus wrote. "The land in Ireland is infinitely more peopled than in England and to give full effect to the natural resources of the country, a great part of the population should be swept from the soil."[51]

The Great Famine wouldn't be the last time British rulers used Malthus's ideas to justify starvation in other nations. The British governor general of India between 1876 and 1880 argued that the Indian population "has a tendency to increase more rapidly than the food it raises from the soil."[52] Later he claimed the "limits of increase of production and of population have been reached."[53]

As tens of thousands of Indians died of hunger, the governor general "lavished monies on the enthronement of Queen Victoria as empress of India," a historian writes. "The famine relief offered to the starving by Lytton's administration was less in terms of calorific intake than that Hitler gave to those interned in Buchenwald concentration camp."[54]

In 1942 and 1943, as India produced food and manufactured goods for the British war effort, food shortages emerged. Food imports could have alleviated the crisis, but Prime Minister Winston Churchill refused to allow it.

Why? "Much of the answer must lie in the Malthusian mentality of Churchill and his key advisors," concludes historian Robert Mayhew. "Indians are breeding like rabbits and being paid a million a day by us for doing nothing about the war," Churchill claimed, falsely.

Partly as a result of his decisions, three million people died in the Bengali famine of 1942 to 1943, which was three times the death toll of the Great Irish Famine.[55]

Adolf Hitler, too, was inspired by Malthus. "The productivity of the soil can only be increased within defined limits and up to a certain point," he wrote in *Mein Kampf*. But, in contrast to Malthus, Hitler believed that those limits could be overcome through invasion of foreign territories.

"Strong and direct connections can be drawn between [Malthus's] work," historian Mayhew concludes, "and some of the most abhorrent moments in twentieth-century history."[56]

In the early twentieth century, the Tennessee Valley region of the United States was a lot like the Congo today. Deforestation was rising. Agricultural yields were declining due to soil erosion. Malaria plagued the region. Few had adequate medical care. Fewer had indoor plumbing or electricity.

World War I brought hope to the region. Congress authorized the construction of a dam on the Tennessee River to power a munitions factory. But the war ended before the dam could be finished. Henry Ford offered to buy the complex for $5 million, but taxpayers had already sunk more than $40 million into the project, leading George Norris, a progressive Republican senator, to oppose Ford's offer.

Norris sat on Congress's agriculture committee and had influence. He was sympathetic to the plight of the poor and regularly visited the Tennessee Valley region. He spent nights in poor farmers' shacks. During the next decade, Norris urged the U.S. government to invest in an agricultural modernization program.

By 1933, Norris had convinced Congress and the newly elected President Franklin D. Roosevelt to create what would be called the Tennessee Valley Authority (TVA). It built dams and fertilizer factories and installed irrigation systems. The TVA trained local farmers to teach others how to increase crop yields. And it planted trees.

There were trade-offs. About twenty thousand families were relocated through eminent domain and condemnation. Nearly seventy thousand individual burial plots were either removed or left in place.[57] Huge areas were flooded with water.

But those sacrifices were a small price to pay for the economic growth that resulted, not just for the local region, but for the national economy as a whole. Cheap electricity and economic growth allowed the natural environment to be restored. Fertilizers replenished the depleted soils. Electricity powered irrigation pumps. Farmers grew more food on less land, and planted trees. Over time, the forests returned.

But a homegrown backlash against industrialization and agricultural modernization had started to brew even before the federal government created the TVA.

In 1930, forty-two-year-old Rhodes scholar and Tennessee poet John Crowe Ransom wrote in the opening essay in a famous collection, *I'll Take My Stand*, "the latter-day societies have been seized—none quite so violently as our American one—with the strange idea that the human destiny is not to secure an honorable peace with nature, but to wage an unrelenting war on nature."[58]

Ransom and the other "Southern Agrarians" disparaged cities and industry for their impact on the environment and on people. They declared farm machinery, paved roads, and indoor plumbing as part of the "disease of modern industrial civilization."

Ransom's perspective as a poet at Vanderbilt University was quite different from the one of the poor sharecroppers. The people of the Tennessee Valley region who suffered from malaria and hunger likely might have disagreed with the view that they had been living at peace with nature. Critics of Ransom and the Southern Agrarians called them "typewriter agrarians" in the same way people sometimes criticize upper-middle-class progressives as "latte liberals."

Ransom, Malthus, and the Malthusians who came after him were socially and politically conservative. Malthus was against birth control, viewing it as against God's plan for humans. He was against social welfare programs for the poor, viewing them as self-defeating. British leaders who justified their policies based on Malthus's thinking were conservatives.

By contrast, socialists and leftists loathed Malthus. Marx and Engels called him a "stain on the human race." Malthus, in their view, had made an avoidable situation look inevitable, or "natural."[59] In his 1879 book, *Progress and Poverty*, the progressive American thinker Henry George attacked Malthus as a defender of inequality. "What gave Malthus his popularity among the ruling classes," George wrote, "was the fact that he furnished a plausible reason for the assumption that some have a better right to existence than others."[60]

But then, after World War II, Malthusianism switched sides and became a left-wing political movement in the form of environmentalism, while anti-Malthusianism became a right-wing political movement in the form of libertarian, pro-business, free market conservatism.

Some opposition to ascendant Malthusianism came from the left. Bayard Rustin, a socialist civil rights leader, told *Time* in 1979 that environmentalists were "self-righteous, elitist, neo-Malthusians who call for slow growth or no growth . . . and who would condemn the black underclass, the slum proletariat, and rural blacks, to permanent poverty."[61]

But most resistance to Malthusianism came from the political right. The most prominent critic of Malthusian alarmists was Julian Simon, an economist who argued "natural resources are not finite," and that children weren't just mouths to feed but rather grow up to be producers, not just consumers.[62] Simon was embraced by conservative and libertarian scholars, think tanks, and media, not left-wing and progressive ones.

Why was that?

5. Lifeboat Ethics

In 1948, a conservationist named William Vogt published a best-selling book, *Road to Survival*, which was translated into nine languages and serialized in *Reader's Digest*.[63] In it, Vogt warned of rampant breeding in poor nations, particularly India. "Before the imposition of Pax Britannica, India had an estimated population of less than 100 million people," Vogt wrote.

It was kept in check by disease, famine, and fighting. Within a remarkably short period the British checked the fighting and contributed considerably to making famines ineffectual by building irrigation works, providing means of food storage, and importing food during periods of starvation. . . . While economic and sanitary conditions were being "improved," the Indians went to their accustomed way, breeding with the irresponsibility of codfish. . . . Sex play is the national sport.[64]

Vogt attacked the medical profession's "duty to keep alive as many people as possible."[65] In truth, he wrote, they were just "increasing misery."[66] He argued that a better model was ancient Greece, which "purposefully reduced" human population through "infanticide, emigration, and colonization."[67]

Vogt felt he had a solution. "International control of resources exploitation, in order to protect technologically retarded nations, is indispensable. . . . The United Nations, and associated organizations, should maintain an ecological commission, cutting across organization lines, to assay the influence of all U.N. activities on the relationship between human beings and their indispensible environment."[68]

American leaders and elites embraced Malthusian ideas just as British elites had. In 1965, in the first televised State of the Union address, President Lyndon Johnson described the "explosion in world population and the growing scarcity in world resources" as the most important issue in the world. He called for "population control."[69]

The New York Times scolded Johnson for not being Malthusian enough. After all, Johnson had suggested economic growth was still possible "in a world where the Malthusian specter, more terrible than Malthus ever conceived, is so near to being a reality," editorialized the *Times*.[70]

That same year, the journal *Science* published an article, "Tragedy of the Commons," by University of California at Santa Barbara biologist Garrett Hardin, which argued that environmental collapse was inevitable because of uncontrolled breeding, and that the only way to avoid the tragedy was "mutual coercion," in which everybody agreed to similar sacrifices.[71]

Many conservation leaders embraced Malthusianism. In 1968, Sierra Club executive director David Brower conceived and edited a book, *The Population Bomb*, by Stanford University biologist Paul Ehrlich, which claimed the world was on the brink of mass starvation. "The battle to feed all of humanity is over. In the 1970s and 1980s hundreds of millions of people will starve to death in spite of any crash programs embarked upon now."[72]

Like Vogt and Malthus before him, Ehrlich was particularly concerned with breeding by poor people in developing nations. During a taxi ride from the Delhi airport to his hotel downtown, Ehrlich described the Indians he looked down upon more contemptuously than a biologist would describe animals: "people eating, people washing, people sleeping. . . . People defecating and urinating. People, people, people, people."[73]

Johnny Carson had Ehrlich on *The Tonight Show* six times, helping Ehrlich sell more than three million copies of *The Population Bomb*.[74]

Malthusianism grew an even harder edge in the 1970s. Hardin, the University of California biologist, published an essay, "Lifeboat Ethics: The Case Against Helping the Poor," in which he argued that "we must recognize the limited capacity of any lifeboat."

The picture Hardin painted was of keeping people out of the lifeboat. Otherwise, the people trying to get in would doom the people in the lifeboat, in addition to themselves.

"However humanitarian our intent," said Hardin, "every Indian life saved through medical or nutritional assistance from abroad diminishes the quality of life for those who remain, and for subsequent generations."[75]

Top academic institutions helped make Malthusian ideas mainstream. In 1972, an NGO called the Club of Rome published *The Limits to Growth*, a report concluding that the planet was on the brink of ecological collapse, which *The New York Times* covered on its front page.

"The most probable result," the report declared, "will be a rather sudden and uncontrollable decline in both population and industrial capacity." The collapse of civilization was "a grim inevitability if society continues its present dedication to growth and 'progress.' "[76]

Like the Southern Agrarians, ecologist Barry Commoner and physicist Amory Lovins argued that industrialization was harmful, and that we needed to protect poor nations from economic development.[77]

The Malthusian Ehrlich and the ostensibly socialist Commoner clashed over population and poverty. Commoner blamed poverty for food crises, where Ehrlich blamed overpopulation. Commoner blamed industrial capitalism for environmental degradation, where Ehrlich blamed too many people.[78]

The clash resolved itself when Malthusians including Ehrlich accepted a redistributive agenda of rich nations assisting poor nations with development aid so long as that money went to charity and not things like infrastructure. This was the seed of what the UN would christen "sustainable development."[79]

Lovins, for his part, married the demand for energy scarcity to a romantic vision of a "soft energy" future that rejected the infrastructure of the rich world. In 1976, *Foreign Affairs* published a thirteen-thousand-word essay by Lovins making the case for small-scale energy production instead of large-scale power plants.[80]

Where electricity from large power stations had been viewed in the New Deal era and before as progressive—liberating people from such tasks as washing clothes by hand and providing a clean alternative to wood-burning stoves—Lovins viewed electricity as authoritarian, disempowering, and alienating. "In an electrical world, your lifeline comes not from an understandable neighborhood technology run by people you know who are at your own social level, but rather from an alien, remote, and perhaps humiliatingly uncontrollable technology run by faraway, bureaucratized, technical elite who have probably never heard of you," he wrote.[81]

The Malthusians significantly modified Malthus. Where Malthus warned that overpopulation would result in a scarcity of food, Malthusians in the 1960s and 1970s warned that energy abundance would result in overpopulation, environmental destruction, and societal collapse.

Ehrlich and Lovins said they opposed nuclear energy because it was abundant. "Even if nuclear power were clean, safe, economic, assured of ample fuel, and socially benign," Lovins said, "it would still be unattractive because of the political implications of the kind of energy economy it would lock us into."[82]

Behind advocacy ostensibly motivated by concerns for the environment lay a very dark view of human beings. "It'd be little short of disastrous for us to discover a source of cheap, clean, and abundant energy," said Lovins, "because of what we would do with it."[83] Ehrlich agreed. "In fact, giving society cheap, abundant energy at this point would be the moral equivalent of giving an idiot child a machine gun."[84]

6. Power Against Progress

So much death and suffering was coming, Ehrlich and Holdren believed, humankind needed to play "triage," and leave some people to die. In the "concept of triage," they wrote, "those in the third groups are those who will die regardless of treatment. . . . The Paddocks [the authors of the 1967 book *Famine 1975!*] felt that India, among others, was probably in this category. Bangladesh is today a more clear-cut example."[85]

Ehrlich and Holdren argued that the world likely did not have enough energy to support the development aspirations of the world's poor. "Most plans for modernizing agriculture in less-developed nations call for introducing energy-intensive practices similar to those used in North America and western Europe—greatly increased use of fertilizers and other farm chemicals, tractors and other machinery, irrigation, and supporting transportation networks—all of which require large inputs of fossil fuels," they noted.[86]

A better way, they said, was "much greater use of human labor and relatively less dependence on heavy machinery and manufactured fertilizers and pesticides." Such labor-intensive farming "causes far less environmental damage than does energy-intensive Western agriculture," they claimed.[87] In other words, the "secret" to "alternative farming methods" was for small farmers in poor nations to remain small farmers.

Malthusians justified their opposition to the extension of cheap energy and agricultural modernization to poor nations by using left-wing and socialist language of redistribution. It wasn't that poor nations needed to develop, it was that rich nations needed to consume less.

Ehrlich and Holdren claimed in their 1977 textbook that the only way to feed seven billion people by the year 2000 was for people in the rich world to eat less meat and dairy—the same recommendation IPCC made in 2019.[88] "Unless food can be transferred from the rich nations on an unprecedented scale," they wrote, "the death rate from starvation in [less-developed countries] seems likely to rise in the next few decades."[89]

Where in 1977, Ehrlich and Holdren proposed international control of the "development, administration, conservation and distribution of all natural resources," many green NGOs and U.N. agencies today similarly seek control over energy and food policies in developing nations in the name of climate change and biodiversity.[90]

By 1980, nearly half a century after President Roosevelt created the TVA, the Democratic Party had reversed itself on the question of abundance versus scarcity. In 1930, Democrats had understood the necessity of cheap energy and food to lifting people out of poverty, but by 1980, President Jimmy Carter's administration had endorsed the "limits to growth" hypothesis.

"If present trends continue," claimed the blue-ribbon authors of *The Global 2000 Report to the President of the United States*, "the world in 2000 will be more . . . vulnerable to disruption than the world we live in now [and] . . . the world's people will be poorer in many ways than they are today."[91]

But fears of overpopulation were declining. Demographers knew that population growth rate had peaked around 1968. In 1972, the editor of *Nature* predicted, "The problems of the 1970s and 1980s will not be famine and starvation but, ironically, problems of how best to dispose of food surpluses." The same editor noted that fear-mongering "seems like patronizing neo-colonialism to people elsewhere."[92]

Others agreed. One demographer said the problem wasn't a population explosion but rather a "nonsense explosion."[93] Danish economist Ester Boserup, working for the FAO, did historical research finding that as the human population increased, people had long found ways to increase food production. Malthus had been wrong even about the preindustrial era, she found.[94] In 1981, the Indian economist Amartya Sen published a book showing that famines occur not because of lack of food but due to war, political oppression, and the collapse of food *distribution*, not production, systems. Sen won the Nobel Prize for economics in 1998.[95]

By 1987, demographers knew the number of humans added annually to the global population had reached its peak. Seven years later, the U.N. held its last Family Planning meeting. Between 1996 and 2006, United Nations family planning spending declined 50 percent.[96]

In 1963, two economists published an influential book called *Scarcity and Growth*. In it, they described how nuclear changed the classical economic assumption of natural resources as scarce and limited. The "notion of [an] absolute limit to natural resource availability is untenable when the definition of resources changes drastically and unpredictably over time."[97]

Policymakers, journalists, conservationists, and other educated elites in the fifties and sixties knew that nuclear was unlimited energy, and that unlimited energy meant unlimited food and water. We could use desalination to convert ocean water into freshwater. We could create fertilizer without fossil fuels, by splitting off nitrogen from the air, and hydrogen from water, and combining them. We could create transportation fuels without fossil fuels, by taking carbon dioxide out of the atmosphere to make an artificial hydrocarbon, or by using water to make pure hydrogen gas.

Some had known this for much longer. In 1909, three decades before scientists would split the atom, an American physicist wrote a best-selling book, inspired by the discovery of radium by Paul and Marie Curie, describing a very similar vision of a nuclear-powered world, and the benefits that would result from such high power densities.[98]

Nuclear energy not only meant infinite fertilizer, freshwater, and food but also zero pollution and a radically reduced environmental footprint. Nuclear energy thus created a serious problem for Malthusians and anyone else who wanted to argue that energy, fertilizer, and food were scarce.

And so some Malthusians argued that the problem with nuclear was that it produced *too much* cheap and abundant energy.

"If a doubling of the state's population in the next twenty years is encour-

aged by providing the power resources for this growth," wrote the Sierra
Club's executive director, opposing Diablo Canyon nuclear plant, "[Califor-
nia's] scenic character will be destroyed." [99]

And this stoked fears of the bomb. The British philosopher Bertrand
Russell argued that "nothing is more likely to lead to an H-bomb war than
the threat of universal destitution through over-population." They called
the growing population in developing nations a "population explosion."
And Erhlich titled his book, *The Population Bomb*. [100]

Ehrlich and Holdren argued that a nuclear energy accident could be worse
than a bomb. "A large reactor's inventory of long-lived radioactivity is more
than one thousand times that of the bomb dropped on Hiroshima," [101] they
wrote, implying that it would have done one thousand times the damage.

The implication was wrong. Nuclear reactors cannot detonate like
bombs. The fuel is not sufficiently "enriched" to do so. But mixing up reactors and bombs was, as we saw, the go-to strategy for Malthusian environmentalists.

And, as would become routine in U.N. reports, including those published
by the IPCC for the next three decades, the United Nations' 1987 report
"Our Common Future" attacked nuclear energy as unsafe and strongly recommended against its expansion. [102]

There is a pattern. Malthusians raise the alarm about resource or environmental problems and then attack the obvious technical solutions. Malthus had
to attack birth control to predict overpopulation. Holdren and Ehrlich had to
claim fossil fuels were scarce to oppose the extension of fertilizers and industrial agriculture to poor nations and to raise the alarm over famine. And climate activists today have to attack natural gas and nuclear energy, the main
drivers of lower carbon emissions, in order to warn of climate apocalypse.

7. The Climate Bomb

As it became clear that the growth in the global birth rate had peaked, Malthusian thinkers started to look to climate change as a replacement apocalypse for overpopulation and resource scarcity.

The influential Stanford University climate scientist Stephen Schneider embraced the Malthusianism of John Holdren and Paul Ehrlich, and he invited them to educate his scientists.

"John did a more credible job of laying out the population-resources-environmental problems than nearly anyone else could have done at the time," wrote Schneider. "That talk helped the [National Center for Atmospheric Research] scientists to see the big picture clearly and early on." [103]

While collaborating with Holdren and George Woodwell, a cofounder of Environmental Defense Fund, at a conference organized by the American Association for the Advancement of Science in 1976, Schneider said he made the case that "humans were multiplying out of control and were using technology and organization in a dangerous, unsustainable way." [104]

Schneider attracted media attention by speaking of climate change in apocalyptic terms. "We were moving beyond the halls onto the world stage." Thanks to the work of friend Paul Ehrlich, he said, "In the summer of 1977 I appeared on four [CBS *The Tonight Show Starring Johnny*] *Carson* shows." [105]

In 1982, a group of economists who called themselves "ecological economists" met in Stockholm, Sweden, and published a manifesto arguing that nature imposes hard limits on human activity.

"Ecological economists distinguished themselves from neo-Malthusian catastrophists by switching the emphasis from resources to systems," notes an environmental historian. "The concern was no longer centered on running out of food, minerals, or energy. Instead, ecological economists drew attention to what they identified as ecological thresholds. The problem lay in overloading systems and causing them to collapse." [106]

Ecological economics, not to be confused with the mainstream environmental economics used by IPCC and other scientific bodies, was popular among philanthropies and environmental leaders in wealthy nations. Pew Charitable Trusts, MacArthur Foundation, and other philanthropies invested heavily in ecological economists. Environmental leaders, including Al Gore, Bill McKibben, and Amory Lovins, promoted its ideas. [107]

McKibben has done more to popularize Malthusian ideas than any other writer. The first book about global warming written for a popular audience was his 1989 book, *The End of Nature*. In it, McKibben argued that human-kind's impact on the planet would require the same Malthusian program developed by Ehrlich and Commoner in the 1970s. Economic growth would have to end. Rich nations must return to farming and transfer wealth to poor nations so they could improve their lives modestly but not industrialize. And the human population would have to shrink to between 100 million and 2 billion.[108]

Where just a few years earlier, Malthusians had demanded limits on energy consumption by claiming fossil fuels were scarce, now they demanded limits by claiming the atmosphere was scarce. "It's not that we're running out of stuff," explained McKibben in 1998; "what we're running out of are what the scientists call 'sinks.' Places to put the by-products of our large appetites. Not garbage dumps—we could go on using Pampers till the end of time and still have empty space left to toss them away. But the atmospheric equivalent of garbage dumps." [109]

By the twenty-first century, Stanford University climate scientist Schneider was as much an activist as a scientist. In a chapter in which he describes his participation in the 2007 IPCC report, Schneider writes, "I was sometimes disgusted how national interests trump planetary interests and the here-and-now overshadows long-term sustainability. I remembered my 'five horsemen of the environmental apocalypse': ignorance, greed, denial, tribalism, and short-term thinking." [110]

Environmentalists used climate change as a fresh reason for opposing hydroelectric dams and flood control even though, as Briscoe, the environmental engineer from South Africa, noted, "Adaptation [to climate change] is 80 percent about water." [111]

Briscoe pointed to evidence that Western environmental groups contributed to food shortages.

"Look at the food crisis last year," said Briscoe in 2011. "There were many voices bemoaning the crisis, with press coverage dominated by NGOs, and aid agencies who immediately called for greater support for agriculture in the developing world. What they did not mention was what their roles had been in precipitating this crisis.[112]

"The NGOs did not reflect on the fact that many NGOs had stridently

lobbied against many irrigation projects and other agricultural modernization projects because these 'were not pro-poor and destroyed the environment,' " Briscoe said. "What the aid agencies did not say was that lending
for agriculture had declined from 20 percent of official development assistance in 1980 to 3 percent in 2005 when the food crisis hit." [113]

One of the groups opposing the construction of the Grand Inga Dam in the
Congo is is a little-known but influential environmental NGO called International Rivers, based in Berkeley, California. While few people have heard
of the organization, it has, since its founding in 1985, helped stop 217 dams
from being built, mostly in poor nations. "If we think of forests as the lungs
of the planet, then rivers most surely are the arteries of the planet," the organization says on its website. [114]

But in 2003, Sebastian Mallaby, a journalist from *The Washington Post*,
discovered International Rivers had severely misrepresented the situation on
the ground in Uganda, where the group was trying to stop a dam. An International Rivers staff person told Mallaby that the Ugandan people near the
proposed dam opposed it. But when he asked her whom he could interview
to verify her assertion, she became evasive and told him that he would get in
trouble with the government if he asked questions. Mallaby did so anyway,
hiring a local sociologist to translate.

"For the next three hours, we interviewed villager after villager, and
found the same story," Mallaby wrote. "The dam people had come and
promised generous financial terms, and the villagers were happy to accept
them and relocate. . . . The only people who objected to the dam were the
ones living just outside its perimeter. They were angry because the project
was *not* going to affect them. They had been offered no generous payout,
and they were jealous of their neighbors." [115]

I found the same thing in my interviews. Not only were Congolese like
Caleb enthusiastic about the Virunga Park dam, Rwandans I interviewed
near hydroelectric dams were ecstatic at the prospects of getting electricity.

Why is International Rivers so opposed to dams? Partly because dams
can make it harder to do recreational rafting. "The Batoka scheme will flood
the gorge and drown the massive rapids that have made Victoria Falls a

prime whitewater rafting location," laments International Rivers about one project. Its allies consist of rafters around the world.[116]

International Rivers and other NGOs routinely work with sympathetic academics to conduct studies that purportedly show why unreliable renewables like solar and wind would be cheaper than reliable ones like hydroelectric dams. In 2018, researchers at the University of California, Berkeley, published a study in which they claimed it would be cheaper to use solar panels, wind turbines, and natural gas than to build the Inga dams.[117]

But the reason so many poor nations begin the process of urbanization, industrialization, and development by building large dams is that they produce inexpensive and reliable power, are simple to build and operate, and can last for a century or longer.[118] Once Congo achieves the security, peace, and good governance it needs to develop, it will likely do the same in building the Inga Dam.

The Inga Dam would have very-high-power densities and thus lower environmental impact than other dams around the world.[119] The proposed Inga Dam would have power densities three times that of dams in Switzerland.[120] And yet International Rivers is not seeking to remove dams in Switzerland nor in California, where for one hundred years they have provided the state with cheap, reliable, and abundant electricity, freshwater for drinking and agriculture, and flood control.[121]

8. Experiments in Poverty

After British tabloids publicized climate-concerned celebrities enjoying their high-energy lives in Sicily, some people chalked up their hypocrisy to ignorance. "This isn't about global warming," said one critic on Twitter. "It's about being seen, rubbing elbows and planning upcoming movies and acting, making it look like they care about environment. . . . Bunch of hypocrites. They would never give up their 'toys.' "[122]

But it's inconceivable the celebrities didn't realize they were behaving hypocritically. It's common knowledge that flying by jet results in significant carbon emissions. Indeed, the celebrities implicitly acknowledged their guilt. Elton John bought carbon offsets to supposedly cancel out Harry and Meghan's emissions, while a spokesperson for Thunberg acknowledged, "It would have been less greenhouse gas emissions if we had not made this departure."[123]

One explanation for the tone-deaf hypocrisy of celebrities who moralize about climate change while jetting around the globe is that they aren't being tone deaf at all. On the contrary, they are flaunting their special status. Hypocrisy is the ultimate power move. It is a way of demonstrating that one plays by a different set of rules from the ones adhered to by common people.

To be sure, the Duke, the Duchess, and Thunberg may not have *consciously* decided to flaunt their status. But neither did Harry and Meghan pledge to never fly private again. Nor did Thunberg cancel her trip.

Were the statements that environmentalists made in support of the right of poor nations to develop mere virtue signaling? Perhaps.

Whereas in January 2019, Thunberg had paid lip service to the need for poor nations to develop, in September she said, "We are in the beginning of a mass extinction and all you can talk about is money and fairy tales of eternal economic growth."[124]

But economic growth was what lifted Suparti out of poverty, saved the whales, and is the hope for Bernadette, once Congo achieves security and peace. Economic growth is necessary for creating the infrastructure required for protecting people from natural disasters, climate-related or not. And economic growth created Sweden's wealth, including that of Thunberg's own family. It is fair to say that without economic growth, the person who is Greta Thunberg would not exist.

In 2016, I traveled to India to give a talk, I arrived early to see some of the country. I interviewed people who live near, and pick through, one of Delhi's largest garbage dumps. I visited a community near a nuclear power plant. And I interviewed rural villagers with Joyashree Roy, a professor of economics at Jadavpur University, Kolkata, in India, and an IPCC coordinating lead author.

I met Joyashree through our common interest in rebound, which is the way in which we might use the money we save from not eating meat or using more energy-efficient appliances to do other things that ultimately result in higher energy usage. She agreed to take me to some of the communities she works with.

Joyashree took me to a poor neighborhood where there was an ongoing experiment in daylighting, which is a way to provide light to people living in dark homes. It is done by cutting a hole in the roof and sticking into it a plastic bottle that refracts sunlight into the home. Daylighting has received widespread positive publicity in Western nations. Joyashree asked me what I thought of it and I told her I found it offensive that Western NGOs viewed sticking a plastic bottle in the roof of a shack as something to celebrate. She seemed to agree.

Criticism of such low-energy experiments had been growing. In 2013, while in Tanzania to promote "Power Africa," a U.S. government program supporting electrification, President Barack Obama dribbled and headed a modified soccer ball known as a Soccket: after you play with it for thirty minutes, it can power an LED light for three hours. "You can imagine this in villages all across the continent," Obama enthused. But with a price tag of $99, the Soccket cost more than a month's wages for the average Congolese. For only $10, one could buy a superior lantern that doesn't require dribbling. More to the point, it wasn't exactly the kind of energy that could industrialize Africa.[125]

Two years later, an Indian village made worldwide headlines after it rebelled against the solar panel and battery "micro-grid" Greenpeace had created as a supposed model of energy leapfrogging for the world's poorest people. The electricity was unreliable and expensive. "We want real electricity," chanted villagers at a state politician, "not fake electricity!" Children held up signs saying the same thing. By "real electricity" they meant reliable grid electricity, which is mostly produced from coal.[126]

Joyashree is in her element in the countryside. Until then, I had interacted with her only in professional environments, where her manner was formal and reserved. In the countryside she was extroverted, open, and warm. In one of the villages, the people were being sickened by drinking water with high levels of arsenic, a naturally occurring toxic metalloid. Joyashree had been working with villagers to develop a water purification system.

In 2018, Joyashree was IPCC's lead coordinating author of the chapter on "Sustainable Development, Poverty Eradication and Reducing Inequalities," in its report on keeping temperatures from rising more than 1.5 degrees above preindustrial temperatures.

I wrote two columns criticizing the chapter for having an antinuclear bias. "Nuclear energy," her chapter claimed, "can increase the risks of proliferation, have negative environmental effects (e.g., for water use), and have mixed effects for human health when replacing fossil fuels." [127]

I didn't think Joyashree was antinuclear, and interviewed her by phone about the chapter in 2019. She said it reflected what was in the peer-reviewed scientific literature. "If you find thousands of articles published on renewables as good things, and only two on nuclear, it suggests there is a huge consensus on renewables, and little on nuclear," she said. "Renewables like solar and wind are not unmixed blessings, but there are very few who write on that." [128]

I pressed Joyashree on why her chapter suggested nations could leapfrog. She defended leapfrogging as a concept, but then expressed frustration with people who advocate energy demand reductions, even among India's poorest. "This is something that I as a social scientist cannot accept," she said.

"Because it's an experiment on the most vulnerable people in the world?" I asked.

"Yeah," she said. And then she said, "Yeah. Yeah. Yeah. Yeah. Yeah," before laughing in a tone that sounded like frustration.

12

FALSE GODS FOR LOST SOULS

1. Parable Bears

Toward the end of 2017, *National Geographic* posted a video, set to sad music, of an emaciated and slow-moving polar bear to YouTube. "This is what climate change looks like," read the caption. A photojournalist and a filmmaker had filmed the bear the previous July. The video has been viewed at least 2.5 billion times.[1]

One of the viewers was student climate activist Greta Thunberg. "I remember when I was younger and in school, our teachers showed us films of plastic in the ocean, starving polar bears, and so on," she recalled in spring 2019. "I cried through all the movies."[2]

Climate change is polar bears' greatest threat, concluded scientists in 2017, because it is melting the arctic ice caps at a rate of 4 percent per year. "But to hear climate denialists tell it, polar bears are doing just fine," reported *The New York Times*. "Predictions about devastating declines in polar bear populations, they say, have failed to materialize."[3]

For forty years, climate denialists funded by fossil fuel companies have misled the public about the science of climate change in the same way that tobacco companies misled the public about the science linking smoking and cancer, argued Harvard historians of science Naomi Oreskes and Erik Conway, in their influential 2010 book, *Merchants of Doubt*. The book is one of the most-cited sources as proof that political inaction on climate change is due to climate science skepticism promoted by right-wing think tanks funded by fossil fuel interests.[4]

A critical moment in the history of climate denial came in 1983, Oreskes and Conway argue, when the National Academy of Sciences published its first major report on climate change.

"The result, *Changing Climate: Report of the Carbon Dioxide Assessment Committee*, was really two reports," they write, "five chapters detailing the likelihood of anthropogenic climate change written by natural scientists, and two chapters on emissions and climate impacts by economists—which presented very different impressions of the problem. The synthesis sided with the economists, not the natural scientists."[5]

Right-wingers had effectively hijacked the National Academy of Sciences, they argue, which the U.S. government created in 1863 to objectively evaluate scientific questions for policymakers. William Nierenberg, who chaired the National Academy of Sciences' Carbon Dioxide Assessment Committee, was a political conservative who trampled over the views of natural scientists like John Perry, a meteorologist, who had concluded, "The problem is already upon us."

Oreskes and Conway write, "Perry would be proven right, but [economist Thomas] Schelling's view would prevail politically. Indeed, it provided the kernel of the emerging . . . idea that climate skeptics would echo for the next three decades . . . that we could continue to burn fossil fuels without restriction and deal with the consequences through migration and adaptation."

Schelling's argument, they said, was "equivalent to arguing that medical researchers shouldn't try to cure cancer, because that would be too expensive, and in any case people in the future might decide that dying from cancer is not so bad."[6]

But what economist Schelling argued about climate change was in no way similar to the tobacco industry's denial of the link between smoking and cancer. Nor was he saying the climate equivalent of "dying from cancer is not so bad." In fact, his argument rested upon acknowledgment that emissions were warming the planet and could be harmful.

Schelling's view was simply that the effects of restricting energy consumption could be worse than the effects of global warming. That view was mainstream back then and remains so today. Indeed, it is at the heart of the discussions of climate mitigation in reports by the Intergovernmental Panel on Climate Change (IPCC), the U.N. Food and Agriculture Organization (FAO), and other scientific bodies.

And rather than being similar to a tobacco industry scientist, Schelling is widely considered one of the most brilliant and humanistic economists of the twentieth century. As a professor at Harvard, Schelling received the Nobel Prize in 2005 for his pioneering work on everything from preventing nuclear war to ending racial discrimination.

The author of the summary, which Oreskes and Conway claim "sided with the economists," was Jesse Ausubel, the expert in energy transitions who worked with Cesare Marchetti in the 1970s at the International Institute for Applied Systems Analysis in Vienna. Ausubel also coauthored an article with Yale University's William Nordhaus, who won the Nobel Prize in 2018 for his work on the economics of climate change.[7]

In other words, there was nothing right-wing or climate denialist at all about *Changing Climate*. *The New York Times* thought so highly of the report that it reprinted its entire four-page summary. Moreover, the framework Ausubel, Nordhaus, and Schelling developed became the basis for much of the scientific and analytical work of the IPCC.[8]

As for the scientist John Perry, he flatly rejected the version of events presented by Oreskes and Conway. "You assert that the members of the committee did not concur with the Synthesis," he wrote in a 2007 email to Oreskes. However, he said, "I would never have permitted the report to go forward if any member had raised explicit objections."[9]

Perry added, "It seems to me that you are evaluating a report of 1983 in terms of the received wisdom of 2007. Nierenberg's strong support for a broad and vigorous research program was very welcome, and I don't recall any bitter complaints about his softness on immediate policy actions."[10]

But Oreskes and Conway were apparently unfazed by Perry's rejection of their narrative because they published their fable unchanged.

What about the polar bears? "The climate denialists" were right: devastating declines in the number of polar bears have indeed failed to materialize, which was something the creators of the starving polar bear footage were forced to admit.

"National Geographic went too far with the caption," said one of them, seeking to shift the blame. But the primary purpose of her expedition was to link dying polar bears to climate change. "Documenting [the effects of climate change] on wildlife hasn't been easy," she added.[11] But the reason it wasn't easy is that there was no evidence for polar bear famine.

Of the nineteen subpopulations, two increased, four decreased, five are stable, and eight have never been counted. There is no discernible overall trend.[12]

It is possible that sea ice's decline is making the polar bear population smaller than it would otherwise be. But even if it were, other factors may be having a larger impact. Hunting, for one, killed 53,500 polar bears between 1963 and 2016, an amount twice as large as today's estimated total population of 26,000.[13]

As for fossil fuel–funded climate denialists misleading people about polar bears, we couldn't find any. The main critic of polar bear exaggerations is a zoologist in Canada named Susan Crockford, who told me that she neither accepts money from fossil fuel interests nor denies that the planet is warming due to human activities.[14]

As such, the misinformation about polar bears perfectly captures the ways in which many of the stories people tell about climate change don't have much to do with science.[15]

2. Climate Politics Takes Its Tol

As a university student in the Netherlands, Richard Tol was a member of both Greenpeace and Friends of the Earth. Concerned about climate change, he earned a PhD in economics in 1997 and delved into the then-emerging field of economics of climate change with gusto, becoming one of the most-cited economists in the world on the topic.

Now a professor at the University of Sussex, in Britain, Tol became involved with the IPCC shortly after it was created, in 1994. He played a major role in all three working groups: science, mitigation, and adaptation.

Tol's reputation came, in part, from being on a team that rigorously established that carbon dioxide and other greenhouse gases are warming the planet. "We were the first to show that in a statistically sound way," he says.[16]

Tol doesn't look like a typical economics professor. When I first met him in London a few years back, his T-shirt was untucked, he was wearing a trench coat, and his hair and beard were long and frizzy, as though he had just stuck a fork in an electric socket.

In 2012, Tol was named the convening lead author of one of the chapters in the IPCC's fifth review of climate change, a prestigious position reflective of his expertise and respect from his peers. He was assigned to the team to draft the IPCC's *Summary for Policymakers*, which is often the only version of the report most journalists, politicians, and informed members of the public read.

The *Summary* authors are charged with distilling key messages into forty-four pages, even though "everyone knows that policy and media will only pick up a few sentences," Tol says.[17] "This leads to a contest between [authors of] chapters. 'My impact is worst, so I will get the headlines.' "

Tol said that the primary message of an earlier draft of the *Summary* was the same one I've emphasized with regards to the Congo: "Many of the more worrying impacts of climate change are really symptoms of mismanagement and underdevelopment."[18]

But representatives from European nations wanted the report's focus to be on emissions reductions, not economic development. "IPCC is partly a scientific organisation and partly a political organisation," Tol explained. As a "political organisation, its job is to justify greenhouse gas emission reduction."

The *Summary* was written during a weeklong meeting in Yokohama, Japan. "Some of these delegates are scholars, others are not," Tol explained. "The Irish delegate, for instance, thinks that unmitigated climate change would put us on a 'highway to hell,' referring, I believe, to an AC/DC song rather than a learned paper."[19]

Two years later, despite Tol's protests, IPCC approved a *Summary for Policymakers* that was far more apocalyptic than the science warranted. "The IPCC shifted from . . . 'Not without risk, but manageable,' to 'We're all gonna die,' " explained Tol. It was a shift, he said, "from what I think is a relatively accurate assessment of recent developments in literature to . . . the Four Horsemen of the Apocalypse. . . . Pestilence, Death, Famine and War were all there." [20]

IPCC's *Summary* left out key information, Tol alleges. The *Summary* "omits that better cultivars and improved irrigation increase crop yields. It shows the impact of sea level rise on the most vulnerable country, but does not mention the average. It emphasizes the impacts of increased heat stress but downplays reduced cold stress. It warns about poverty traps, violent conflict and mass migration without much support in the literature. The media, of course, exaggerated further." [21]

It wasn't the first time IPCC had exaggerated climate change's impact in a *Summary*. In 2010, an IPCC *Summary* falsely claimed climate change would result in the melting of the Himalayan glaciers by 2035. This was a serious case of alarmism given that 800 million people depend on the glaciers for irrigation and drinking water.

Shortly after, four scientists published a letter in *Science* pointing out the error, with one of them calling it "extremely embarrassing and damaging," adding, "These errors could have been avoided had the norms of scientific publication including peer review and concentration upon peer-reviewed work, been respected." [22]

Roger Pielke, the University of Colorado expert on climate change and natural disasters, similarly found instances where IPCC authors were exaggerating or misrepresenting the science for effect.

"What does Pielke think about this?" an external IPCC reviewer at one point asked about a claim being made about climate change and natural disasters. The official IPCC response was, "I believe Pielke agrees." But he didn't. Indeed, he had never been consulted. "The IPCC included misleading information in its report," Pielke said, "and then fabricated a response to a reviewer, who identified the misleading information, to justify keeping that material in the report." [23]

Thanks in part to the criticisms of IPCC by Pielke and other scientists, environment ministers from around the world demanded an independent review of IPCC policies and procedures by the InterAcademy Council, the international organization of national science academies. The InterAcademy

Council made recommendations for improving research quality that IPCC adopted, such as better practices for including research that had not yet been published in peer-reviewed journals.[24]

But the IPCC kept publishing apocalyptic summaries and press releases, and IPCC contributors and lead authors kept making apocalyptic claims, such as that sea level rise will be "unmanageable" and that "the potential risk of multi-breadbasket failure is increasing."[25] And, as Tol noted, journalists exaggerated further. The system appeared biased toward exaggeration.

Ausubel was one of the first to recognize the politicization of climate science. After pioneering ways to forecast energy demand, transitions, and emissions for *Changing Climate*, he helped create the IPCC. "And then the expected happened," Ausubel said. "Opportunists flowed in. By 1992, I stopped wanting to go to climate meetings."[26]

In response to the IPCC's decision to let the exaggerators write the *Summary for Policymakers*, Tol resigned. "I simply thought it was incredible," he said. "I told Chris Field, the chairman, about this, and I quietly withdrew."[27]

This book began with a defense of the science assembled by IPCC and others against those who claim climate change will be apocalyptic. We saw how the scientific consensus, as reflected in IPCC reports, supports Tol's view that, "Many of the more worrying impacts of climate change are really symptoms of mismanagement and underdevelopment."[28]

Now we must address the question of how so many people, myself included, came to believe that climate change threatened not only the end of polar bears but the end of humanity.

The answer is, in part, that while the IPCC's science is broadly sound, its *Summary for Policymakers*, press releases, and authors' statements betray ideological motivations, a tendency toward exaggeration, and an absence of important context.

As we saw, IPCC authors and press statements have claimed sea level rise will be "unmanageable," world food supplies are in jeopardy, vegetarianism would significantly reduce emissions, poor nations can grow rich with renewables, and nuclear energy is relatively dangerous.

The news media also deserves blame for having misrepresented climate change and other environmental problems as apocalyptic, and for having failed to put them in their global, historical, and economic context. Leading media companies have been exaggerating climate change at least since the 1980s. And, as we have seen, elite publications like *The New York Times* and *The New Yorker* have frequently and uncritically repeated debunked Malthusian dogma for well more than a half century.

IPCC and other scientific organizations are most misleading in what their summaries and press releases don't say, or at least not clearly. They don't clearly say that the death toll from natural disasters has radically declined and should decline further with continued adaptation. They don't clearly say that wood fuel build-up and constructing homes near forests matters more than climate change in determining the severity and impact of fires in much of the world. And they don't clearly say that fertilizer, tractors, and irrigation matter more than climate change to crop yields.

3. Who Framed Roger Pielke?

In early 2015, Raúl Grijalva, a U.S. congressman from Arizona, sent a letter to the University of Colorado's president suggesting that Roger Pielke might have taken money from the fossil fuel industry.

Grijalva offered no evidence Pielke had taken fossil fuel money but noted that a different scientist, one who was skeptical of climate change, had received funding from an electric utility that operates fossil fuel power plants, and hinted that Pielke might have done the same.

Grijalva asked the university's president to hand over any and all of Pielke's correspondence relating to his recent congressional testimony as well as all government testimony he had ever given, including in draft form. And he demanded documents showing Pielke's funding sources.[29]

Meanwhile, ExxonMobil, Grijalva wrote, "may have provided false or misleading information" about its support for a scientist in the past. "If true, these may not be isolated incidents," wrote Grijalva, who posted the letter to his website and distributed it to reporters.[30]

The result was widespread media coverage of Grijalva's implied accusation that Pielke might be secretly taking money from Exxon. Grijalva sent similar letters to six other university presidents whose schools employed scientists who testified in front of Congress about climate change.

It was hardly the first time progressives had attacked Pielke. Starting in 2008, seven separate authors with Center for American Progress (CAP), the large progressive think tank, had published more than 150 blog posts attacking Pielke as "the uber-denier" and a "trickster and a careerist."[31]

"Roger Pielke is the single most disputed and debunked person in the entire realm of people who publish regularly on disasters and climate change," claimed a CAP spokesperson.[32]

Center for American Progress was able to persuade many in the news media that Pielke was a climate skeptic. In 2010, the magazine *Foreign Policy* included Pielke in an article titled, "The FP Guide to Climate Skeptics," and wrote, "For his work questioning certain graphs presented in IPCC reports, Pielke has been accused by some of being a climate change 'denier.' "[33]

After learning of the investigation, Pielke responded with a blog post on his website: "I am under 'investigation,' " reads the headline. "Before continuing, let me make one point abundantly clear: I have no funding, declared or undeclared, with any fossil fuel company or interest. I never have."[34]

In January 2014, John Podesta, the founder of the Center for American Progress—who oversaw aspects of the 2009 green stimulus and the effort to pass cap-and-trade legislation—took a position in the White House where he oversaw climate change policy and communications. A few days after Podesta's appointment, John Holdren, a senior advisor and director of the White House Office of Science and Technology Policy under President Barack Obama, emailed him to say, "I'm looking forward to working with you on the President's climate-change agenda."[35]

In February 2014, Holdren appeared in front of the same Senate committee that Pielke had testified before a year earlier. A senator asked Holdren about Pielke's earlier testimony. Holdren replied that Pielke's research was "not representative of the mainstream scientific opinion."

A few days later, Holdren published a three-thousand-word article on the White House website criticizing Pielke's testimony before Congress as "misleading" and "not representative of mainstream views."[36]

But Grijalva's attack on Pielke had crossed a line, even for most activist scientists and journalists. When "politicians seek to probe beyond possible sources of external influence on published work and attempt to expose internal discussions that they find inconvenient," editorialized the British journal *Nature* about the investigation, "that sends a chilling message to all academics and to the wider public." [37]

"Publicly singling out specific researchers based on perspectives they have expressed and implying a failure to appropriately disclose funding sources—and thereby questioning their scientific integrity," wrote the American Meteorological Society in a public statement, "sends a chilling message to all academic researchers." [38]

Eventually, Grijalva backed down—at least partially. He admitted to a reporter that his request for Pielke's correspondence was "overreach." He said, "As long as we get a response as to the funding sources, I think everything else is secondary and not necessary." [39]

While other university presidents ignored Grijalva's request, the then–University of Colorado president cooperated and investigated Pielke. The resulting investigation confirmed that Pielke had never received any funding from fossil fuel companies. [40]

The effort to delegitimize Pielke was one of the most audacious and effective attacks by a fossil fuel–funded think tank on a climate expert in history.

It can be partly explained by the financial interests of CAP's donors. As we saw, renewable energy and natural gas interests funded CAP during the period when CAP was overseeing both Obama's green stimulus program and the administration's effort to pass cap-and-trade climate legislation in Congress, between 2009 and 2010.

It can also be partly explained by politics. Some people told me that the campaign of character assassination against Pielke was due to the widespread belief among Democratic, progressive, and environmental leaders that they needed to be able to claim that climate change's impacts are happening right now, and are catastrophic, in order to pass legislation to subsidize renewables and tax fossil fuels, and recruit, mobilize, and win over swing voters.

But having lived through the period I perceived that his persecution seemed to go well beyond money and political power. It felt like a witch hunt, a term we apply to campaigns like Senator Joseph McCarthy's persecution of scientists and artists as alleged communists in the 1950s. The scapegoating of Pielke, like apocalyptic environmentalism more broadly, had an undeniably religious quality to it.

4. False Gods

In 2019, Bill McKibben published a new book, *Falter*, which argued that climate change is the "greatest challenge humans have ever faced."[41]

It was a strong claim by America's leading environment writer and most influential climate leader. McKibben writes for *The New Yorker* and *The New York Times* and his organization, 350.org, has a nearly $20 million annual budget. He is respected by other journalists, members of Congress, presidential candidates, and millions of Americans. As such, it is worth taking his claim seriously.[42]

As we have seen, the death toll and damage from extreme events have declined 90 percent during the last century, including in poor nations. For McKibben's argument to be true, that long, salutary trend will need to reverse itself, and quickly.[43]

And, for McKibben's claim to be true, climate change must prove to be a greater challenge than coping with the Black Death, which killed about half of all Europeans, about fifty million people; the control of infectious diseases, which killed hundreds of millions; the great wars of Europe and the Holocaust, which killed more than 100 million people; postwar independence movements and the spread of nuclear weapons; and Africa's World Wars, which killed millions of people.

And climate change must prove to be greater than all other contemporary challenges, from the monumental task of lifting one billion souls out of extreme poverty in a world where manufacturing is playing a smaller role in economic development, to the battles and wars that killed tens of thousands of people last year in the Middle East and Africa.

McKibben's apocalyptic vision didn't start with climate change. In 1971, when Bill McKibben was eleven years old, police arrested his father for defending the right of Vietnam Veterans Against the War to gather on the town green in Lexington, Massachusetts. McKibben's mother says her son was "furious that he wasn't allowed to be arrested with his father."[44] Later, McKibben protested nuclear power himself, and he was tear-gassed for it.

After Harvard, McKibben says his "leftism grew more righteous."[45] He was influenced by *New Yorker* writer Jonathan Schell, who wrote books claiming humankind had to get rid of nuclear weapons or risk extinction. In his 1989 *The End of Nature*, he described climate change as an apocalyptic threat, like nuclear war.

The underlying problem, said McKibben, was spiritual. Through capitalist industrialization, humankind had lost its connection to nature. "We can no longer imagine that we are part of something larger than ourselves," he wrote in *The End of Nature*. "That is what this all boils down to."[46]

In the early twentieth century, the American scholar William James defined religion as the belief in "an unseen order, and that our supreme good lies in adjusting ourselves thereto."[47] The scholar Paul Tillich defined religion more broadly to include belief systems and moral frameworks. For environmentalists, the unseen order we need to adjust ourselves to is nature.

Throughout this book we have seen environmental support for various behaviors, technologies, and policies motivated not by what the science tells us but by intuitive views of nature. These intuitive views rest on the appeal-to-nature fallacy.

The appeal-to-nature fallacy holds that "natural" things, e.g., tortoiseshell, ivory, wild fish, organic fertilizer, wood fuel, and solar farms, are better for people and the environment than "artificial" things, e.g., plastics from fossil fuels, farmed fish, chemical fertilizer, and nuclear plants.

It is fallacious for two reasons. First, the artificial things are as natural as the natural things. They are simply newer. Second, the older, "natural" things are "bad," not good, if "good" is defined as protecting sea turtles, elephants, and wild fish.

This background and largely unconscious idea of nature is, in my experience, very strong. I have seen many environmentalists dismiss evidence demonstrating, for example, the larger impact of renewable energy and organic farming on landscapes. "Natural" things must, by definition, be better for the environment.

Irrational ideas about nature repeatedly creep into the environmental sciences. In the 1940s, scientists attempted to create a science of nature, ecology, which was based on cybernetics, the science of self-regulating systems, which had been usefully applied in World War II for guiding antiaircraft missiles. Cybernetics also applies to systems like thermostats, which turn on the furnace when it gets too cold and turn it off when it gets too hot.

The assumption was that nature, when left alone, achieves a kind of harmony or equilibrium. Like the thermostat switching on in the cold, nature gracefully, gradually self-regulates species and environments in and out of existence when it senses things are out of proportion. It's an interwoven system that works as a whole unless humans interfere with its operations.

But "nature" doesn't operate like a self-regulating system. In reality, different natural environments change constantly. Species come and go. There is no whole or "system" to collapse. There's just a changing mix of plants, animals, and other organisms over time. We might prefer one version of that mixture, like the Amazon rainforest, but there is nothing in the mixture telling us that it is better or worse than some other combination, like a farm or desert.[48] The same is true for what we call "the climate." The climate simply refers to the weather over a period of time, say thirty years, either in a particular place or the entire planet.

In reality, much of what we see in the climate and Earth systems are "dials," like higher temperatures and rising sea levels, not "switches," like sudden ice sheets melting or forests burning out of control. Apocalyptic scientists and activists list various changes, such as melting ice sheets, changing ocean circulations, and deforestation, and suggest that they will add up to an apocalyptic sum greater than their parts. But they cannot offer a clear mechanism, amid so much complexity and uncertainty, for how such an apocalyptic scenario could occur.

While scientists borrowed from cybernetics to describe nature as a self-regulating system, the notion of nature existing in a delicate balance is

Neoplatonism, and ungrounded in empirical reality. "The commonplaces of modern ecology, such as 'everything connects,' " writes environmental philosopher Mark Sagoff, "recalls the neoplatonic view of nature as an integrated mechanism into which every species fits." [49]

Some ecological scientists recognized that they had inadvertently and unconsciously imposed a fundamentally religious idea onto science. "I am convinced that modern ecological theory, so important in our attitudes towards nature and man's interference with it," admitted one, "owes its origin to the [Judeo-Christian intelligent] design argument. The wisdom of the creator is self-evident. . . . no living thing is useless, and all are related one to the other." [50]

The flip side of nature's interconnectedness was collapse, which E.O. Wilson adopted as the core assumption of his apocalyptic *species area model*. What the scientists had re-created in both the species area model, and the prediction that extinctions would result in the extinction of humankind, were the first and last book of the Bible. Genesis tells the story of Eden, and humankind's fall from it, as a consequence of its hubris. The book of Revelation tells the story of the end of the world, the ultimate consequence of humankind's fall.

Environmentalism today is the dominant secular religion of the educated, upper-middle-class elite in most developed and many developing nations. It provides a new story about our collective and individual purpose. It designates good guys and bad guys, heroes and villains. And it does so in the language of science, which provides it with legitimacy. [51]

On the one hand, environmentalism and its sister religion, vegetarianism, appear to be a radical break from the Judeo-Christian religious tradition. For starters, environmentalists themselves do not tend to be believers, or strong believers, in Judeo-Christianity. In particular, environmentalists reject the view that humans have, or should have, dominion, or control, over Earth. [52]

On the other hand, apocalyptic environmentalism is a kind of new Judeo-Christian religion, one that has replaced God with nature. In the Judeo-Christian tradition, human problems stem from our failure to adjust ourselves to God. In the apocalyptic environmental tradition, human problems stem from our failure to adjust ourselves to nature. In some Judeo-Christian traditions, priests play the role of interpreting God's will or laws, including discerning right from wrong. In the apocalyptic environmentalist tradition, scientists play that role. "I want you to listen to the scientists," Thunberg and others repeat.[53]

Most environmentalists are unaware that they are repeating Judeo-Christian myths, concludes a scholar who studied the phenomenon closely. Because Judeo-Christian myths and morals are prevalent in our culture, environmentalists know them subconsciously and repeat them unintentionally, albeit in the ostensibly secular language of science and nature.[54]

Having first experienced and then studied the phenomenon for fifteen years, I believe that secular people are attracted to apocalyptic environmentalism because it meets some of the same psychological and spiritual needs as Judeo-Christianity and other religions. Apocalyptic environmentalism gives people a purpose: to save the world from climate change, or some other environmental disaster. It provides people with a story that casts them as heroes, which some scholars, as we will see, believe we need in order to find meaning in our lives.

At the same time, apocalyptic environmentalism does all of this while retaining the illusion among its adherents that they are people of science and reason, not superstition and fantasy. "For the many people skeptical of institutional Christian religion," wrote a leading scholar, "but seeking greater religious meaning in their lives, that is no doubt part of the attraction of secular religion."[55]

There is nothing wrong with religions and often a great deal right about them. They have long provided people with the meaning and purpose they need, particularly in order to survive life's many challenges. Religions can be a guide to positive, prosocial, and ethical behavior.

"Whether or not God exists (and as an atheist I personally doubt it)," noted psychologist Jonathan Haidt, ". . . religious believers in the United States are happier, healthier, longer-lived, and more generous to charity and to each other than are secular people."[56]

The trouble with the new environmental religion is that it has become increasingly apocalyptic, destructive, and self-defeating. It leads its adherents to demonize their opponents, often hypocritically. It drives them to seek to restrict power and prosperity at home and abroad. And it spreads anxiety and depression without meeting the deeper psychological, existential, and spiritual needs its ostensibly secular devotees seek.

5. Apocalypse Angst

To believe in imminent apocalypse is, according to one scholar, to believe that "the accepted texture of reality is about to undergo a staggering transformation, in which the long-established institutions and ways of life will be destroyed."[57]

And, to some extent, that is what is happening, and has been happening since the end of the Cold War, World War II, and the rise of the scientific and industrial revolutions.

For thousands of years, religion sought to constrain what we today call science, efforts to understand the world, including ourselves. Even Sir Francis Bacon, the father of modern physics, wrote in 1605 that the pursuit of knowledge outside of morality was dangerous.

After the rediscovery of classical texts including Plato and Aristotle in the Middle Ages, Western thinkers initially concentrated on reconciling classical philosophy with Christianity. Over time, however, the focus shifted to making sense of the natural world, leading to what is called the scientific revolution. Although most early scientists professed that they did their science in service of God, they pursued knowledge without knowing whether it would lead to good or ill. And privately they saw that their experiments worked the same whether or not they were pious.

During the Enlightenment, philosophers tried to apply the same rational approach to morality and politics in the form of "secular humanism." It borrowed from Judeo-Christianity the idea that humans were special, but it emphasized the use of science and reason in the pursuit of virtue and did not require belief in God, the afterlife, or other core aspects of religion. Politically, secular humanists believed in the fundamental equality of all human beings.

But it quickly became clear there was no "objective" basis for morality. In the 1800s, philosophers pointed out all the ways in which what we believe to be "good" is based on our own selfish needs and changing historical and social contexts. Morality, it turned out, was relative to your time, place, and social position. By the 1920s, European philosophers argued that moral judgements could not be justified empirically and were simply expressions of emotion, which have no specific content and are thus meaningless. We couldn't start deciding what was right or wrong based on what made particular people sad or happy at any given moment.[58]

After World War II, many leading scholars and universities in Europe and the United States rejected the teaching of morality and virtue as unscientific and thus without value. "Reason reveals life to be without purpose or meaning," was the intellectual consensus, a historian notes. "Science is the only legitimate exercise of the intellect, but that leads inevitably to technology and, ultimately, to the bomb.[59]

"From humanists we learned that science threatened civilization," the historian added. "From the scientists we learned that science cannot be stopped. Taken together, they implied there is no hope." The result, he argued, was a "culture of despair."[60]

Apocalyptic environmentalism emerged from this crisis of faith and, over the decades, became pronounced during moments of global change and crisis. In 1970, when fears of overpopulation were at their peak, Earth Day was held in the midst of national turmoil over the Vietnam War. In 1983, during heightened Cold War tensions, more than 300,000 people protested in London's Hyde Park against nuclear weapons. And in the early 1990s, climate change emerged as the new apocalyptic threat at the end of the Cold War.

After the dissolution of the Soviet Union, people in the West no longer had an external enemy against which to direct their negative energy and define themselves. "Being the sole winner in a conflict means concentrating on oneself all the criticism that could earlier be deflected onto others," observed Pascal Bruckner in *The Fanaticism of the Apocalypse*.[61]

In the wake of the 2016 elections in Britain and the United States, where voters rejected, in one way or another, the established global order, climate alarmism grew more extreme.

Where environmentalism had until recently offered the prospects of utopia in the form of a return to low-energy and renewable-powered agrarian societies, it is striking the extent to which apocalyptic environmentalist leaders have deemphasized that vision for an emphasis on climate Armageddon.

Green utopianism is still there. Apocalyptic environmentalists in Europe and the United States advocate a Green New Deal not just to reduce carbon emissions but also to create good jobs with high pay, reduce economic inequality, and improve community life.

But negativity has triumphed over positivity. In place of love, forgiveness, kindness, and the kingdom of heaven, today's apocalyptic environmentalism offers fear, anger, and the narrow prospects of avoiding extinction.

I happened to be in London to witness Extinction Rebellion's forcible closure of central London, which was about a week before the Tube protest. I stayed in a hotel a few blocks from Trafalgar Square, where hundreds of activists camped out for two weeks.

The mostly white, educated, upper-middle-class Extinction Rebellion activists were strikingly similar in socioeconomic status, ideology, and behavior to the Earth First! activists I'd met while working to save the last unprotected ancient redwood forest in northern California in the late 1990s.

But Extinction Rebellion was far more death-obsessed than Earth First! At London Fashion Week, there were Extinction Rebellion activists carrying coffins; there were large banners with the word DEATH on them; there were women wearing black mourning veils; and there were dead-silent activists wearing blood-red gowns who had painted their faces ghost-white save for their red lips.

Extinction Rebellion's heavy focus on death bothered me and so, after I returned to the United States, I reached out to my friend Richard Rhodes, the author of *The Making of the Atomic Bomb*. Dick and I had talked about death earlier in the year, and so I was curious to hear his reaction to the ubiquity of death symbolism in Extinction Rebellion's protests.

"You know Becker's book *The Denial of Death?*" Rhodes asked. "It won the Pulitzer Prize."

I said I did. According to the anthropologist Ernest Becker, all humans, not just religious people, need to believe, consciously or unconsciously, that we are, in one way or another, immortal, that some part of us will never die.

Humans are unique, Becker believed, in that we realize that we are going to die from a very young age. Our deaths rightly frighten us; we are all born with strong survival instincts. But since too much fear of death gets in the way of living, healthy individuals repress their fears, making them mostly unconscious.[62]

To defend ourselves psychologically against this low-level fear we create what Becker calls an "immortality project," a way of feeling that some part of us will live on after our deaths. Many people feel immortal by having children and grandchildren. Others feel immortal creating art, businesses, writing, or communities that will last after they are gone.

We subconsciously cast ourselves as the heroes of our immortality projects. "It doesn't matter whether the cultural hero is magical, religious, and primitive, or secular, scientific, and civilized," Becker wrote. "It is still a mythical hero-system in which people serve in order to earn a feeling of primary value of cosmic specialness, of ultimate usefulness to creation, of unshakable meaning."[63]

Such appear to be the benefits of climate activism. "Extinction Rebellion," Sarah Lunnon told me, "offered a way of being courageous." Zion Lights of Extinction Rebellion pointed to research showing "children who engage with climate activism have better mental health than kids who know about climate change but don't do anything about it." And Thunberg's climate activism allowed her to escape depression. "It is like day and night," said her father. "It is an incredible transformation."[64]

There is much to admire in the idealism of environmentalism and vegetarianism, which are popular among adolescents and young adults. They represent "the will to give a future to the entire planet, including its animals," conclude the Italian psychologists who studied vegetarian beliefs.[65]

But vegetarianism also appears to stem from above-average existential angst. The "fear of death may be recognized as inherent not only to single

individuals," write the Italians, "but as the connotation of an ecological ethics which expresses itself through vegetarianism."[66]

For Becker, an exaggerated fear of death reveals a deep and often subconscious dissatisfaction with one's life. What we really fear when we obsess
over our death is that we aren't making the most of our lives. We feel stuck
in bad relationships, unsupportive communities, or oppressive careers.

That was certainly the case for me. I was drawn toward the apocalyptic
view of climate change twenty years ago. I can see now that my heightened
anxiety about climate change reflected underlying anxiety and unhappiness
in my own life that had little to do with climate change or the state of the
natural environment.

Perhaps it is a coincidence, but it is notable that the spike in environmental alarmism comes at a time when anxiety, depression, and suicide are
rising within the general population, especially among adolescents, in both
the United States and Europe.[67] Seventy percent of American teenagers call
anxiety and depression a major problem.[68]

Because addressing our personal lives is painful and difficult, suggests
Becker, we often look for external demons to conquer. Doing so makes us
feel heroic, and creates a feeling of immortality through the recognition,
validation, and love we receive from others.

6. Lost Souls

Preaching climate apocalypse is different from fearing death. Where most
of us don't want to think about our deaths, some people seem excited when
they talk about the end of the world, whether from the Mayan apocalypse,
Y2K, or climate change.

"That deep sense that the world might come to an end is a strange mix
of fear and desire," Dick said. "Everyone would like to see their enemies
destroyed in horrible ways, wouldn't they?" he chuckled.

Dick admitted he fantasized about seeing the atomic destruction of his
hometown, where he had been starved and abused as a boy, something he
wrote about in his affecting 1990 memoir, *A Hole in the World.*[69]

"Remember when the TV movie, *The Day After* [about nuclear apoca-lypse], was shown [in 1983]? It was located in and around Kansas City, Mis-souri, my hometown. I lived there at the time. I remember watching it when it was first shown," he said.

"When those missiles came over the horizon and were targeted at Kansas City, I remember having mixed feelings. 'Oh wow! They're going to destroy Kansas City!' But also, 'Oh my god, how horrible!' " Dick laughed at this. "You know, both." [70]

Few apocalyptic environmentalists have as understandable a reason to fantasize of their hometowns being annihilated as Dick Rhodes. But could a similar hatred of human civilization, and perhaps humanity itself, be behind claims of environmental apocalypse?

"There's a strain of the environmental movement that is Calvinistic," Dick said. "In the sense that the world is an evil place and it would be better if it were destroyed and turned back over to the natural kingdom."

If the climate apocalypse is a kind of subconscious fantasy for people who dislike civilization, it might help explain why the people who are the most alarmist about environmental problems are also the most opposed to the technologies capable of addressing them, from fertilizer and flood control to natural gas and nuclear power.

After two weeks of Extinction Rebellion parading coffins through the street, blocking traffic, and stopping Tube trains, many Britons had had enough. "Let us no longer beat around the bush about these people," wrote one British columnist. "This is an upper-middle-class death cult." [71]

"I want you to panic," said Thunberg to a gathering of world leaders in Davos, Switzerland, in January 2019.[72] "If standing up against the climate and ecological breakdown and for humanity is against the rules then the rules must be broken," she tweeted in October.[73]

Two days later, two male Extinction Rebellion protesters stood on top of a train, to block it from moving forward, in the London Tube. Angry com-muters kicked and beat one of the young protesters and another young man filming the event.

I was frightened by the scene and watched it several times. My thought each time was that the commuters could quite easily have killed the two men. Many in the crowd were gripped by a sudden, uncontrollable fear and wildly unthinking behavior. In other words, they were in a panic.

When I complained to Lunnon by phone about the Tube incident, her response was that, thanks to Extinction Rebellion, climate change had been in the news extensively in the run-up to the British elections.

She offered the same justification for the Tube protest on British television. A cohost on *This Morning* asked Lunnon if her group's decision to cancel further protests was "sort of an admission that you got it wrong this time?"[74]

"I don't think we've got it wrong necessarily," said Lunnon, "because today, I'm sitting in this studio and we've been bumped three times from *This Morning*'s studios in the last ten days, and it's taken us being this disruptive to get on your program—"

"Is that why you did it?" asked one of the hosts.

"That man glued himself onto an electric train," asked the other host, angrier now, "so you could get on *This Morning*?"

"Not so we can get on *This Morning*, obviously," said Lunnon, backtracking. "But unless we do very, very disruptive actions, people do not want to talk to us."

On BBC's *Newsnight*, as host Emma Barnett was wrapping up the program, Lunnon practically shouts, "No! We waited! We waited for thirty years for capitalism to work and it hasn't worked!

"And, Emma," says Lunnon, "when you go home, you will look at your children and think, '*I've got a 50-50 chance of it staying in their world that we know as it is today.*'" Lunnon pivoted to another camera and says, "And the people looking at home: you as well."[75]

Thunberg displayed a similar anger in her speech to the United Nations in September 2019. "This is all wrong," she says. "I shouldn't be standing here. I should be back in school on the other side of the ocean. Yet you all come to us young people for hope?" she said, practically shouting. "How dare you!"[76]

The late British philosopher Roger Scruton thought deeply about the politics of anger. On the one hand, he believed it was valuable and necessary. "Resentment is to the body politic what pain is to the body," he wrote. "Bad to feel it, but good to be capable of feeling it, since without the ability to feel it we will not survive."[77]

The problem, Scruton argued, is when "resentment loses the specificity of its target and becomes directed to society as a whole," Scruton concludes. At that point resentment becomes "an existential posture" adopted not "to negotiate within existing structures, but to gain total power, so as to abolish the structures themselves. . . . That posture is, in my view, the core of a serious social disorder." Another word for what Scruton is describing is nihilism.[78]

Young people learning about climate change for the first time might understandably believe, upon listening to Lunnon and Thunberg, that climate change is the result of deliberate, malevolent actions. In reality, it is the opposite. Emissions are a by-product of energy consumption, which has been necessary for people to lift themselves, their families, and their societies out of poverty, and achieve human dignity. Given that's what climate activists have been taught to believe, it's understandable that so many of them would be so angry.

But such anger can worsen loneliness. "When you educate yourself about the climate you do tend to separate yourself from your friends and family," Lunnon said on Sky News. "It's very hard to separate yourself from your friends and your family who may well be catching flights and . . . eating roast dinners every Sunday."[79]

The problem is that people tend to feel put down by climate activists who condemn economic growth, eating meat, flying, and driving. "Why would people listen to you," asked Lunnon, "when, you know, you're some kind of new age Puritan?"[80]

As such, while some climate leaders and activists may derive psychological benefits from climate alarmism, the evidence suggests that many more people are harmed by it, including the alarmists themselves.

Lauren Jeffrey, the seventeen-year-old in Britain, noticed heightened anxiety among her peers after Extinction Rebellion's protests. "In October I was hearing people my age saying things I found quite disturbing," says Jeffrey. " 'It's too late to do anything.' 'There is no future anymore.' 'We're basically doomed.' 'We should give up.' " [81]

Thunberg and her mother say that watching videos about plastic waste, polar bears, and climate change contributed to her depression and eating disorder. "So when I was eleven, I became ill. I fell into depression. I stopped talking and I stopped eating. In two months, I lost about ten kilos of weight. Later on, I was diagnosed with Asperger syndrome, OCD [obsessive-compulsive disorder], and selective mutism." [82]

Twenty years ago, I discovered that the more apocalyptic environmentalist books and articles I read, the sadder and more anxious I felt. This was in sharp contrast to how I felt after reading histories of the civil rights movement, whose leaders were committed to an ethos, and politics, of love, not anger.

It was, in part, my awareness of the impact that reading about climate and the environment had on my mood that led me to doubt whether environmentalism could be successful. It was only several years later that I started to question environmentalism's claims about energy, technology, and the natural environment.

Now that I have, I can see that much of my sadness over environmental problems was a projection, and misplaced. There is more reason for optimism than pessimism.

Conventional air pollution peaked fifty years ago in developed nations and carbon emissions have peaked or will soon peak in most others.

The amount of land we use for meat production is declining. Forests in rich nations are growing back and wildlife are returning.

There is no reason poor nations can't develop and adapt to climate change. Deaths from extreme events should keep declining.

Cruelty to animals in meat production has declined and should continue to decline, and, if we embrace technology, habitats available for endangered species, including for gorillas, and penguins, should keep growing in size.

None of that means there isn't work to do. There is plenty. But much if not most of it has to do with accelerating those existing, positive trends, not trying to reverse them in a bid to return to low-energy agrarian societies.

And so, while I can empathize with the sadness and loneliness behind the anger and fearmongering about climate change, deforestation, and species extinction, I can see that much of it is wrong, based on unaddressed anxieties, disempowering ideologies, and misrepresentations of the evidence.

7. Environmental Humanism

The answer from many rational environmentalists, including myself, who are alarmed by the religious fanaticism of apocalyptic environmentalism, has been that we need to better maintain the divide between science and religion, just as scientists need to maintain the divide between their personal values and the facts they study.

Others, like Scruton, urge us to aim for a world "where conflicts are resolved according to a shared conception of justice" and the "building and governance of institutions, and to the thousand ways in which people enrich their lives through corporations, traditions and spheres of accountability."[83]

But Scruton himself doubted that such a rationalist project could succeed against the apocalyptic tendencies of the regressive left. "Clearly we are dealing with the religious need, a need planted in our 'species being,' " he writes. "There is a longing for membership that no amount of rational thought, no proof of the absolute loneliness or humanity or of the unredeemed nature of our sufferings, can ever eradicate."[84]

Attempts to affirm the boundary between science and religion will thus likely not work so long as apocalyptic environmentalists speak to deep human needs for meaning and purpose and environmental rationalists don't.

As such, we need to go beyond rationalism and re-embrace humanism, which affirms humankind's specialness, against Malthusian and apocalyptic environmentalists who condemn human civilization and humanity itself. As environmental humanists, whether scientists, journalists, or activists, we must ground ourselves first in our commitment to the transcendent moral purpose of universal human flourishing and environmental progress, and then in rationality.

When Sir Francis Bacon urged that the quest for truth be balanced by morality, he quoted the apostle Paul: "knowledge bloweth up, but charity buildeth up." Bacon didn't call for hard limits on knowledge so much as a constant orientation of scientists around the emotion behind the best moralities. The "corrective spice" to science, said Bacon, was "charity (or love)."[85]

As such, when we hear activists, journalists, IPCC scientists, and others claim climate change will be apocalyptic unless we make immediate, radical changes, including massive reductions in energy consumption, we might consider whether they are motivated by love for humanity or something closer to its opposite.

As I write these words, Bernadette remains a refugee in her own country. She and her husband remain at risk of being kidnapped or worse. And she must spend her days working someone else's land, which she understandably experiences as a profound loss of dignity.

I have kept Bernadette in the foreground throughout *Apocalypse Never* to remind us, particularly those of us in developed nations, to feel gratitude for the civilization we take for granted, put claims of climate apocalypse in perspective, and inspire empathy and solidarity for those who do not yet enjoy the fruits of prosperity.

The stories we tell matter. The picture promoted by apocalyptic environmentalists is inaccurate and dehumanizing. Humans are not unthinkingly destroying nature. Climate change, deforestation, plastic waste, and species extinction are not, fundamentally, consequences of greed and hubris but rather side effects of economic development motivated by a humanistic desire to improve people's lives.

A core ethic of environmental humanism is that rich nations must support, not deny, development to poor nations. Specifically, rich nations should lift the various restrictions on development aid for energy production in poor and developing nations. It is hypocritical and unethical to demand that poor nations follow a more expensive and thus slower path to prosperity than the West followed. As the last nations to develop, it is already going to be harder for them to industrialize.

The good news is that in many nations, including African ones, cheap hydroelectricity and natural gas will likely be available. But if coal is the best option for poor and developing nations, then rich nations in the West must support that option.

As the cradle of humankind's evolution, the Albertine Rift region is of great significance to people everywhere. The presence of the endangered mountain gorillas, our evolutionary cousins, magnifies its importance. Despite the success of gorilla protection, much of the conservation activities in the region have failed in the absence of military security and economic development.

Cheap hydroelectricity from the Grand Inga Dam could power the southern African region, including the factories, which remain the only way we know how to transform large numbers of unskilled subsistence farmers into city people. The main opposition to the dam outside of the Congo comes from apocalyptic environmentalists. Environmental humanists should stand up to them.

The West owes Congo a significant debt. The palm oil it provided the world helped save the whales, a magnificent achievement. But its people continue to suffer. Belgians colonized and created the nation but then abandoned it in the early 1960s, leaving the country in disarray. Neither the United States nor the Soviet Union covered itself with glory there during the Cold War.

Since the end of the Cold War, the United Nations and Western nations have failed to end the worsening violence in Eastern Congo, which most experts believe stems from Rwanda's desire to maintain chaos as cover for continued mineral extraction. Many of the experts I interviewed believe Western governments are afraid to demand change from Rwanda out of some mix of guilt for having failed to stop the 1994 genocide and fear that the region will return to civil war.

Until then, conservationists and environmental humanists have an interest in advocating for a resolution to the ongoing insecurity, which has traumatized generations of people with no end in sight.

In early 2020, I chatted through Facebook, using Google Translate, with Suparti, in Indonesia.

"Yes, I still live in the same house," she said. In 2016, she married a man she met through Facebook who works for Sharp Semiconductor and quickly became pregnant. "My daughter was supposed to be born on January 18 but there were complications. I eventually underwent a caesarean section on January 30."

The birth of her daughter sounded harrowing. "I had to go to four hospitals and most of them had no doctor at that time because it was the end of the year," she said. "Finally, I found a hospital with a doctor. We went into surgery right away."

One hundred years ago, Suparti could easily have suffered a far worse outcome.

"Thanks to God," said Suparti, "she was born healthy and safe. I went back to work forty days later."

Suparti still works for the chocolate factory and has seen her salary rise and her responsibilities grow. She is now involved in the factory's computer systems, having risen through the ranks of packaging and labeling from pouring chocolate back in 2015.

"I want four children, two boys and two girls," she said. "Thanks to God, my parents are all healthy, my siblings are all good, and I am happy. I am grateful to have a husband who is very kind and patient."

Suparti sent me two photos through Facebook. One was of her husband and her, and the other was of her daughter wearing a bright pink hijab. A big smile stretched across her tiny face.

News media, editors, and journalists might consider whether their constant sensationalizing of environmental problems is consistent with their professional commitment to fairness and accuracy, and their personal commitment to being a positive force in the world. While I am skeptical that stealth environmental activists working as journalists are likely to change how they do their reporting, I am hopeful that competition from outside traditional news media institutions, made possible by social media, will inject new competitiveness into environmental journalism and raise standards.

Improving environmental journalism requires coming to grips with some fundamentals. Power density determines environmental impact. As such, coal is good when it replaces wood and bad when it replaces natural gas or nuclear. Natural gas is good when it replaces coal and bad when it replaces uranium. Only nuclear energy can power our high-energy human civilization while reducing humankind's environmental footprint. Power-dense farming, including of fish, creates the prospect of shrinking humankind's largest environmental impact.

We need to correct our misunderstanding of nuclear energy. It was born from good intentions, not bad ones, nor from some mindless accident of science. Nuclear weapons were created to prevent war and end war, and that is all they have been used for and all they will ever be good for. The United States and other developed nations should renew the commitments they made under "atoms for peace" in the 1950s in the form of a Green Nuclear Deal, for reasons including but going beyond climate change.

That will require recognizing that nuclear weapons, like nuclear energy, are here to stay. We can't get rid of them, even if we wanted to, for reasons experts have understood since 1945. And trying to do so has contributed to decades of tension and conflict, culminating in the unnecessary and disastrous U.S. and British invasion of Iraq in 2003.

The continued existence of nuclear weapons reminds us that they hold the potential to create, if not apocalypse, then at the very least, the destruction of cities and perhaps even civilizations. Such weapons should, if we are sentient human beings, stimulate some amount of anxiety. Call it existential angst. We need to find a better way to manage, and channel, those anxieties. Confronting them directly as objects of death, and even as symbols of the apocalypse, may help.

When Dick Rhodes and I talked about Extinction Rebellion's protests in London, it wasn't the first time we had talked about death. A few months earlier, over lunch, I told him I thought the continued existence of nuclear weapons should remind us to be happy to be alive.

"You mean like a *memento mori*," said Dick.

"Yes!" I said, after he reminded me that *memento mori*, which means "remember you will die" in Latin, is something people use to remind them of their mortality and thus feel gratitude toward life.

The classic *memento mori* is the skull, like those seen in still-life paintings by European artists in the Middle Ages. They appear to have become more popular after the Black Death.[86]

What if we treated objects of mass destruction as *memento mori*?

Experiments inspired by Ernest Becker's work suggest that doing so could help with our angst. When psychologists encourage people to think of their eventual deaths, they tend to feel anxious about how they are living their lives. But when they encourage people to imagine they are dying, and to look back on their lives, they tend to do so with gratitude, appreciation, and greater love toward those around them.[87]

The same has been true for me after visiting poor and developing nations, whether as a researcher or tourist. I believe that if environmental scientists, journalists, and activists spent more time talking to people about their daily struggles in places like Indonesia and the Congo, they would be less likely to see the end of the world, and reason to panic, in every new environmental problem.

8. Love > Science

In early 2017, I leased retail space in Berkeley, California, for my new research organization, Environmental Progress, which I had created the year before. The previous tenant had sold fast-fashion that was no longer fashionable.

I loved the idea of being in street-level retail space in Berkeley, home to the inventors of nuclear energy, its opponents, and later, in the form of Will Siri, its advocates. We would be picking up where Siri left off. And I felt the desire to be more grounded, to actually advocate something, not just research and write about it.

Most of the space was a single large room with high ceilings and windows. There was an open upstairs balcony, about five feet wide, that hovered over the room. Wood and tarnished copper bars ringed the balcony, from which protruded tacky wood beams. The floor was concrete.

We knew the existence of an openly pronuclear environmental organization in downtown Berkeley might trigger people, so we sought to make the space calming and inviting.

To open up the space, we cut off the beams and painted the whole room white, except for the windows and copper bars, which we polished. Helen painted, stained, and sealed the floor, resulting in a rich, marbled, and calming ocean of sky blue. It worked. Most of the people who come in off the street remain calm and even open-minded, upon learning that we are pronuclear.

Because the space is zoned retail, we are required to sell items, and so we started purchasing and reselling nuclear memorabilia from the fifties and sixties, including uranium glass jewelry and "Atoms for Peace" stamps.

We turned the upstairs balcony into a gallery. I created the exhibit with photos I'd taken around the world that tell the story of environmental progress through people's lives. There is a photo of Bernadette, half-angry and half-proud, standing where the baboons ate her sweet potatoes. There is a photo of one of the mountain gorilla orphans with a Mona Lisa smile and a flower in her mouth. There is a photo of Caleb, smiling, and holding the hand of Daniel, the Spanish engineer, as we walk to the Virunga dam. There is a photo of Suparti standing next to the sewing machine she doesn't use. And there is a photo, the only one that's not mine, of the Diablo Canyon nuclear plant.

In my upstairs office, which overlooks the balcony and the main room, Helen hung a poster from the World's Fair in Chicago from 1933 to 1934. It was made during some of the darkest years of the Great Depression, and yet it expresses a far more optimistic view of the future than the one offered by most environmentalists today.

The Fair's theme was technological innovation, and the name of the exposition was "A Century of Progress," which celebrated Chicago's 100th anniversary. The tagline was "Science Finds, Industry Applies, Man Adapts"—a perfect motto for mitigating and adapting to climate change. The Fair celebrated "technology and progress, a utopia, or perfect world, founded on democracy and manufacturing"—a perfect motto for achieving prosperity for all.[88]

Nature and prosperity for all determine how we do our research, not the other way around. No amount of science can prove that you should support our mission, just as no amount of science can make you want to protect mountain gorillas. All we can do is show you photos and tell you stories about Bernadette and Suparti and hope they make you care, in the same way they have made us care.

When my staff and I meet, these images of a high-energy, prosperous world with flourishing wildlife frame our thinking. They represent our commitment to goals that are human and natural, rational and moral, and physical and spiritual. Research should be, we believe, in service of some ultimate value. Ours are the love of humanity, and the love of nature.

By the time we were finished renovating the office, I realized we had created a shrine, of sorts, to a vision that could be fairly called spiritual. Nature and prosperity for all is our transcendent moral purpose. Environmental Progress is our immortality project.

In 2015, a few days after Bernadette showed me where baboons had eaten her sweet potatoes, Helen and I viewed the endangered mountain gorillas up-close and in-person.

We stayed at Merode's resort in Virunga Park. In the heart of a war zone we ate delicious food and slept in the rainforest, surrounded by colobus monkeys and chimpanzees. It was Eden inside the apocalypse.

To see the gorillas, Virunga Park rangers, armed with machine guns, took us by Jeep, and then by foot, through the cattle pastures that were once habitat for the primates. We passed through communities of farmers who came out of their homes to wave and smile at us. We got out and hiked for a bit through cattle ranches that run next to the edge of the park boundary, a reminder of the pent-up demand for land and the pressure of a growing human population.

We smelled the gorillas before we saw them. Their smell was unique, a pungent mix of very strong body odor, musky perfume, and skunk. We stared at them and they stared back. We gasped at our first sight of a silverback. He grunted disapprovingly at the clacking of my camera's shutter. He wanted to be left alone. Our guide motioned for us to give him space. The hour we had with the gorillas flew by.

Helen and I resolved on the spot to see them again, which we did in Uganda one week later. As in Congo, reaching the gorillas required walking through village after village of poor farmers pressed up against the park boundary and trying to cope with animals eating their crops. The first gorilla we saw had violated the invisible park border and was eating from a tree on a farmer's land. The second gorilla we saw was a magnificent silverback, one that was much more tolerant of our noisy cameras and our group's oohing and ahh-ing than the one in Congo had been.

Finally we were treated to the sight of an infant gorilla. Its mother lay a few feet away. She smiled at us as her baby played near her. It felt like we were two individuals in a single primate community. Scientists warn against the anthropomorphization of animals, but it is impossible not to see gorillas as kin, particularly when they are smiling at you.

People who see gorillas in the wild feel awe and wonder, a mix of happiness, surprise, and a bit of fear. "This wonderful and frightful production of nature walks upright like a man," wrote a sea captain after he and his men had stumbled upon gorillas on the African coast in 1774. It was a story the man likely recounted for years afterward with emotion and excitement. [89]

Scientists have long named self-interest as a reason for why humans should care about endangered species like the mountain gorilla. But if the mountain gorillas were ever to go extinct, humankind would become spiritually, not materially, poorer.

Happily, nobody saves mountain gorillas, yellow-eyed penguins, and sea turtles because they believe human civilization depends on it. We save them for a simpler reason: we love them. [90]

EPILOGUE

Few things make one feel more immortal than saving the life of a nuclear plant. Maybe that's because nuclear energy itself could be said to be immortal. One thousand years from now, future humans might still be producing nuclear power from the same locations they do so today. Today's nuclear plants can easily operate for eighty years and might run for one hundred if they are well maintained and parts are replaced when necessary. Like Cesar Marchetti's typewriters, each successive plant will be better until we arrive at fusion and start the process all over again.

And so, between 2016 and 2020, I worked with environmental humanists around the world to save nuclear power plants. It's had an impact. Where nuclear energy was viewed just a few years ago as optional, it is today increasingly viewed as essential for dealing with climate change.

In 2019, my colleagues and I organized pronuclear demonstrations in more than thirty cities around the world. We were inspired by Greta Thunberg's Student Strike for the Climate, which started with just one or two people, like herself, in many places. We asked people to take the simplest of tasks: "Stand-up for Nuclear." In German town squares, outside train stations in South Korea, and on university campuses in the United States, pronuclear activists stood at tables, handing out information, showing videos, and debunking nuclear myths. Many offered balloons and face-painting for children.

In December 2019, 120 people from Germany, Poland, Switzerland, Austria, the Netherlands, and the Czech Republic rallied near the Philippsburg nuclear power plant, forty-five minutes from the French border, which the German government had forced to close prematurely. "Public opinion towards nuclear energy is changing," said Bjorn Peters, a German pronuclear leader, in late 2019.

Saving nuclear has been two steps forward and one step back. California and New York are moving forward with plans to prematurely close Diablo Canyon and Indian Point nuclear plants, which provide reliable and low-cost carbon-free power to roughly six million people, thanks to the influence of groups like NRDC, EDF, Sierra Club, and 350.org. At the same time, a growing number of people appear to recognize the foolishness of doing so, and are changing their minds about the technology.

In researching and writing this book I was pleased to find support for nuclear energy from unexpected quarters, including from people with whom I disagreed, at least on some questions. Christine Figgener, the German sea turtle scientist, who started a worldwide discussion on straws and plastics, told me she supports nuclear. "I think the world is not black and white," she said. And Zion Lights of Extinction Rebellion told me she had changed her mind after a scientist friend told her it was safe. "I said, 'That's not what I've been told.' And he said, 'Don't just listen to what people tell you.' And so I looked it up and he was right. The data shows it is safe. And I realized solar panels and batteries are not going to meet demand." [1]

Some appear to be moderating their views of climate change, as well. In December 2019, David Wallace-Wells, the author of the apocalyptic 2019 book *The Uninhabitable Earth,* which claimed climate change was "much, much worse than you think," wrote that, "For once, the climate news might be better than you thought. It's certainly better than I've thought." [2] Wallace-Wells pointed to research by Pielke and others showing that the IPCC's high–coal use scenario, known by its technical name, RCP 8.5, was highly improbable, and temperatures were likely to peak at below 3 degrees centigrade above pre-industrial levels. [3]

In December 2019, I received an email from the IPCC inviting me to be an expert reviewer of its next assessment report. It wasn't the first time IPCC had reached out to me. After I criticized the IPCC in 2018 for its biased treatment of nuclear and renewables, a lead author of an IPCC report encouraged me to become a reviewer. I hadn't followed up on that invitation, but this time I did.

I also received an invitation to testify before Congress on the state of climate science, which I did in January 2020. For more than twenty years, I noted, the national conversation about climate change has been polarized between those who deny it and those who exaggerate it. Happily, it appeared that some scientists, journalists, and activists were finally pushing back against the extremes on both sides. None of the members of the House Committee on Science, Space, and Technology denied climate change. Most said they thought something should be done to address it.

Environmental humanism will eventually triumph over apocalyptic environmentalism, I believe, because the vast majority of people in the world want both prosperity and nature, not nature without prosperity. They are just confused about how to achieve both. For while some environmentalists claim their agenda will also deliver a greener prosperity, the evidence shows that an organic, low-energy, and renewable-powered world would be worse, not better, for most people and for the natural environment.

While environmental alarmism may be a permanent feature of public life, it need not be so loud. The global system is changing. While that brings new risks, it also brings new opportunities. Confronting new challenges requires the opposite of panic. With care, persistence, and, I dare say, love, I believe we can moderate the extremes and deepen understanding and respect in the process. In so trying, I believe we will bring ourselves closer to the transcendent moral purpose most people, perhaps even some currently apocalyptic environmentalists, share: nature and prosperity for all.

ACKNOWLEDGMENTS

I am grateful to my family for its love and support. I have dedicated *Apocalypse Never* to my children, Joaquin and Kestrel, who bring me great joy. My parents, Robert Shellenberger and Judith Green, and Nancy and Don O'Brian, taught me to love my fellow humans, care for our planet, and think for myself. They helped me appreciate nature and instilled in me a sense of justice and responsibility. Helen Jeehyun Lee is my best friend and loving wife, and the loyal skeptic I needed. My siblings and in-laws, Kim Shellenberger and Rich Balaban, Julie O'Brian and Jeff Oxenford, Mark and Gina O'Brian, provided me with affection and encouragement.

I have benefited from a supportive community of colleagues. Eric Nelson at HarperCollins helped me see the book I wanted to write. Rob Sartori and Nena Madonia Oshman were persistent and effective champions. I am indebted to Madison Czerwinski, Mark Nelson, Paris Wines, Alexandra Gates, Emmet Penney, Jemin Desai, Sid Bagga, and Gabrielle Haigh from Environmental Progress, and Mark Schlabach for being intelligent, critical, and hardworking researchers, fact-checkers, and editors. I could not have written *Apocalypse Never* without them.

I am grateful that, since our founding in 2016, Environmental Progress's loyal supporters and board members, Frank Batten, Bill Budinger, John Crary, Paul Davis, Kate Heaton, Steve Kirsch, Ross Koningstein, Michael Pelizzari, Jim Swartz, Barrett Walker, Matt Winkler, and Kristin Zaitz, allowed us to follow the facts wherever they led.

Finally, I am grateful to the very many people who gave their time to make *Apocalypse Never* possible: Syarifah Nur Aida, Kevin Anderson, Jesse Ausubel, Pascal Bruckner, Chuck Casto, Hinh Dinh, Kerry Emanuel, Christine Figgener, Chris Helman, Ana Paula Henkel, Laura Jeffrey, Caleb Kabanda, Michael Kavanagh, John Kelly, Julie Kelly, Annette Lanjouw, Claire Lehmann, Timothy Lenton, Michael Lind, Zion Lights, Lisa

Linowes, Sarah Lunnon, Alastair McNeilage, Andrew McAfee, Sudip Mukhopadhyay, Michael Oppenheimer, Andrew Plumptre, Roger Pielke, Helga Rainer, Richard Rhodes, Joyashree Roy, Mark Sagoff, Sarah Sawyer, Laura Seay, Mamy Bernadette Semutaga, Mouad Srifi, Suparti, Nobuo Tanaka, Geraldine Thomas, and Tom Wigley.

NOTES

Introduction

1. Damien Gayle, "Avoid London for Days, Police Warn Motorists, amid 'Swarming Protests,'" *The Guardian*, November 21, 2018, https://www.theguardian.com.
2. Savannah Lovelock and Sarah Lunnon, interviewed by Sophie Ridge, *Sophie Ridge on Sunday*, Sky News, October 6, 2019, https://www.youtube.com/watch?v=ArO_-xH5Vm8.
3. Cameron Brick and Ben Kenward, "Analysis of Public Opinion in Response to the Extinction Rebellion Actions in London," Ben Kenward (website), April 22, 2019, http://www.benkenward.com/XRSurvey/AnalysisOfPublicOpinionInResponseToTheExtinctionRebellionActionsInLondonV2.pdf.
4. "International Poll: Most Expect to Feel Impact of Climate Change, Many Think It Will Make Us Extinct," YouGov, September 14, 2019, https://yougov.co.uk.
5. Ben Kenward, "Analysis of Public Opinion in Response to the Extinction Rebellion Actions in London, April 2019," April 20, 2019, https://drive.google.com/file/d/1dfKFPVghcb4-b9peTyCgR3pu84_8OjGR/view.
6. Mattha Busby, "Extinction Rebellion Protesters Spray Fake Blood on to Treasury," *The Guardian*, October 3, 2019, https://www.theguardian.com.
7. "Provisional UK Greenhouse Gas Emissions Statistics 2018," Department for Business, Energy & Industrial Strategy, March 28, 2019, https://www.gov.uk/government/statistics/provisional-uk-greenhouse-gas-emissions-national-statistics-2018. "Historical Electricity Data: 1920 to 2018," Department for Business, Energy & Industrial Strategy, July 25, 2019, https://www.gov.uk/government/statistical-data-sets/historical-electricity-data.

The carbon intensity of electricity from the United Kingdom's power stations fell by 50 percent between 2000 and 2018, based on emissions and electricity production data from the UK government.

8. Sarah Lunnon, interviewed by Phillip Schofield and Holly Willoughby, *This Morning*, ITV, October 17, 2019, https://www.youtube.com/watch?v=ACL mQsPocNs.

9. Ibid.

10. Ibid.

11. Ibid.

12. "Protesters Dragged off DLR Train as Extinction Rebellion Delay Commuters in London," *The Guardian*, October 17, 2019, https://www.theguardian .com. *The Telegraph*, "Commuters Turn on Extinction Rebellion as Protesters Target the Tube," YouTube, October 17, 2019, https://www.youtube.com /watch?v=9P1UXYS6Bmg.

1: It's Not the End of the World

1. Coral Davenport, "Major Climate Report Describes a Strong Risk of Crisis as Early as 2040," *New York Times*, October 7, 2018, https://www.nytimes .com.

2. Chris Mooney and Brady Dennis, "The World Has Just over a Decade to Get Climate Change Under Control, U.N. Scientists Say," *Washington Post*, October 7, 2018, https://www.washingtonpost.com.

3. Christopher Flavelle, "Climate Change Threatens the World's Food Supply, United Nations Warns," *New York Times*, August 8, 2019, https://www.ny times.com.

4. Ibid.

5. Gaia Vince, "The Heat Is On over the Climate Crisis. Only Radical Measures Will Work," *The Guardian*, May 18, 2019, https://www.theguardian.com.

6. Robinson Meyer, "The Oceans We Know Won't Survive Climate Change," *The Atlantic*, September 25, 2019, https://www.theatlantic.com.

7. Stefan Rahmstorf, Jason E. Box, Georg Feulner et al., "Exceptional Twentieth-Century Slowdown in Atlantic Ocean Overturning Circulation," *Nature Climate Change* 5 (2015): 475–80, https://doi.org/10.1038/nclimate 2554.

8. S. E. Chadburn, E. J. Burke, P. M. Cox et al., "An Observation-Based Constraint on Permafrost Loss as a Function of Global Warming," *Nature Climate Change* 7 (2017): 340–44, https://doi.org/10.1038/nclimate3262.

9. Robinson Meyer, "The Oceans We Know Won't Survive Climate Change," *The Atlantic*, September 25, 2019, https://www.theatlantic.com.

10. Richard W. Spinrad and Ian Boyd, "Our Deadened, Carbon-Soaked Seas," *New York Times*, October 15, 2015, https://www.nytimes.com. John Ross, "Ex-judge to Investigate Controversial Marine Research," *Times Higher Education*, January 8, 2020.

11. "Top 20 Most Destructive California Fires," California Department of Forestry and Fire Protection, August 8, 2019, https://www.fire.ca.gov/media/5511/top20_destruction.pdf.

12. Natacha Larnaud, " 'This Will Only Get Worse in the Future': Experts See Direct Line Between California Wildfires and Climate Change," CBS News, October 30, 2019, https://www.cbsnews.com.

13. Leonardo DiCaprio (@LeoDiCaprio), "The reason these wildfires have worsened is because of climate change and a historic drought. Helping victims and fire relief efforts in our state should not be a partisan issue," Twitter, November 10, 2018, 3:32 p.m., https://twitter.com/leodicaprio/status/1061401158856687616.

14. Alexandria Ocasio-Cortez (@AOC), "This is what climate change looks like. The GOP like to mock scientific warnings about climate change as exaggeration. But just look around: it's already starting. We have 10 years to cut carbon emissions in half. If we don't, scenes like this can get much worse. #GreenNewDeal," Twitter, October 28, 2019, 6:09 a.m., https://twitter.com/aoc/status/1188805012631310336.

15. Farhad Manjoo, "It's the End of California as We Know It," *New York Times*, October 30, 2019, https://www.nytimes.com.

16. Scott Neuman, "Enormous 'Megafire' in Australia Engulfs 1.5 Million Acres," NPR, January 10, 2020, accessed January 15, 2020, https://www.npr.org. "Fathers, son, newlyweds: Nation mourns 29 lives lost to bushfires," *Daily Telegraph*, January 22, 2020, https://www.dailytelegraph.com.au.

17. David Wallace-Wells, *The Uninhabitable Earth: Life After Warming* (New York: Crown Publishing Group, 2019), 16.

18. Pat Bradley, "Leading Environmentalist Reacts to Latest IPCC Report," WAMC, October 9, 2018, https://www.wamc.org/post/leading-environmentalist-reacts-latest-ipcc-report.

19. Coral Davenport, "Major Climate Report Describes a Strong Risk of Crisis as Early as 2040," *New York Times*, October 7, 2018, https://www.nytimes.com.

20. Guardian News, " 'I Want You to Panic': 16-Year-Old Issues Climate Warning at Davos," YouTube, January 25, 2019, https://www.youtube.com/watch?v=RjsLm5PCdVQ.

21. William Cummings, " 'The World Is Going to End in 12 Years if We Don't Address Climate Change,' Ocasio-Cortez says," *USA Today*, January 22, 2019, https://www.usatoday.com.

22. Andrew Freedman, "Climate Scientists Refute 12-Year Deadline to Curb Global Warming," Axios, January 22, 2019, https://www.axios.com.

23. Ibid.

24. V. Masson-Delmonte, Panmao Zhai, Hans-Otto Pörtner et al., eds, *Global Warming of 1.5°C. An IPCC Special Report on the the Impacts of Global Warming of 1.5°C above Pre-industrial Levels and Related Global Greenhouse Gas Emission Pathways, in the Context of Strengthening the Global Response to the Threat of Climate Change, Sustainable Development, and Efforts to Eradicate Poverty*, Intergovernmental Panel on Climate Change, 2018, https://www.ipcc.ch/site/assets/uploads/sites/2/2019/06/SR15_Full_Report_High_Res.pdf.

25. Liz Kalaugher, "Scientist or Climate Activist—Where's the Line?" *Physics World*, September 20, 2019, https://physicsworld.com/a/climate-scientist-or-climate-activist-wheres-the-line.

26. Kerry Emanuel (climate scientist, MIT) in discussion with the author, November 15, 2019.

27. Andrew Freedman, "Climate Scientists Refute 12-Year Deadline to Curb Global Warming," Axios, January 22, 2019, https://www.axios.com.

28. Hannah Ritchie and Max Roser, "Global Deaths from Natural Disasters," Our World in Data, accessed October 25, 2019, https://ourworldindata.org/natural-disasters. Data published by EMDAT (2019): OFDA/CRED International Disaster Database, Université Catholique de Louvain—Brussels—Belgium. Data for individual years are summed over ten-year intervals from first to last year of each calendar decade.

29. Giuseppe Formetta and Luc Feyen, "Empirical Evidence of Declining Global Vulnerability to Climate-Related Hazards," *Global Environmental Change* 57 (July 2019): article 101920, https://doi.org/10.1016/j.gloenvcha.2019.05.004.

30. "Sea Level Change: Scientific Understanding and Uncertainties," in *Climate Change 2013: The Physical Science Basis*, edited by Thomas F. Stocker, Dahe Quin, Gian-Kasper Plattner et al., Intergovernmental Panel on Climate Change, 2013, https://www.ipcc.ch/site/assets/uploads/2018/03/WG1AR5_SummaryVolume_FINAL.pdf, 47–59.

31. The Zuidplaspolder in the western Netherlands is 6.76m below sea level. The IPCC in its Medium scenario (RCP4.5) predicts a 0.39m median sea level rise from a 1990 baseline through year 2200.

32. "Bangladesh Delta Plan 2100," Dutch Water Sector, May 20, 2019, https://www.dutchwatersector.com/news/bangladesh-delta-plan-2100; "Deltaplan Bangladesh," Deltares, https://www.deltares.nl/en/projects/deltaplan-bangladesh-2.

33. Jon. E. Keeley (U.S. Geological Survey scientist), in discussion with the author, November 4, 2019; Jon. E. Keeley and Alexandra Syphard, "Different historical fire–climate patterns in California,"*International Journal of Wildland Fire* 26, no. 4 (January 2017): 253, https://doi.org/10.1071/WF16102.

34. Alexandra D. Syphard, John E. Keeley, Anne H. Pfaff, and Ken Ferschweiler, "Human Presence Diminishes the Importance of Climate in Driving Fire Activity Across the United States," *Proceedings of the National Academy of Sciences of the United States of America* 114, no. 52 (December 2017): 13750–55, https://doi.org/10.1073/pnas.1713885114.

35. Alexandria Symonds, "Amazon Rainforest Fires: Here's What's Really Happening," *New York Times*, August 23, 2019, https://www.nytimes.com.

36. Timothy D. Clark, Graham D. Raby, Dominique G. Roche et al., "Ocean Acidification Does Not Impair the Behaviour of Coral Reef Fishes," *Nature* 557 (2020): 370–75, https://doi.org/10.1038/s41586-019-1903-y. John Ross, "Ex-judge to Investigate Controversial Marine Research," *Times Higher Education*, January 8, 2020.

37. *The Future of Food and Agriculture: Alternative Pathways to 2050*, Food and Agriculture Organization of the United Nations, 2018, http://www.fao.org/3/I8429EN/i8429en.pdf, 76–77.

38. Eric Holt-Giménez, Annie Shattuck, Miguel Altieri, et al., "We Already Grow Enough Food for 10 Billion People . . . and Still Can't End Hunger," *Journal of Sustainable Agriculture* 36, no. 6 (2012): 595–98, http://dx.doi.org/10.1080/10440046.2012.695331. *The Future of Food and Agriculture: Alternative Pathways to 2050*, 82.

39. *The Future of Food and Agriculture: Alternative Pathways to 2050*, 76–77.

40. Keywan Riahi, Arnulf Grübler, and Nebojsa Nakicenovic, "Scenarios of Long-Term Socio-economic and Environmental Development Under Climate Stabilization," *Technological Forecasting and Social Change* 74, no. 7 (September 2007): 887–935, https://doi.org/10.1016/j.techfore.2006.05.026.

41. Though some experts call the country "Democratic Republic of the Congo," in part because there is a different, smaller nation called "Republic of the Congo," many in the DRC refer to it simply as "the Congo."

42. For more information on the Great African War and its impact on the Congo, see David Van Reybrouck, *Congo: The Epic History of a People* (New York: HarperCollins, 2010) and Gérard Prunier, *Africa's World Wars: Congo, the Rwandan Genocide, and the Making of a Continental Catastrophe* (Oxford, UK: Oxford University Press, 2010), 338.

43. "Democratic Republic of the Congo Travel Advisory," U.S. Department of State, Bureau of Consular Affairs, January 2, 2020, https://travel.state.gov /content/travel/en/traveladvisories/traveladvisories/democratic-republic -of-the-congo-travel-advisory.html.

44. "Access to Electricity (% of Population): Congo, Rep., Congo, Dem. Rep," World Bank Group, https://data.worldbank.org.

45. *Africa Energy Outlook: A Focus on Energy Prospects in Sub-Saharan Africa* (Paris: International Energy Agency, 2014), 32.

46. Pauline Bax and Williams Clowes, "Three Gorges Has Nothing on China-Backed Dam to Power Africa," Bloomberg, August 5, 2019, https://www .bloomberg.com.

47. Katharine J. Mach, Caroline M. Kraan, W. Neil Adger et al., "Climate as a Risk Factor for Armed Conflict," *Nature* 571 (2019): 193–97, https://doi .org/10.1038/s41586-019-1300-6.

48. "Globally, medium to high levels of land degradation are related to increased conflict, as are very high levels of water scarcity, but the relative increases in risk are quite small. Increasing levels of land degradation increase the risk of conflict from a baseline of 1 percent to between 2–4 percent." Hedrik Urdal, "People vs. Malthus: Population Pressure, Environmental Degradation and Armed Conflict Revisited," *Journal of Peace Research* 42 (2005): 417–34, https://doi.org/10.1177/0022343305054089. Clionadh Raleigh and Henrik Urdal, "Climate Change, Demography, Environmental Degradation and Armed Conflict," *Environmental Change and Security Programme* 13 (2008–2009): 27–33, https://doi.org/10.1016/j.polgeo.2007.06.005.

49. Michael Oppenheimer, Bruce C. Glavovic, Jochen Hinkel et al., "Sea Level Rise and Implications for Low-Lying Islands, Coasts and Communities," in *IPCC Special Report on the Ocean and Cryosphere in a Changing Climate*, Intergovernmental Panel on Climate Change, 2019, https://www.ipcc.ch/site/as sets/uploads/sites/3/2019/11/08_SROCC_Ch04_FINAL.pdf, 321–445.

50. Clionadh Raleigh and Henrik Urdal, "Climate Change, Demography, Environmental Degradation and Armed Conflict," *Environmental Change and Security Programme* 13 (2008–2009): 27–33, https://doi.org/10.1016/j.pol geo.2007.06.005.

51. David Van Reybrouk, *Congo: An Epic History of a People* (New York: Harper-Collins, 2014).

52. World Bank, "GDP (current US$)-Congo, Dem. Rep.," World Bank, accessed January 15, 2020, https://data.worldbank.org.

53. Ibid.

54. Peter Jones and David Smith, "UN Report on Rwanda Fuelling Congo Conflict 'Blocked by US,' " *The Guardian*, June 20, 2012, https://www.theguard ian.com.

55. Tom Wilson, David Blood, and David Pilling, "Congo Voting Data Reveal Huge Fraud in Poll to Replace Kabila," *Financial Times*, January 15, 2019, https://www.ft.com.

56. Myles Allen and Sarah Lunnon, interviewed by Emma Barrett, *Newsnight*, BBC, aired October 10, 2019, on BBC.

57. Sarah Lunnon (Extinction Rebellion spokesperson) in discussion with the author, November 26, 2019.

58. Gaia Vince, "The Heat Is On over the Climate Crisis. Only Radical Measures Will Work," *The Guardian*, May 18, 2019, https://www.theguardian.com.

59. Sarah Lunnon (Extinction Rebellion spokesperson) in discussion with the author, November 26, 2019.

60. Johan Rockström (director of the Potsdam Institute for Climate Impact Research) in discussion with the author, November 27, 2019.

61. Ibid.

62. Ibid.

63. Ibid.

64. Hans van Meijl et al., "Comparing impacts of climate change and mitigation on global agriculture by 2050," *Environmental Research Letters* 13, no. 6 (2018), https://iopscience.iop.org/article/10.1088/1748–9326/aabdc4/pdf.

65. "This occurs because . . . land-based mitigation leads to less land availability for food production, potentially lower food supply, and therefore food price increases." Cheikh Mbow et al., "Chapter Five: Food Security," in V. Masson-Delmotte et al., eds., *Climate Change and Land: An IPCC special report on climate change, desertification, land degradation, sustainable land management, food security, and greenhouse gas fluxes in terrestrial ecosystems* (IPCC,

2019), https://www.ipcc.ch/site/assets/uploads/2019/11/SRCCL-Full
-Report-Compiled-191128.pdf.

66. FAO, *The Future of Food and Agriculture–Alternative Pathways to 2050* (Rome: United Nations, 2018), accessed December 16, 2019, http://www.fao.org /global-perspectives-studies/food-agriculture-projections-to-2050/en.

67. Roger Pielke, Jr., "I Am Under Investigation," *The Climate Fix* (blog), February 25, 2015, https://theclimatefix.wordpress.com/2015/02/25/i-am-under -investigation.

68. Roger Pielke, Jr., *The Climate Fix: What Scientists and Politicians Won't Tell You About Global Warming* (New York: Basic Books, 2011), 162.

69. Laurens M. Bouwer, Ryan P. Crompton, Eberhard Faust et al., "Confronting Disaster Losses," *Science* 318 (December 2007): 753, https://www.research gate.net/publication/5871449_Disaster_management_Confronting_disas ter_losses. Peter Höppe and Roger Pielke, Jr., eds., "Workshop on Climate Change and Disaster Losses: Understanding and Attributing Trends and Projections," May 25–26, 2006, Hohenkammer, Germany, https://sciencepolicy .colorado.edu/research_areas/sparc/research/projects/extreme_events /munich_workshop/ccdl_workshop_brochure.pdf.

70. Roger Pielke, Jr., *The Climate Fix: What Scientists and Politicians Won't Tell You About Global Warming*, 171.

71. Jessica Weinkle, Chris Landsea, Douglas Collins, et al., "Normalized Hurricane Damage in the Continental United States 1900–2017," *Nature Sustainability* 1 (2018): 808–813, https://doi.org/10.1038/s41893-018-0165-2; Roger Pielke, Jr., *The Climate Fix: What Scientists and Politicians Won't Tell You About Global Warming*, 171.

72. Roger Pielke, Jr., *The Climate Fix: What Scientists and Politicians Won't Tell You About Global Warming*, 171.

73. Pielke cited two studies for this claim: Thomas Knutson, Suzana J. Camargo, Johnny C. L. Chan et al., "Tropical Cyclones and Climate Change Assessment: Part I: Detection and Attribution," *American Meteorological Society*, October 2019, https://journals.ametsoc.org/doi/pdf/10.1175 /BAMS-D-18-0189.1, and "2019 Atlantic Hurricane Season," National Hurricane Center and Central Pacific Hurricane Center, https://www.nhc .noaa.gov/data/tcr. Roger Pielke, Jr., "When Is Climate Change Just Weather? What Hurricane Dorian Coverage Mixes Up, on Purpose," *Forbes*, September 4, 2019, https://www.forbes.com.

74. Roger Pielke, Jr., *The Climate Fix: What Scientists and Politicians Won't Tell You About Global Warming*, 170–172.

75. Roger Pielke, Jr., "My Unhappy Life as a Climate Heretic," *Wall Street Journal*, December 2, 2016, https://www.wsj.com.

76. Christopher B. Field, Vicente Barros, Thomas F. Stocker et al., eds., *Managing the Risks of Extreme Events and Disasters to Advance Climate Change Adaptation: Special Report of the Intergovernmental Panel on Climate Change*, Intergovernmental Panel on Climate Change, https://www.ipcc.ch/site/assets/uploads/2018/03/SREX_Full_Report-1.pdf, 9.

77. Roger Pielke, Jr., *The Climate Fix: What Scientists and Politicians Won't Tell You About Global Warming*, 175.

78. Roger Pielke, Jr., "Disasters Cost More than Ever—but Not Because of Climate Change," FiveThirtyEight, March 19, 2014, https://fivethirtyeight.com.

79. Roger Pielke, Jr., *The Climate Fix: What Scientists and Politicians Won't Tell You About Global Warming*, 174.

80. Roger Pielke, Jr., "Disasters Cost More than Ever—but Not Because of Climate Change."

81. Ibid.

82. Robinson Meyer, "The Oceans We Know Won't Survive Climate Change," *Atlantic*, September 25, 2019, https://www.theatlantic.com.

83. Michael Oppenheimer (IPCC author) in discussion with the author, December 26, 2019.

84. Ibid.

85. Ibid.

86. Ibid.

87. Ibid.

88. "Project Information Document (PID) Concept Stage: DRC—Growth with Governance in the Mineral Sector," Report no. AB3834, World Bank, March 24, 2009, http://documents.worldbank.org/curated/en/341011468234300132/pdf/Project0Inform1cument1Concept0Stage.pdf. Ernest Mpararo, "Democratic Republic of the Congo," Transparency International, 2018, https://www.transparency.org/country/COD. For an overview of Congo's history and the role of colonialism and decolonization, see Van Reybrouck, *Congo*. Congo ranks 161st out of 180 (the highest) in terms of perceived corruption.

89. "Top 20 Deadliest California Wildfires," California Department of Forestry and Fire Protection, September 27, 2019, https://www.fire.ca.gov /media/5512/top20_deadliest.pdf.

90. Jon E. Keeley and Alexandra D. Syphard, "Twenty-first Century California, USA, Wildfires: Fuel-Dominated vs. Wind-Dominated Fires," *Fire Ecology* 15, article no. 24, July 18, 2019, https://doi.org/10.1186/s42408-019-004 1-0.

91. Jon E. Keeley and Alexandra D. Syphard, "Historical Patterns of Wildfire Ignition Sources in California Ecosystems," *International Journal of Wildland Fire* 27, no. 12 (2018): 781–99, https://doi.org/10.1071/WF18026.

92. Jon. E. Keeley (U.S. Geological Survey scientist) in discussion with the author, November 4, 2019.

93. Ibid.

94. Ibid.

95. Ibid.

96. Ibid.

97. David Packham, interviewed by Andrew Bolt, *The Bolt Report*, Sky News Australia, November 11, 2019, https://www.youtube.com/watch?v=E6Rrg Brb6R8.

98. S. Ellis, P. Kanowski, and R.J. Whelan, "National Inquiry on Bushfire Mitigation and Management," University of Wollongong Research Online, March 31, 2004, https://ro.uow.edu.au.

Luke Henrique-Gomes, "Bushfires death toll rises to 33 after body found in burnt-out house near Moruya," *Guardian*, January 24, 2020, https:// www.theguardian.com. "Bushfire—Southeast Victoria," Australia Institute for Disaster Resilience, https://knowledge.aidr.org.au/resources/bush fire-south-east-victoria. Bernard Teague, Ronald McLeod, and Susan Pascoe, *Final Report Summary*, Victorian Brushfires Royal Commission, 2009. http:// royalcommission.vic.gov.au/finaldocuments/summary/PF/VBRC_Sum mary_PF.pdf.

"Black Friday bushfires, 1939," Australia Institute for Disaster Resilience, https://knowledge.aidr.org.au/resources/bushfire-black-friday-victoria -1939.

99. Lauren Jeffrey (British YouTuber) in discussion with the author, December 3, 2019.

100. Dan McDougall, " 'Ecological grief': Greenland residents traumatised by climate emergency," *Guardian*, August 12, 2019, https://www.theguardian

.com; Mary Ward, "Climate anxiety is real, and young people are feeling it," *Sydney Morning Herald*, September 20, 2019, https://www.smh.com.au.

101. Susan Clayton et al., "Mental Health and Our Changing Climate: Impacts, Implications, and Guidance," American Psychological Association and ecoAmerica, March 2017, https://www.apa.org/news/press/releases/2017/03/mental-health-climate.pdf.

102. Reuters, "One in five UK children report nightmares about climate change," March 2, 2020.

103. Sonia Elks, "Suffering Eco-anxiety over Climate Change, Say Psychologists," Reuters, September 19, 2019, https://www.reuters.com.

104. Rupert Read, "How I Talk with Children About Climate Breakdown," YouTube (video), August 18, 2019, https://youtu.be/6Lt0jCDtYSY.

105. Zion Lights, interviewed by Andrew Neil, *The Andrew Neil Show*, BBC, aired October 10, 2019, on BBC, https://www.youtube.com/watch?time_continue=7&v=pO1TTcETyuU&feature=emb_logo.

106. Lauren Jeffrey, "An Open Letter to Extinction Rebellion," YouTube (video), October 21, 2019, https://www.youtube.com/watch?v=iyYPLkWV3l0.

107. Peter James Spielmann, "UN Predicts Disaster if Global Warming Not Checked," Associated Press, June 29, 1989, https://apnews.com.

108. Roger Pielke, Jr., "The Uninhabitable Earth—Future Imperfect," *Financial Times*, March 8, 2019, https://www.ft.com.

109. Allyson Chiu, " 'It Snuck up on Us': Scientists Stunned by 'City-Killer' Asteroid That Just Missed Earth," *Washington Post*, July 26, 2019, www.washingtonpost.com.

110. Nick Perry, "Death toll from New Zealand volcano rises to 21 as victim dies from injuries nearly two months later," *Independent*, January 29, 2020, https://www.independent.co.uk.

111. " 'Stealth Transmission' Fuels Fast Spread of Coronavirus Outbreak," Mailman School of Public Health at Columbia University, February 26, 2020, https://www.mailman.columbia.edu/public-health-now/news/stealth-transmission-fuels-fast-spread-coronavirus-outbreak.

112. Kerry Emanuel (climate scientist, MIT) in discussion with the author, November 15, 2019.

113. International Energy Agency, "Global CO_2 Emissions in 2019," February 11, 2020. European Commission, "Progress Made in Cutting Emissions," accessed March 2, 2020. Carbon Brief, "Analysis: Why US carbon emissions have fallen 14% since 2005," August 15, 2017. BP Energy Data, 2019.

114. USA emissions from EIA, "US Carbon dioxide emissions from energy consumption (from 1973)," EIA, accessed February 1, 2020, https://www.eia.gov/environment/data.php.

UK emissions from UK Department for Business, Energy and Industrial Strategy, "National Statistics: Provisional UK Greenhouse Gas Emissions National Statistics 2018," UK government website, March 28, 2019, https://www.gov.uk/government/statistics/provisional-uk-greenhouse-gas-emissions-national-statistics-2018.

115. Justin Ritchie (@jritch), "With this in mind, it is fair to say that the global energy system today, as modeled by IEA, is tracking much closer to 2° of warming this century than previously thought. Thread: 5/11," Twitter, November 18, 2019, 9:37 a.m., https://twitter.com/jritch/status/1196482584710004736.

2: Earth's Lungs Aren't Burning

1. Leonardo DiCaprio (@leonardodicaprio), "#Regram #RG @rainforest alliance: The lungs of the Earth are in flames . . . ," Instagram photo, August 22, 2019, https://www.instagram.com/p/B1eBsWDlfF1/?utm_source=ig_web_button_share_sheet. Cristiano Ronaldo (@cristiano), "The Amazon Rainforest produces more than 20% of the world's oxygen and its been burning for the past 3 weeks. It's our responsibility to help to save our planet. #prayforamazonia," Twitter, August 22, 2019, https://twitter.com/Cristiano/status/1164588606436106240?s=20. Journalists, too, described the Amazon as the lungs of the world. "The Amazon remains a net source of oxygen today," said a journalist on CNN. Susan Scutti, "Here's What We Know About the Fires in the Amazon Rainforest," CNN, August 24, 2019, https://www.cnn.com.

2. Alexandria Symonds, "Amazon Rainforest Fires: Here's What's Really Happening," New York Times, August 23, 2019, https://www.nytimes.com.

3. Max Fisher, " 'It's Really Close': How the Amazon Rainforest Could Self-Destruct," New York Times, August 30, 2019, https://www.nytimes.com.

4. Symonds, "Amazon Rainforest Fires: Here's What's Really Happening."

5. Franklin Foer, "The Amazon Fires Are More Dangerous than WMDs," The Atlantic, August 24, 2019. Alexander Zaitchik, "Rainforest on Fire," The Intercept, July 6, 2019, https://theclimatecenter.org/rainforest-on-fire.

6. Emmanuel Macron, "Our house is burning. Literally. The Amazon rainforest—the lungs which produces 20% of our planet's oxygen—is on fire. It is an international crisis. Members of the G7 Summit, let's discuss this emergency first order in two days! #ActForTheAmazon," Twitter, August 22, 2019, 12:15 p.m., https://twitter.com/emmanuelmacron/status/1164617008962527232.

7. Kendra Pierre-Louis, "The Amazon, Siberia, Indonesia: A World of Fire," *New York Times*, August 28, 2019, https://www.nytimes.com.

8. "Communication to the Committee on the Rights of the Child in the Case of *Sacchi et al. v. Argentina et al.*," September 23, 2019, https://childrenvsclimatecrisis.org/wp-content/uploads/2019/09/2019.09.23-CRC-communication-Sacchi-et-al-v.-Argentina-et-al-2.pdf.

9. "Land Use," Food and Agriculture Organization of the United Nations, http://www.fao.org/faostat/en/#data/RL.

10. Jane Brody, "Concern for the Rainforest Has Begun to Blossom," *New York Times*, October 13, 1987, https://timesmachine.nytimes.com.

11. Michael Shellenberger, "An Interview with Founder of Earth Innovation, Dan Nepstad," Environmental Progress, August 25, 2019, http://environmentalprogress.org/big-news/2019/8/29/an-interview-with-founder-of-earth-innovation-dan-nepstad.

12. Yadvinder Malhi, "Does the Amazon Provide 20% of Our Oxygen?" August 24, 2019, http://www.yadvindermalhi.org/blog/does-the-amazon-provide-20-of-our-oxygen.

13. Doyle Rice, "What Would the Earth Be Like Without the Amazon Rainforest?," *USA Today*, August 28, 2019, https://www.usatoday.com.

14. Valerie Richardson, "Celebrities Get Fact-Checked After Sharing Fake Photos of Amazon Rainforest Fire," *Washington Times*, August 26, 2019, https://www.washingtontimes.com.

15. "Aerial View of the Taim Ecological Station on Fire," *Baltimore Sun*, March 29, 2013, http://darkroom.baltimoresun.com.

16. Niraj Chokshi, "As Amazon Fires Spread, So Do the Misleading Photos," *New York Times*, August 23, 2019, https://www.nytimes.com.

17. "PRODES-Amazônia," Instituto Nacional de Pesquisas Espaciais, November 18, 2019, accessed December 16, 2019, http://www.obt.inpe.br/OBT/assuntos/programas/amazonia/prodes.

18. "Monitoramento dos Focos Ativos por Países," Queimadas Instituto Nacional De Pesquisas Espaciais, January 1, 2020, http://queimadas.dgi.inpe.br/queimadas/portal-static/estatisticas_paises.

19. Michael Shellenberger, "An Interview with Founder of Earth Innovation, Dan Nepstad," Environmental Progress, August 25, 2019, http://environmental progress.org/big-news/2019/8/29/an-interview-with-founder-of-earth-in novation-dan-nepstad.

20. Luke Gibson, Tien Ming Lee, Lian Pin Koh et al., "Primary Forests Are Irreplaceable for Sustaining Tropical Biodiversity," *Nature*, no. 478 (2011): 378–381, https://doi.org/10.1038/nature10425.

21. *Years of Living Dangerously*, season 2, episode 4, "Fueling the Fire," directed by Jon Meyersohn and Jonathan Schienberg, National Geographic, November 16, 2016.

22. Michael Williams, *Deforesting the Earth: From Prehistory to Global Crisis, an Abridgment* (Chicago: University of Chicago, 2006), 87, 106.

23. Germany's per capita carbon emissions in 2018 were 10.0 tons per capita as 82.8 million Germans emitted 830 million tonnes of CO_2 equivalent. Brazil's 211 million people emitted 2000 million tons of CO_2 in 2018 including from Amazon deforestation and fires, a rate of 9.5 tonnes per capita. German emission data: German Federal Environmental Agency, "Indicator: Greenhouse gas emissions," Umwelt Bundesamt, https://www.umweltbundesamt.de /en/indicator-greenhouse-gas-emissions. Brazil carbon emission data: "Total Emissions," SEEG Brazil, accessed February 2, 2020, http://plata forma.seeg.eco.br/total_emission.

24. N. Andela, D. C. Morton, L. Giglio et al., "A human-driven decline in global burned area," *Science* 356, no. 6345 (June 30, 2017): 1356–1362, https://doi .org/10.1126/science.aal4108. Xiao Peng Song, M. C. Hansen, S. V. Stehman et al., "Global land change from 1982 to 2016," *Nature*, no. 560 (August 8, 2018): 639–643, https://doi.org/10.1038/s41586-018-0411-9.

25. Food and Agriculture Organization of the United Nations. FAO, "FAOSTAT Statistical Database," FAOSTAT, accessed January 15, 2020, http://www .fao.org/faostat/en/#data. Between 1995 and 2015, forested area in Europe increased by over 17 million hectares. Belgium, the Netherlands, Switzerland, and Denmark are a combined 15.6 million hectares.

26. Alex Gray, "Sweden's forests have doubled in size over the last 100 years," *World Economic Forum*, December 13, 2018, https://www.weforum.org /agenda/2018/12/swedens-forests-have-been-growing-for-100-years.

27. Jing M. Chen, Weimin Ju, Philippe Ciais et al., "Vegetation Structural Change Since 1981 Significantly Enhanced the Terrestrial Carbon Sink," *Na-*

ture Communications 10, no. 4259 (October 2019): 1–7, https://www.nature
.com/articles/s41467-019-12257-8.pdf.

28. "The significant reduction in deforestation that has taken place in recent
    years, despite rising food commodity prices, indicates that policies put in place
    to curb conversion of native vegetation to agriculture land might be effective.
    This can improve the prospects for protecting native vegetation by investing
    in agricultural intensification." Alberto G. O. P. Barretto, Göran Berndes,
    Gerd Sparovek, and Stefan Wirsenius, "Agricultural Intensification in Brazil
    and Its Effects on Land-Use Patterns: An Analysis of the 1975–2006 Period,"
    Global Change Biology 19, no. 6 (2013): 1804–1815, https://doi.org/10.1111
    /gcb.12174.

29. Jing M. Chen et al., "Vegetation structural change since 1981 significantly en-
    hanced the terrestrial carbon sink," *Nature Communications* 10, no. 4259 (Octo-
    ber 2019): 1–7, https://www.nature.com/articles/s41467-019-12257–8.pdf.

30. Ibid.

31. Jon Lloyd and Graham D. Farquhar, "Effects of Rising Temperatures and
    Carbon Dioxide on the Physiology of Tropical Forest Trees," *Philosophical
    Transactions of the Royal Society* 363, no. 1498 (February 2008): 1811–1817,
    https://doi.org/10.1098/rstb.2007.0032.

32. Sean M. McMahon, Geoffrey G. Parker, and Dawn R. Miller, "Evidence for
    a Recent Increase in Forest Growth," *Proceedings of the National Academy of
    Sciences of the United States of America* 107, no. 8 (February 2010): 3611–3615,
    https://www.ncbi.nlm.nih.gov/pmc/articles/PMC2840472.

33. Alex Gray, "Sweden's Forests Have Doubled in Size over the Last 100 Years,"
    World Economic Forum, December 13, 2018, https://www.weforum.org
    /agenda/2018/12/swedens-forests-have-been-growing-for-100-years.

34. A major study of 111 nations found a negative relationship between tem-
    perature and labor productivity that was statistically significant. In fact,
    researchers found that a nation's temperature level is the second-most con-
    tributing factor to explaining labor productivity overall. The greatest con-
    tributing factor was simply already being a highly developed nation. Kemal
    Yildirim, Cuneyt Koyuncu, and Julide Koyuncu, "Does Temperature Af-
    fect Labor Productivity: Cross-Country Evidence," *Applied Economet-
    rics and International Development* 9, no. 1 (2009): 29–38, https://www.re
    searchgate.net/profile/Cuneyt_Koyuncu/publication/227410116_Does
    _Temperature_Affect_Labor_Productivity_Cross-Country_Evidence

/links/0a85e53467d19369e8000000/Does-Temperature-Affect-Labor-Pro
ductivity-Cross-Country-Evidence.pdf.

35. Pedro Renaux, "Poverty Grows and Poor Population in 2017 Amounts to
54.8 Million," Agência IBGE, December 6, 2018, https://agenciadenoticias
.ibge.gov.br/en/agencia-news/2184-news-agency/news/23316-poverty
-grows-and-poor-population-in-2017-amounts-to-54-8-million.

36. "Amazon Tribes," Survival International, accessed January 2, 2020, https://
www.survivalinternational.org/about/amazontribes.

37. "Brazilian Indians," Survival International, accessed January 2, 2020,
https://www.survivalinternational.org/tribes/brazilian.

38. Linda Rabben, "Kayapo Choices: Short-Term Gain vs. Long-Term Damage," *Cultural Survival Quarterly Magazine*, June 1995, https://www.cultural
survival.org/publications/cultural-survival-quarterly/kayapo-choices
-short-term-gain-vs-long-term-damage.

39. Michael Williams, *Deforesting the Earth: From Prehistory to Global Crisis, an
Abridgment*, Ibid., 21–23.

40. Ibid., 19.

41. Christopher Sandom et al., "Global late Quaternary megafauna extinctions linked to humans, not climate change," *Proceedings of the Royal Society B, Biological Sciences* 281, no. 1787 (2014), https://doi.org/10.1098/rs
pb.2013.3254/.

42. Michael Williams, *Deforesting the Earth: From Prehistory to Global Crisis, an
Abridgment*, Ibid., 24–26.

43. Ibid., 25–29.

44. J. A. J. Gowlett, "Discovery of Fire by Humans: A Long and Convoluted
Process," *Philosophical Transactions of the Royal Society B* 371, no. 1696 (June
2016), http://dx.doi.org/10.1098/rstb.2015.0164.

45. Ibid.

46. Michael Williams, *Deforesting the Earth: From Prehistory to Global Crisis, an
Abridgment*, 146.

47. Keith Thomas, *Man and the Natural World: Changing Attitudes in England,
1500–1800* (Oxford University Press, 1983), 192. Animals in England, writes
a historian, "had been divided into the wild, to be tamed or eliminated, the
domestic, to be exploited for useful purposes, and the pet, to be cherished for
emotional satisfaction."

48. Michael Shellenberger, "An Interview with Founder of Earth Innovation, Dan
Nepstad," Environmental Progress, August 25, 2019, http://environmental

progress.org/big-news/2019/8/29/an-interview-with-founder-of-earth-in novation-dan-nepstad.

49. "Brazil and the Amazon Forest," Greenpeace, accessed January 20, 2020, https://www.greenpeace.org/usa/issues/brazil-and-the-amazon-forest. Michael Shellenberger, "An Interview with Founder of Earth Innovation, Dan Nepstad," Environmental Progress, August 25, 2019, http://environmental progress.org/big-news/2019/8/29/an-interview-with-founder-of-earth-in novation-dan-nepstad.

50. Michael Shellenberger, "An Interview with Founder of Earth Innovation, Dan Nepstad."

51. Ibid.

52. John Briscoe, "Infrastructure First? Water Policy, Wealth, and Well-Being," Belfer Center, January 28, 2012, https://www.belfercenter.org/publication /infrastructure-first-water-policy-wealth-and-well-being.

53. John Briscoe, "Invited Opinion Interview: Two Decades at the Center for World Water Policy," *Water Policy* 13, no. 2 (February 2011): 151, https:// doi.org/10.2166/wp.2010.000.

54. John Briscoe, "Infrastructure First? Water Policy, Wealth, and Well-Being."

55. Rhett Butler, "Greenpeace Accuses McDonald's of Destroying the Amazon," Mongabay, April 7, 2006, https://news.mongabay.com.

56. Michael Shellenberger, "An Interview with Founder of Earth Innovation, Dan Nepstad."

57. Ibid.

58. Ibid.

59. Ibid.

60. Ibid.

61. Ibid.

62. David P. Edwards et al., "Wildlife-Friendly Oil Palm Plantations Fail to Protect Biodiversity Effectively," *Conservation Letters* 3 (2010): 236–42, https:// doi.org/10.1111/j.1755-263X.2010.00107.x.

63. Michael Shellenberger, "An Interview with Founder of Earth Innovation, Dan Nepstad."

64. Dave Keating, "Macron's Mercosur Veto—Are Amazon Fires Being Used as a Smokescreen for Protectionism?," *Forbes*, August 23, 2019, https://www .forbes.com.

65. Ibid.

66. Ibid.

67. "Jair Bolsonaro to Merkel: Reforest Germany, Not Amazon," Deutsche Welle, August 15, 2019, https://www.dw.com/en/jair-bolsonaro-to-merkel-reforest-germany-not-amazon/a-50032213.

68. "Brazil's Lula Blasts Rich Nations on Climate," Reuters, February 6, 2007, https://www.reuters.com.

69. "The analyses show that in agriculturally consolidated areas (mainly southern and southeastern Brazil), land-use intensification (both on cropland and pastures) coincided with either contraction of both cropland and pasture areas, or cropland expansion at the expense of pastures, both cases resulting in farmland stability or contraction. In contrast, in agricultural frontier areas (i.e., the deforestation zones in central and northern Brazil), land-use intensification coincided with expansion of agricultural lands. These observations provide support for the thesis that (i) technological improvements create incentives for expansion in agricultural frontier areas; and (ii) farmers are likely to reduce their managed acreage only if land becomes a scarce resource. The spatially explicit examination of land-use transitions since 1960 reveals an expansion and gradual movement of the agricultural frontier toward the interior (center-western Cerrado) of Brazil." Barretto et al., "Agricultural Intensification in Brazil and Its Effects on Land-Use Patterns: An Analysis of the 1975–2006 Period."

70. Bernardo B. N. Strassburg, Agnieszka E. Latawiec, Luis G. Barioni et al., "When Enough Should Be Enough: Improving the Use of Current Agricultural Lands Could Meet Production Demands and Spare Natural Habitats in Brazil," *Global Environmental Change* 28 (September 2014): 84–97, https://doi.org/10.1016/j.gloenvcha.2014.06.001.

71. Philip K. Thornton and Mario Herrero, "Potential for Reduced Methane and Carbon Dioxide Emissions from Livestock and Pasture Management in the Tropics," *Proceedings of the National Academy of Sciences of the United States of America* 107, no. 46 (November 2010): 19667–72, https://doi.org/10.1073/pnas.0912890107.

72. Michael Shellenberger, "An Interview with Founder of Earth Innovation, Dan Nepstad."

73. Juliana Gil, Rachael Garrett, and Thomas Berger, "Determinants of Crop-Livestock Integration in Brazil: Evidence from the Household and Regional Levels," *Land Use Policy* 59, no. 31 (December 2016): 557–568, https://doi.org/10.1016/j.landusepol.2016.09.022.

74. LaMont C. Cole, "Man's Ecosystem," *BioScience* 16, no. 4 (April 1966):

243–48, https://doi.org/10.2307/1293563/. Wallace S. Broecker, "Man's Oxygen Reserves," *Science* 168, no. 3939 (June 1970), 1537–38, https://doi .org/10.1126/science.168.3939.1537. Broecker ended his article "Hopefully the popular press will bury the bogeyman it created." I am grateful to Mark Sagoff for bringing this history to my attention.

3: Enough with the Plastic Straws

1. Christine Figgener, "Sea Turtle with Straw up Its Nostril—'NO' TO PLASTIC STRAWS," YouTube, August 10, 2015, https://www.youtube.com /watch?v=4wH878t78bw.
2. Ibid.
3. Michael A. Lindenberger, "How a Texas A&M Scientist's Video of a Sea Turtle Soured Americans on Drinking Straws," *Dallas Morning News*, July 19, 2018, https://www.dallasnews.com.
4. Hilary Brueck, "The Real Reason Why So Many Cities and Businesses Are Banning Plastic Straws Has Nothing to Do with Straws at All," *Business Insider*, October 22, 2018, https://www.businessinsider.com.
5. Sophia Rosenbaum, "She Recorded That Heartbreaking Turtle Video. Here's What She Wants Companies like Starbucks to Know About Plastic Straws," *Time*, July 17, 2018, https://time.com.
6. Jenna R. Jambeck, Roland Geyer, Chris Wilcox et al., "Plastic Waste Inputs from Land into the Ocean," *Science* 347, no. 6223 (February 2015): 768–771, http://doi.org/10.1126/science.1260352.
7. Christine Figgener (sea turtle biologist) in conversation with the author, November 6, 2019.
8. Chris Wilcox, Melody Puckridge, Gamar A. Schuyler et al., "A Quantitative Analysis Linking Sea Turtle Mortality and Plastic Debris Ingestion," *Scientific Reports* 8 (September 2018): article no. 12536, https://www.nature.com /articles/s41598-018-30038-z.pdf. Chris Wilcox, Nicholas J. Mallos, George H. Leonard et al., "Using Expert Elicitation to Estimate the Impacts of Plastic Pollution on Marine Wildlife," *Marine Policy* 65 (March 2016): 107–14, https://doi.org/10.1016/j.marpol.2015.10.014.
9. Christine Figgener (sea turtle biologist) in conversation with the author, November 6, 2019/.
10. Leandro Bugoni, L'igia Krause, and Maria Virgínia Petry, "Marine Debris

and Human Impacts on Sea Turtles in Southern Brazil," *Marine Pollution Bulletin* 42, no. 12 (December 2001): 1330–1334, https://doi.org/10.1016/S0025-326X(01)00147-3.

11. Chris Wilcox, Melody Puckridge, Gamar A. Schuyler et al., "A Quantitative Analysis Linking Sea Turtle Mortality and Plastic Debris Ingestion."

12. Iliana Magra, "Whale Is Found Dead in Italy with 48 Pounds of Plastic in Its Stomach," *New York Times*, April 2, 2019, https://www.nytimes.com. Matthew Haag, "64 Pounds of Trash Killed a Sperm Whale in Spain, Scientists Say," *New York Times*, April 12, 2018, https://nytimes.com. Daniel Victor, "Dead Whale Found With 88 Pounds of Plastic Inside Body in the Philippines," *New York Times*, March 18, 2019.

13. Seth Borenstein, "Science Says: Amount of Straws, Plastic Pollution Is Huge," Associated Press, April 21, 2018, https://apnews.com.

14. Michelle Paleczny, Edd Hammill, Vasiliki Karpouzi, and Daniel Pauly, "Population Trend of the World's Monitored Seabirds, 1950–2010," *PLOS ONE* 10, no. 6 (June 2015): e0129342, https://journals.plos.org/plosone/article/file?id=10.1371/journal.pone.0129342&type=printable.

15. Laura Parker, "Nearly Every Seabird on Earth Is Eating Plastic," *National Geographic*, September 2, 2015, https://www.nationalgeographic.com.

16. Seth Borenstein, "Science Says: Amount of Straws, Plastic Pollution Is Huge."

17. Chris Wilcox, Erik Van Sebille, and Britta Denise Hardesty, "Threat of Plastic Pollution to Seabirds Is Global, Pervasive, and Increasing," *Proceedings of the National Academy of Sciences of the United States of America* 112, no. 38 (August 2015): 11899–904, https://doi.org/10.1073/pnas.1502108112.

18. Claudia Giacovelli, Anna Zamparo, Andrea Wehrli et al., *Single-Use Plastics: A Roadmap for Sustainability*, United Nations Environment Programme, 2018, https://wedocs.unep.org/bitstream/handle/20.500.11822/25496/singleUsePlastic_sustainability.pdf?sequence=1&isAllowed=y, 12.

19. Susan Freinkel, *Plastics: A Toxic Love Story* (New York: Houghton Mifflin Harcourt, 2011), 7–8.

20. Roland Geyer, Jenna R. Jambeck, and Kara Lavender Law, "Production, Use, and Fate of All Plastics Ever Made," *Science Advances* 3, no. 7 (July 19, 2017): e1700782, http://advances.sciencemag.org/content/3/7/e1700782.

21. Jambeck et al., "Plastic Waste Inputs from Land into the Ocean."

22. Ibid.

23. W. C. Li, H. F. Tse, and L. Fok, "Plastic Waste in the Marine Environ-

ment: A Review of Sources, Occurrence and Effects," *Science of the Total Environment* 566–567 (2016): 333–49, http://dx.doi.org/10.1016/j.scito tenv.2016.05.084.

24. L. Lebreton, B. Slat, F. Ferrari et al., "Evidence That the Great Pacific Garbage Patch Is Rapidly Accumulating Plastic," *Scientific Reports*, 2018, no. 8 (March 22, 2018), article no. 4666, https://doi.org/10.1038/s41598-018 -22939-w.

25. Christine Figgener (sea turtle biologist) in conversation with the author, November 6, 2019.

26. Ibid.

27. "Facts and Figures About Materials, Waste and Recycling," Environmental Protection Agency, October 30, 2019, accessed January 2, 2019, https:// www.epa.gov/facts-and-figures-about-materials-waste-and-recycling/plas tics-material-specific-data#PlasticsTableandGraph.

28. Changing the Way We Use Plastics, European Commission, 2018, https:// ec.europa.eu/environment/waste/pdf/pan-european-factsheet.pdf.

29. Christine Figgener (sea turtle biologist) in conversation with the author, November 6, 2019.

30. Mike Ives, "China Limits Waste. 'Cardboard Grannies' and Texas Recyclers Scramble," *New York Times*, November 25, 2017, https://www.nytimes.com.

31. Mike Ives, "Recyclers Cringe as Southeast Asia Says It's Sick of the West's Trash," *New York Times*, June 7, 2019, https://www.nytimes.com.

32. Ibid.

33. Motoko Rich, "Cleansing Plastic from Oceans: Big Ask for a Country That Loves Wrap," *New York Times*, June 27, 2019, https://www.nytimes.com.

34. Roland Geyer et al., "Production, use, and fate of all plastics ever made," *Science Advances* 3, no. 7 (July 19, 2017), http://advances.sciencemag.org/con tent/3/7/e1700782.

35. Christine Figgener (sea turtle biologist) in conversation with the author, November 6, 2019.

36. Harvey Black, "Rethinking Recycling," *Environmental Health Perspectives* 103, no. 11 (1995): 1006–1009, https://www.ncbi.nlm.nih.gov/pmc/articles /PMC1519181/pdf/envhper00359-0034-color.pdf.

37. Benjamin Brooks, Kristen Hays, and Luke Milner, *Plastics recycling: PET and Europe lead the way, Petrochemicals special report* (S&P Global Platts, September 2019), https://www.spglobal.com/platts/plattscontent/_assets/_files /en/specialreports/petrochemicals/plastic-recycling-pet-europe.pdf.

38. Daniel Hoornweg and Perinaz Bhada-Tata, *What a Waste: A Global Review of Solid Waste Management*, World Bank Urban Development Series, Knowledge Papers no. 15, March 2012, https://openknowledge.worldbank.org /bitstream/handle/10986/17388/68135.pdf?sequence=8&isAllowed=y/, 46.

39. Jambeck et al., "Plastic Waste Inputs from Land into the Ocean."

40. Marcus Eriksen, Laurent C. M. Lebreton, Henry S. Carson et al., "Plastic Pollution in the World's Oceans: More than 5 Trillion Plastic Pieces Weighing over 250,000 Tons Afloat at Sea," *PLOS ONE* 9, no. 12 (December 10, 2014): e111913, https://journals.plos.org/plosone/article/file?id=10.1371 /journal.pone.0111913&type=printable. As indicated in the title, the authors' final estimate of the total number of plastic pieces in the ocean came in at 5 trillion particles—both macroplastic and microplastic—weighing 269,000 tons.

41. Ibid.

42. Ibid.

43. Ibid.

44. Collin P. Ward, Cassia J. Armstrong, Anna N. Walsh, Julia H. Wash, and Christopher M. Reddy, "Sunlight Converts Polystyrene to Carbon Dioxide and Dissolved Organic Carbon," *Environmental Science Technology Letters* 6, no. 11 (October 10, 2019): 669–674, https://doi.org/10.1021/acs .estlett.9b00532.

45. "Sunlight Degrades Polystyrene Faster than Expected," National Science Foundation, October 18, 2019, https://www.nsf.gov/discoveries/disc _summ.jsp?cntn_id=299408&org=NSF&from=news.

46. Collin P. Ward, "Sunlight Converts Polystyrene to Carbon Dioxide and Dissolved Organic Carbon."

47. Ibid.

48. Emily A. Miller, Loren McClenachan, Roshikazu Uni et al., "The Historical Development of Complex Global Trafficking Networks for Marine Wildlife," *Science Advances* 5, no. 3 (March 2019): eaav5948, http://dx.doi.org/10.1126 /sciadv.aav5948.

49. Martha Chaiklin, "Imports and Autarky: Tortoiseshell in Early Modern Japan," in *Luxury and Global Perspective: Objects and Practices, 1600–2000*, edited by Bernd-Stefan Grewe and Karen Hoffmeester (Cambridge, UK: Cambridge University Press, 2016), 218–21, 230, 236.

50. Ibid.

51. Stephanie E. Hornbeck, "Elephant Ivory: An Overview of Changes to Its

Stringent Regulation and Considerations for Its Identification," *AIC Objects Specialty Group Postprints* 22 (2015): 101–22, http://resources.conserva tion-us.org/osg-postprints/wp-content/uploads/sites/8/2015/05/osg022 -08vII.pdf.

52. Ibid.

53. "Ivory: Where It Comes From, Its Uses and the Modes of Working It," *New York Times*, August 14, 1866, https://timesmachine.nytimes.com.

54. Ibid.

55. Ibid.

56. "The World's Ivory Trade," *New York Times*, July 23, 1882, https://timesma chine.nytimes.com.

57. Ibid.

58. Ibid.

59. Chaiklin, "Imports and Autarky: Tortoiseshell in Early Modern Japan."

60. Freinkel, *Plastics: A Toxic Love Story.*

61. Terri Byrne, "Ivoryton's Keys Are Musical," *New York Times*, December 25, 1977, https://timesmachine.nytimes.com.

62. Susan Freinkel, "A Brief History of Plastic's Conquest of the World," *Scientific American*, May 29, 2011, https://www.scientificamerican.com.

63. Christine Figgener (sea turtle biologist) in discussion with the author, November 6, 2019.

64. According to the IUCN, "Disease appears to be a problem in some populations in some years, with diphtheritic stomatitis (caused by the bacteria *Corynebacterium* spp.) and the blood parasite *Leucocytozoon tawaki*, formerly only known from Fiordland penguins, causes of mortality for chicks. . . . Human disturbance, particularly from unregulated tourists at breeding areas, negatively affects energy budgets, fledgling weight and probability of survival." "Yellow-Eyed Penguin, *Megadyptes antipodes*," IUCN Red List of Threatened Species 2018, BirdLife International, 2018, http://dx.doi .org/10.2305/IUCN.UK.2018-2.RLTS.T22697800A132603494.en.

65. "Southern Royal Albatross, *Diomedea epomophora*," IUCN Red List of Threatened Species 2018, BirdLife International, 2018, http://dx.doi.org /10.2305/IUCN.UK.2018-2.RLTS.T22698314A132641187.en.

66. Thomas Mattern, "New Zealand's Mainland Yellow-Eyed Penguins Face Extinction Unless Urgent Action Taken," University of Otago, May 17, 2017, https://www.otago.ac.nz/news/news/otago648034.html.

67. P. E. Michael, R. Thomson, C. Barbraud et al., "Illegal Fishing Bycatch

Overshadows Climate as a Driver of Albatross Population Decline," *Marine Ecology Progress Series* 579 (September 2017): 185–99, http://dx.doi .org/10.3354/meps12248.

68. Bugoni et al., "Marine Debris and Human Impacts on Sea Turtles in Southern Brazil."

69. Christine Figgener (sea turtle biologist) in conversation with the author, November 6, 2019.

70. A. Abreu-Grobois and P. Plotkin (IUCN SSC Marine Turtle Specialist Group), "Olive Ridley, *Lepidochelys olivacea*," IUCN Red List of Threatened Species 2008, BirdLife International, http://dx.doi.org/10.2305/IUCN .UK.2008.RLTS.T11534A3292503.en.

71. Nathaniel L. Bindoff, William W. L. Cheung, James G. Kairo et al., "Changing Ocean, Marine Ecosystems, and Dependent Communities," in *IPCC Special Report on the Ocean and Cryosphere in a Changing Climate*, Intergovernmental Panel on Climate Change, 2019, https://www.ipcc.ch/site/assets/up loads/sites/3/2019/11/09_SROCC_Ch05_FINAL-1.pdf.

72. *The State of World Fisheries and Aquaculture: Meeting the Sustainable Development Goals*, Food and Agriculture Organization of the United Nations, 2018, http://www.fao.org/3/I9540EN/i9540en.pdf, vii.

73. "Sharks," IUCN Red List of Threatened Species, https://www.iucnred list.org/search?query=sharkspercent20&searchType=species. Sandra Altherr and Nicola Hodgins, *Small Cetaceans, Big Problems: A Global Review of the Impacts of Hunting on Small Whales, Dolphins and Porpoises*, Pro Wildlife, Animal Welfare Institute, and Whale and Dolphin Conservation, November 2018, https://awionline.org/sites/default/files/publication/digital_down load/AWI-ML-Small-Cetaceans-Report.pdf.

74. Christine Figgener (sea turtle biologist) in conversation with the author, November 6, 2019.

75. Sophia Rosenbaum, "She Recorded That Heartbreaking Turtle Video. Here's What She Wants Companies like Starbucks to Know About Plastic Straws," *Time*, July 17, 2018, https://time.com.

76. Christine Figgener (sea turtle biologist) in conversation with the author, November 6, 2019.

77. Rebecca L. C. Taylor, "Bag Leakage: The Effect of Disposable Carryout Bag Regulations on Unregulated Bags," *Journal of Environmental Economics and Management* 93 (January 2019): 254–71, https://doi.org/10.1016/j .jeem.2019.01.001.

78. Bjørn Lomborg, "Sorry, Banning Plastic Bags Won't Save Our Planet," *The Globe and Mail*, June 20, 2019, https://www.theglobeandmail.com.

79. Eriksen et al., "Plastic Pollution in the World's Oceans: More than 5 Trillion plastic Pieces Weighing over 250,000 Tons Afloat at Sea."

80. For the specific case of carbonated drinks, see Franklin Associates, *Life Cycle Inventory of Three Single-Serving Soft Drink Containers*, report prepared for PET Resin Association, 2009, http://www.petresin.org/pdf/FranklinLCI SodaContainers2009.pdf. For a general review, see Pan Demetrakakes, "This Material, or That?," *Packaging Digest*, March 11, 2015, www.packaging digest.com/beverage-packaging/material-or.

81. Christine Figgener (sea turtle biologist) in conversation with the author, November 6, 2019.

82. Frida Røyne and Johanna Berlin, "The Importance of Including Service Life in the Climate Impact Comparison of Bioplastcs and Fossil-Based Plastics," Research Institutes of Sweden, Report no. 23, 2018, http://ri.diva-portal .org/smash/get/diva2:1191391/FULLTEXT01.pdf.

83. A lifecycle assessment found that an engine component storage box made from conventional plastic could last six times as long as a bioplastics box under development. Ibid.

84. Christine Figgener (sea turtle biologist) in conversation with the author, November 6, 2019.

85. "Composting PLA and TPS results in higher impacts than landfilling in seven categories: smog, acidification, carcinogenics, non-carcinogenics, respiratory effects, ecotoxicity, and fossil fuel depletion." Troy A. Hottle, Melissa M. Bilec, and Amy E. Landis, "Biopolymer Production and End of Life Comparisons Using Life Cycle Assessment," *Resources, Conservation and Recycling* 122 (July 2017): 295–306, https://doi.org/10.1016/j.resconrec.2017.03.002.

86. Kunnika Changwichan, Thapat Silalertruksa, and Shabbir H. Gheewala, "Eco-Efficiency Assessment of Bioplastics Production Systems and End-of-Life Options," *Sustainability* 10, no. 4 (March 2018): 952, https://doi .org/10.3390/su10040952.

87. Daniel Posen, Paulina Jaramillo, Amy E. Landis, and W. Michael Griffin, "Greenhouse Gas Mitigation for U.S. Plastics Production: Energy First, Feedstocks Later," *Environmental Research Letters* 12, no. 3 (December 2017), https://doi.org/10.1088/1748-9326/aa60a7.

88. Christine Figgener (sea turtle biologist) in conversation with the author, November 6, 2019.

89. Marydele Donnelly, "Trade Routes for Tortoiseshell," State of the World's Sea Turtles (SWOT), Report Volume 3, February 1, 2008, https://www .seaturtlestatus.org/articles/2008/trade-routes-for-tortoiseshell. Associated Press, "Japan Agrees to End Endangered Hawksbill Turtle Imports After '92," *Los Angeles Times*, June 19, 1991, https://www.latimes.com.

90. Tina Deines, "Endangered hawksbill turtle shell trade is much bigger than scientists ever suspected," *National Geographic*, March 27, 2019, https://www .nationalgeographic.com.

91. Christine Figgener (sea turtle biologist) in conversation with the author, November 6, 2019.

92. "World's Biggest Producer of Plastic to Curtail Its Use," *Bloomberg News*, January 19, 2020, https://www.bloomberg.com.

4: The Sixth Extinction Is Canceled

1. Elizabeth Kolbert, *The Sixth Extinction: An Unnatural History* (New York: Henry Holt & Company, 2014), 266–267.

2. "World Is 'on Notice' as Major UN Report Shows One Million Species Face Extinction," UN News, May 6, 2019, https://news.un.org/en /story/2019/05/1037941. "UN Report: Nature's Dangerous Decline 'Unprecedented'; Species Extinction Rates 'Accelerating,' " Sustainable Development Goals, May 6, 2019, https://www.un.org/sustainabledevelopment /blog/2019/05/nature-decline-unprecedented-report.

3. "UN Report: Nature's Dangerous Decline 'Unprecedented.' "

4. Ibid.

5. Kolbert, *The Sixth Extinction*.

6. Luke J. Harmon and Susan Harrison, "Species Diversity Is Dynamic and Unbounded at Local and Continental Scales," *American Naturalist* 185, no. 5 (2015): 584–93, https://doi.org/10.1086/680859. See also Thomas J. Stohlgren, John D. Barnett, and John T. Kartesz, "The Rich Get Richer: Patterns of Plant Invasions in the United States," *Frontiers in Ecology and the Environment* 1, no. 1 (2003): 11–14, https://doi.org/10.2307/3867959. Scientists have known for decades that the evidence "overwhelmingly supports the openness of communities to new species, even at the small spatial scales where species interact and the influences of competition and resource supply should be strongest." For a recent discussion, see Rubén G. Mateo, Karel Mokany, and Antoine Guisan,

"Biodiversity Models: What If Unsaturation Is the Rule?," *Trends in Ecology & Evolution* 32, no. 8 (2017): 556–66, https://doi.org/10.1016/j.tree.2017.05.003.

7. Fangliang He and Stephen P. Hubbell, "Species–Area Relationships Always Overestimate Extinction Rates from Habitat Loss," *Nature* 473 (2011): 368–371, https://doi.org/10.1038/nature09985.

8. Dov F. Sax, Steven D. Gaines, and James Brown, "Species Invasions Exceed Extinctions on Islands Worldwide: A Comparative Study of Plants and Birds," *The American Naturalist* 160, no. 6 (2002): 766–783, https://doi.org/10.1086/343877.

9. Chris D. Thomas, "Rapid Acceleration of Plant Speciation During the Anthropocene," *Trends in Ecology & Evolution* 30, no. 8 (2015): 448–455, https://doi.org/10.1016/j.tree.2015.05.009.

10. Kolbert, *The Sixth Extinction*, 186.

11. Ibid., 186–187.

12. Mark Sagoff, "Welcome to the Narcisscene," *Breakthrough Journal* no. 9 (Summer 2018), https://thebreakthrough.org/journal/no-9-summer-2018/welcome-to-the-narcisscene.

13. The IUCN Red List of Endangered Species, https://www.iucnredlist.org.

14. Ibid.

15. The number of marina animal genera increased from 2,000 to 5,500 over 100 million years. Genus (p. genera) is the taxonomic rank above species. Marine animal fossils are hardy and easier to study, so scientists use their fossil record to approximate overall extinctions and growth in Earth's geological history. J. J. Sepkoski, "A Compendium of Fossil Marine Animal Genera," *Bulletins of American Paleontology* 363 (2002): 1–560.

16. Ibid.

17. Peter Brannen, "Earth Is Not in the Midst of a Sixth Mass Extinction," *The Atlantic*, June 13, 2017, https://www.theatlantic.com.

18. *Protected Planet Report 2018*, United Nations Environment Programme, 2018, https://livereport.protectedplanet.net/pdf/Protected_Planet_Report_2018.pdf.

19. Marine Deguignet, Diego Juffe-Bignoli, Jerry Harrison et al., *2014 United Nations List of Protected Areas*, United Nations Environment Programme, 2014, www.unep-wcmc.org.

20. A. J. Plumptre, S. Ayebare, D. Segan et al., "Conservation Action Plan for the Albertine Rift," Wildlife Conservation Society and Its Partners, 2016, http://conservationcorridor.org/cpb/Plumptre_et_al_2016.pdf, 12.

21. "Living Planet Index," 2018, Zoological Society of London and WWF, www
.livingplanetindex.org.

22. Hannah Behrendt, Carole Megevand, and Klas Sander, "Deforestation Trends
in the Congo Basin: Reconciling Economic Growth and Forest Protection,"
Working Paper 5, "Wood-Based Biomass Energy," Regional Commission in
Charge of Forestry in Central Africa, April 2013, https://www.profor.info
/sites/profor.info/files/Biomass%20Energy_Sectoral%20Report_Final
%5Bweb%5D_may13_0.pdf.

23. Mark Jenkins, "Who Murdered the Virunga Gorillas?," *National Geographic*,
July 2008, www.nationalgeographic.com.

24. Ibid.

25. Ibid.

26. Holly Dranginis, "Congo's Charcoal Cartel," *Foreign Affairs*, May 12, 2016,
https://www.foreignaffairs.com.

27. Behrendt et al., "Deforestation Trends in the Congo Basin," 1.

28. Sophie Lewisohn, "Virunga: Preserving Africa's National Parks Through
People-Centred Development," Capacity4dev, European Union, April 3,
2018, https://europa.eu/capacity4dev/articles/virunga-preserving-africas
-national-parks-through-people-centred-development. Amy Yee, "The
Power Plants That May Save a Park, and Aid a Country," *New York Times*,
August 30, 2017, https://www.nytimes.com.

29. Andrew Plumptre (senior scientist, Africa Program, Wildlife Conservation
Society) in discussion with the author, February 10, 2015, and November 6,
2019.

30. Michael J. Kavanagh (journalist) in discussion with the author, November 29,
2014.

31. Abe Streep, "The Belgian Prince Taking Bullets to Save the World's Most
Threatened Park," *Outside*, November 5, 2014, https://www.outsideonline
.com.

32. Ibid.

33. Jeffrey Gettlemen, "Oil Dispute Takes a Page from Congo's Bloody Past,"
New York Times, November 15, 2014, https://www.nytimes.com.

34. George Schaller, *The Year of the Gorilla* (Chicago: University of Chicago
Press, 1988), 3.

35. Ibid., 8.

36. Paul Raffaele, "Gorillas in Their Midst," *Smithsonian*, October 2007, https://
www.smithsonianmag.com.

37. Andrew J. Plumptre et al., "Conservation Action Plan for the Albertine Rift" (unpublished report for Wildlife Conservation Society and its partners, 2016), 5, 7.

38. "What I was hearing in the mid-90s and early 2000s while working for IGCP was that the conflict in the DRC was all about greed and people wanting to exploit the minerals. Others said it was all about grievances and the Rwandan conflict. Doing my PhD I came to the conclusion that both aspects are at play, but the causes of the conflict stem from grievances." Michael Shellenberger, "Violence, the Virungas, and Gorillas: An Interview with Conservationist Helga Rainer," Breakthrough Institute, November 20, 2014, https://the breakthrough.org/issues/conservation/violence-the-virungas-and-gorillas.

39. Mark Dowie, *Conservation Refugees: The Hundred-Year Conflict Between Global Conservation and Native Peoples* (Cambridge, MA: MIT Press, 2009), xxi.

40. Mahesh Rangarajan and Ghazala Shahabuddin, "Displacement and Relocation from Protected Areas," *Conservation and Society* 4, no. 3 (September 2006): 359, https://www.conservationandsociety.org.

41. Mark Dowie, *Conservation Refugees: The Hundred-Year Conflict Between Global Conservation and Native Peoples.*

42. Ibid., xxvi.

43. Sammy Zahran, Jeffrey G. Snodgrass, David G. Maranon et al., "Stress and Telomere Shortening Among Central Indian Conservation Refugees," *Proceedings of the National Academy of Sciences of the United States of America* 112, no. 9 (March 3, 2015): E928–E936, https://doi.org/10.1073/pnas.1411902112.

44. A. J. Plumptre, A. Kayitare, H. Rainer et al., "The Socio-economic Status of People Living near Protected Areas in the Central Albertine Rift," *Albertine Rift Technical Reports* 4 (2004), https://www.researchgate.net/publication/235945000_Socioeconomic_status_of_people_in_the_Central_Albertine_Rift, 28.

45. Alastair McNeilage (primatologist, Wildlife Conservation Society) in discussion with the author, February 5, 2015.

46. Plumptre et al., "The Socio-economic Status of People Living near Protected Areas in the Central Albertine Rift," 98.

47. Michael Shellenberger, "Postcolonial Gorilla Conservation: An Interview with Ecologist Sarah Sawyer," Breakthrough Institute, November 19, 2014, https://thebreakthrough.org/issues/conservation/postcolonial-gorilla-conservation.

48. Ibid.

49. Andrew Plumptre (senior scientist, Africa Program, Wildlife Conservation Society) in discussion with the author, November 6, 2019.

50. "2019–2020 Gorilla Tracking Permit Availability Uganda/Rwanda," Kisoro Tours Uganda, https://kisorotoursuganda.com/2019-2020-gorilla-tracking-permit-availability-uganda-rwanda. Uganda remains a relative bargain at just $600.

51. Michael Shellenberger, "Postcolonial Gorilla Conservation: An Interview with Ecologist Sarah Sawyer," https://thebreakthrough.org/issues/conservation/violence-the-virungas-and-gorillas.

52. Plumptre et al., "The Socio-economic Status of People Living near Protected Areas in the Central Albertine Rift," *Albertine Rift Technical Reports* 4 (2004): 116, https://albertinerift.wcs.org.

53. Alastair McNeilage (primatologist, Wildlife Conservation Society) in discussion with the author, February 5, 2015.

54. Andrew Plumptre (senior scientist, Africa Program, Wildlife Conservation Society) in discussion with the author, February 10, 2015, and November 6, 2019.

55. Michael Shellenberger, "Violence, the Virungas, and Gorillas: An Interview with Conservationist Helga Rainer."

56. Andrew Plumptre et al., "The Socio-economic Status of People Living near Protected Areas in the Central Albertine Rift," 25.

57. Michael Shellenberger, "Postcolonial Gorilla Conservation: An Interview with Ecologist Sarah Sawyer."

58. Andrew Plumptre (senior scientist, Africa Program, Wildlife Conservation Society) in discussion with the author, February 10, 2015.

59. Andrew Plumptre (senior scientist, Africa Program, Wildlife Conservation Society) in discussion with the author, November 6, 2019.

60. Ibid.

61. Ibid.

62. "Once accustomed to harvesting game with traditional weapons for their own community's use, expelled natives often buy rifles, re-enter their former hunting grounds, and begin poaching larger numbers of the same game for the growing 'bush meat,' or the meat from wild animals, trade, which like almost everything else has gone global," noted the environmental journalist. In Cameroon in 2003, "impoverished and embittered refugees invaded both reserves and plundered their natural resources." Dowie, *Conservation Refugees*, xxvi–xxvii.

63. Andrew Plumptre (senior scientist, Africa Program, Wildlife Conservation Society) in discussion with the author, February 10, 2015.

64. Andrew Plumptre (senior scientist, Africa Program, Wildlife Conservation Society) in discussion with the author, November 6, 2019.

65. Shannon Sims, "After Violence, Congo's Virunga National Park Closes for the Year," *New York Times*, June 14, 2018, https://www.nytimes.com.

66. Jason Burke, "Ranger Killed Weeks After Reopening of Virunga National Park," *The Guardian*, March 8, 2019, www.theguardian.com.

67. Mamy Bernadette Semutaga, interviewed by Caleb Kabanda, December 11, 2019.

68. Sunil Nautiyal and Harald Kaechele, "Fuel Switching from Wood to LPG Can Benefit the Environment," *Environmental Impact Assessment Review* 28, no. 8 (November 2008): 523–32, https://doi.org/10.1016/j.eiar.2008.02.004.

69. Alastair McNeilage (primatologist, Wildlife Conservation Society) in discussion with the author, February 5, 2015.

70. Ibid.

71. Ibid.

72. Ibid.

73. Ibid.

74. Ibid.

75. Ibid.

76. Michael J. Kavanagh (journalist) in discussion with the author, November 29, 2014.

77. Amy Yee, "The Power Plants That May Save a Park, and Aid a Country," *New York Times*, August 30, 2017, https://www.nytimes.com.

78. Ibid.

79. *Sub-Saharan Africa: Macro Poverty Outlook*, World Bank Group, October 2019, http://pubdocs.worldbank.org/en/720441492455091991/mpo-ssa.pdf, 222.

80. "History of Hoover," Arizona Power Authority, accessed January 20, 2020, http://www.powerauthority.org/about-us/history-of-hoover.

5: Sweatshops Save the Planet

1. Elizabeth Paton, "Extinction Rebellion Takes Aim at Fashion," *New York Times*, October 6, 2019, https://www.nytimes.com.

2. Olivia Petter, "Extinction Rebellion: 'Why We're Targeting London Fashion Week," *The Independent*, September 12, 2019, https://www.independent.co.uk.

3. Elizabeth Paton, "Extinction Rebellion Takes Aim at Fashion," *New York Times*, October 6, 2019, https://www.nytimes.com.

4. Ibid.

5. XR Boycott Fashion, "On Friday 26th of July, we wrote to the British Fashion Council begging them to cancel London Fashion Week," Facebook, August 4, 2019, https://www.facebook.com/XRBoycottFashion/posts/113234513352496?__tn__=K-R.

6. "Pulse of the Fashion Industry," Global Fashion Agenda, Boston Consulting Group and Sustainable Apparel Coalition, 2019, accessed October 26, 2019, https://www.globalfashionagenda.com/pulse-2019-update.

7. Elizabeth Paton, "Extinction Rebellion Takes Aim at Fashion."

8. Sarah Anne Hughes, "Greenpeace Protests Barbie at Mattel Headquarters," *Washington Post*, June 8, 2011, https://www.washingtonpost.com.

9. Ibid.

10. Steven Greenhouse, "Nike Supports Women in Its Ads but Not Its Factories, Groups Say," *New York Times*, October 26, 1997, https://www.nytimes.com.

11. Jeff Ballinger, "Nike, Sexual Harassment and the 'Corporate Social Responsibility' Racket: How the Company Shut Down the *New York Times*," *Washington Babylon*, November 9, 2017, https://washingtonbabylon.com.

12. Ibid.

13. "Statement on 2019 Copenhagen Fashion Summit," Union of Concerned Researchers in Fashion, May 5, 2019, accessed October 26, 2019, http://www.concernedresearchers.org/ucrf-on-2019-copenhagen-fashion-summit.

    See also, "The changes we are seeing from some brands remain extremely superficial," said one XR activist. Paton, "Extinction Rebellion Takes Aim at Fashion."

14. Syarifah Nur Aida (journalist, Ipeh) in discussion with the author, June 8, 2015.

15. Suparti (factory worker) in discussion with the author, June 8–9, 2015.

16. Ibid.

17. Xiaoping Liu, Guohua Hu, Yimin Chen et al., "High-Resolution Multitemporal Mapping of Global Urban Land Using Landsat Images Based on the Google Earth Engine Platform," *Remote Sensing of Environment* 209 (May 2018): 227–39, https://doi.org/10.1016/j.rse.2018.02.055.

18. Christopher D. Elvidge, Benjamin T. Tuttle, Paul C. Sutton et al., "Global Distribution and Density of Constructed Impervious Surfaces," *Sensors* 7, no. 9 (2007): 1962-79, https://dx.doi.org/10.3390%2Fs7091962.

19. FAO finds reforestation in Europe, Asia, North America, and the Caribbean. Central America, South America, Africa, and Oceania are still deforesting. The global rate of deforestation has been cut by over half since 1990, from 7.3 million to 3.3 million hectares per year as reforestation accelerated. "Data," FAO.

20. "Data," Food and Agriculture Organization of the United Nations, http://www.fao.org/faostat/en/#data. Russell Warman, "Global Wood Production from Natural Forests Has Peaked," *Biodiversity and Conservation* 23, no. 5 (2014): 1063–78, https://doi.org/10.1007/s10531-014-0633-6.

21. FAO projects that the amount of arable land and permanent crop area will stay nearly flat through 2050, as detailed in its report on the subject. Nikos Alexandratos and Jelle Bruinsma, "World Agriculture Towards 2030/2050: The 2012 Revision," ESA Working Paper no. 12-03, Agricultural Development Economics Division, Food and Agriculture Organization of the United Nations, June 2012, http://www.fao.org/3/a-ap106e.pdf.

22. Per FAO, global per capita kilocalorie production was 2,196 in 1961 and 2,884 in 2013. Along with the global population increase from 3.1 billion to 7.2 billion between 1961 and 2013, global food production has tripled. The amount of global land used for agriculture increased from 4.5 billion to 4.8 billion hectares over the same period. "Data," FAO.

23. Ibid.

24. Ibid.

25. *Changes in Farm Production and Efficiency: A Summary Report*, United States Department of Agriculture, Statistical Bulletin 233 (Washington, DC: USDA, 1959), 12–13.

26. A. Bala, "Nigeria," Global Yield Gap and Water Productivity Atlas, http://www.yieldgap.org/en/web/guest/nigeria. Nikolai Beilharz, "New Zealand farmer sets new world record for wheat yield," ABC News, April 3, 2017, https://www.abc.net.au. Matthew B. Espe, Haishun Yang, Kenneth G. Cassman et al., "Estimating Yield Potential in Temperate High-Yielding, Direct-Seeded US Rice Production Systems," *Field Crops Research* 193 (July 2016): 123–32, https://doi.org/10.1016/j.fcr.2016.04.003. Though average yields for some crops such as wheat have plateaued, there is still more room for them to increase. In 2017, a farmer in New Zealand produced an aston-

ishing eight times as much wheat as the Australian average and five times as much as the global average.

27. FAO, *The future of food and agriculture—Alternative pathways to 2050* (Rome: Food and Agriculture Organization of the United Nations, 2018), 76–77.

28. Nathaniel D. Mueller, James S. Gerber, Matt Johnston et al., "Closing Yield Gaps Through Nutrient and Water Management," *Nature* 490 (2012): 254–57, https://doi.org/10.1038/nature11420.

29. Deepak K. Ray, "Increasing Global Crop Harvest Frequency: Recent Trends and Future Directions," *Environmental Research Letters* 8 (2013), https://iop science.iop.org/article/10.1088/1748-9326/8/4/044041/pdf.

30. Luis Lassaletta, Gilles Billen, Bruna Grizzetti et al., "50 Year Trends in Nitrogen Use Efficiency of World Cropping Systems: The Relationship between Yield and Nitrogen Input to Cropland," *Environmental Research Letters* 9, no. 10 (October 2014), https://doi.org/10.1088/1748-9326/9/10/105011.

31. Ibid.

32. "Changes in Erosion 1982–1997," U.S. Department of Agriculture, January 4, 2001, https://www.nrcs.usda.gov/wps/portal/nrcs/detail/soils/ref /?cid=nrcs143_013911. FAO data on crop yields show almost every major crop increasing in yield in the United States between 1982 and 1997. "Data," FAO.

33. Suparti (factory worker) in discussion with the author, June 8–9, 2015.

34. James C. Riley, "Estimates of Regional and Global Life Expectancy, 1800–2001," *Population and Development Review* 31, no. 3 (2005), 537–543, accessed January 16, 2020, www.jstor.org/stable/3401478, "World Population Prospects 2019: Highlights," United Nations, accessed January 14, 2020, https:// www.un.org/development/desa/publications/world-population-pros pects-2019-highlights.html.

35. Max Roser, Hannah Ritchie, and Bernadeta Dadonaite, "Child and Infant Mortality," Our World in Data, November 2019, https://ourworldindata .org/child-mortality. The world series for 1800 to 1960 was calculated by the authors on the basis of the Gapminder estimates of child mortality and the Gapminder series on population by country. For each estimate in that period a population-weighted global average was calculated. The 2017 child mortality rate was taken from the 2019 update of World Bank data.

36. Steven Pinker, *Enlightenment Now: The Case for Reason, Science, Humanism, and Progress* (New York: Penguin Publishing Group, 2019), 86–87.

37. "PovcalNet: An Online Analysis Tool for Global Poverty Monitoring,"

World Bank Group, accessed October 29, 2019, http://iresearch.world bank.org/PovcalNet/home.aspx.

38. Max Roser, Hannah Ritchie, and Esteban Ortiz-Ospina, "World Population Growth," Our World In Data, May 2019, accessed January 16, 2020, https://ourworldindata.org/world-population-growth.

39. Ibid.

40. For 1990 data: U.N. Food and Agriculture Organization, "Undernourishment around the world," *The State of Food Insecurity in the World, 2006* (Rome: FAO, 2006), http://www.fao.org/3/a0750e/a0750e02.pdf.

    For 2018 data: FAO, "Suite of Food Security Indicators," FAOSTAT, accessed January 28, 2020, http://www.fao.org/faostat/en/#data/FS.

41. Emanuela Cardia, "Household Technology: Was It the Engine of Liberation?," Society for Economic Dynamics, *2008 Meeting Papers* no. 826, https://EconPapers.repec.org/RePEc:red:sed008:826.

42. Benjamin Friedman, *The Moral Consequences of Economic Growth* (New York: Vintage, 2005). Steven Pinker, *The Better Angels of Our Nature: Why Violence Has Declined* (New York: Penguin, 2012).

43. Syarifah Nur Aida (Ipeh) (journalist) in discussion with the author, June 8, 2015.

44. Adam Smith, *The Wealth of Nations* (New York: Penguin, 1982), 109–110.

45. Michael Lind, *Land of Promise: An Economic History of the United States* (New York: Harper, 2012).

46. Ibid.

47. Margaret McMillan, Dani Rodrik, and Claudia Sepulveda, "Structural Change, Fundamentals, and Growth: A Framework and Case Studies," NBER Working Paper 23378, National Bureau of Economic Research, May 2017, https://www.nber.org/papers/w23378.pdf, 10.

48. Dani Rodrik, "New Technologies, Global Value Chains, and the Developing Economies," Pathways for Prosperity Commission Background Paper 1, September 2018, https://drodrik.scholar.harvard.edu/files/dani-rodrik /files/new_technologies_global_value_chains_developing_economies. pdf, 7.

49. Robert Gellatelly and Ben Kiernan, *The Specter of Genocide: Mass Murder in Historical Perspective* (Cambridge, UK: Cambridge University Press, 2003), 290–91.

50. "Corruption Perceptions Index 2019," Transparency International, https://www.transparency.org/cpi2019.

51. "GDP per Capita (Current US$)—Indonesia," World Bank Group, https:// data.worldbank.org/indicator/NY.GDP.PCAP.CD?locations=ID.

52. Art Kleiner, "The Man Who Saw the Future," *Strategy + Business*, February 12, 2003, https://www.strategy-business.com.

53. Michael Shellenberger, "Leapfrog or Backfire?" interview with Arthur van Benthem, The Breakthrough Institute, November 18, 2014, https://thebreak through.org.

54. Ibid.

55. Ibid.

56. Ibid.

57. Roger Fouquet and Peter Pearson, "The Long Run Demand for Lighting: Elasticities and Rebound Effects in Different Phases of Economic Development," *Economics of Energy and Environmental Policy* 1, no. 1 (2012): 83–100, https://doi.org/10.5547/2160-5890.1.1.8.

58. Shellenberger, "Leapfrog or Backfire?"

59. William D. Nordhaus, "Do Real-Output and Real-Wage Measures Capture Reality? The History of Lighting Suggests Not," in *The Economics of New Goods*, edited by Timothy F. Bresnahan and Robert J. Gordon (Chicago: University of Chicago Press, 1996), 27–70, https://www.nber.org/chapters /c6064.pdf. Roger Fouquet and Peter J. G. Pearson, "Seven Centuries of Energy Services: The Price and Use of Light in the United Kingdom (1300–2000)," *The Energy Journal* 27, no. 1 (January 2006): 139–77, https://doi .org/10.2307/23296980.

60. Per capita energy data are from the International Energy Agency. Total primary energy supply (TPES) per capita of DRC and Indonesia refer to IEA data from 2017, while US data are for 2018.

61. "Household Air Pollution and Health," World Health Organization, May 8, 2018, https://www.who.int/news-room/fact-sheets/detail/household-air -pollution-and-health.

62. Eric Johnson, "Charcoal versus LPG grilling: A Carbon-Footprint Comparison," *Environmental Impact Assessment Review* 29 (November 2009): 370–78, https://doi.org/10.1016/j.eiar.2009.02.004.

63. *The Potential Air Quality Impacts from Biomass Combustion*, Air Quality Expert Group, 2017, https://uk-air.defra.gov.uk/assets/documents/reports /cat11/1708081027_170807_AQEG_Biomass_report.pdf, 37, 38.

64. Vaclav Smil, *Power Density: A Key to Understanding Energy Sources and Uses* (Cambridge, MA: MIT Press, 2015), 403.

65. "Energy Sources Have Changed Throughout the History of the United States," Today in Energy, U.S. Energy Information Administration, July 3, 2013, https://www.eia.gov/todayinenergy/detail.php?id=11951.

66. Vaclav Smil, *Power Density*.

67. FAO, "Data," FAOSTAT, accessed January 27, 2020, http://www.fao.org /faostat/en/#data. For hydro, fossil, and nuclear plant area, see Vaclav Smil, *Power Density*, 210. For land used for agriculture, FAOSTAT reports that 37 percent of the Earth's landmass, or 41 percent of ice-free landmass, is used for agriculture.

68. Wyoming and Australia produce 15,000 watts of power per square meter compared to just 0.6 watts per square meter for wood fuel from temperate forests. Vaclav Smil, *Power Density*.

69. The power density of English forests was 0.2 W/m^2, of crop residues is 0.05 W/m^2, and of Wyoming coal mines is 11,000 W/m^2. Vaclav Smil, *Power Density*.

70. Vaclav Smil, *Power Density*.

71. Ibid.

72. Jennifer Lee, "When Horses Posed a Public Health Hazard," *City Room* (blog), *New York Times*, June 9, 2008, https://cityroom.blogs.nytimes .com/2008/06/09/when-horses-posed-a-public-health-hazard.

73. Vaclav Smil, *Power Density*.

74. Xiliang Hong et al., "Exergoenvironmental Evaluation for a Coal-Fired Power Plant of Near-Zero Air Pollutant Emission," *Applied Thermal Engineering* 128 (January 5, 2018), https://doi.org/10.1016/j.applther maleng.2017.08.068. Danilo H. D. Rocha et al., "Exergoenvironmental analysis of a ultra-supercritical coal-fired power plant," *Journal of Cleaner Production* 231 (September 10, 2019): 671–682, https://doi.org/10.1016/j .jclepro.2019.05.214. Jingwei Chen et al. "Thermodynamic and environmental analysis of integrated supercritical water gasification of coal for power and hydrogen production," *Energy Conversion and Management* 198 (October 15, 2019): 1–12, https://doi.org/10.1016/j.encon man.2019.111927.

75. David Hosanky, s.v. "Flue gas treatment," *Encyclopedia Britannica*, June 27, 2014, accessed January 16, 2020, https://www.britannica.com/technology /flue-gas-treatment.

76. United States Environmental Protection Agency, "Air Quality—National Summary," July 8, 2019, accessed March 25, 2020, www.epa.gov.

77. Hannah Ritchie and Max Roser, "Energy," Our World in Data, July 2018, https://ourworldindata.org/energy.

78. Dani Rodrik, "An African Growth Miracle?," *Journal of African Economies* 27, no. 1 (December 9, 2016), 10–27, https://doi.org/10.1093/jae/ejw027.

79. Aisha Salaudeen, "Ethiopia's Garment Workers Make Clothes for Some of the World's Largest Clothing Brands but Get Paid the Lowest," CNN, May 13, 2019, https://www.cnn.com.

80. Rodrik, "An African Growth Miracle?"

81. Ibid.

82. Ibid.

83. Hinh Dinh (former World Bank economist) in discussion with the author, February 21, 2016.

84. Kerry Emanuel (MIT climate scientist) in discussion with the author, November 15, 2019.

85. Suparti (factory worker) in discussion with the author, June 8–9, 2015.

6: Greed Saved the Whales, Not Greenpeace

1. John P. Rafferty, "Titanosaurs: 8 of the World's Biggest Dinosaurs," *Encyclopaedia Britannica*, https://www.britannica.com/list/titanosaurs-8-of-the-worlds-biggest-dinosaurs.

2. Lance E. Davis, Robert E. Gallman, and Karin Gleiter, *In Pursuit of Leviathan: Technology, Institutions, Productivity, and Profits in American Whaling* (Chicago: University of Chicago Press, 1997), 21.

3. Neil Pyenson, *Spying on Whales: The Past, Present, and Future of Earth's Most Awesome Creatures* (New York: Penguin, 2018), 176.

4. Nguyen Van Vinh, "Vietnamese Fisherman Worship Whale for Safety," Reuters, June 22, 2007, https://www.reuters.com.

5. "Vietnamese Fishermen Drag Giant Whale Carcass Inland for Burial," Newsflare, May 29, 2019, https://www.newsflare.com/video/296461/animals/vietnamese-fishermen-drag-giant-whale-carcass-inland-for-burial.

6. Andrew Darby, *Harpoon: Into the Heart of Whaling* (Cambridge, MA: Hachette Books, 2009), 106.

7. George Francis Dow, *Whale Ships and Whaling: A Pictorial History* (New York: Dover Publications, 2013), 7.

8. John E. Worth, "Fontaneda Revisited: Five Descriptions of Sixteenth-

Century Florida," *The Florida Historical Quarterly* 73, no. 3 (January 1995): 339–352, https://www.jstor.org/stable/30150454.

9. Davis et al., *In Pursuit of Leviathan.*

10. Johan Nicolay Tønnessen and Arne Odd Johnsen, *The History of Modern Whaling* (Berkeley: University of California Press, 1982), 129.

11. Taylor Kate Brown, "Hunting Whales with Rowing Boats and Spears," BBC, April 26, 2015, https://www.bbc.com.

12. Glover Morrill Allen, *The Whalebone Whales of New England* (Boston: Printed for the Society with Aid from the Gurdon Saltonstall Fund, 1916), 161.

13. Walter S. Tower, *A History of the American Whale Fishery* (Philadelphia: University of Pennsylvania, 1907), 50.

14. Davis et al., *In Pursuit of Leviathan.*

15. Johan Nicolay Tønnessen and Arne Odd Johnsen, *The History of Modern Whaling* (Berkeley: University of California Press, 1982), 129.

16. "Development of the Pennsylvania Oil Industry," National Historic Chemical Landmark, American Chemical Society, 2009, https://www.acs.org/content/acs/en/education/whatischemistry/landmarks/pennsylvaniaoilindustry.html.

17. "Samuel Kier," Pennsylvania Center for the Book, 2007, accessed December 5, 2019, https://www.pabook.libraries.psu.edu/literary-cultural-heritage-map-pa/bios/Kier__Samuel_Martin.

18. In 1845. W. S. Tower, *A History of the American Whale Fishery* (Philadelphia, PA: University of Pennsylvania, 1907), 127.

19. "Development of the Pennsylvania Oil Industry."

20. Davis et al., *In Pursuit of Leviathan*, 355–63. Had petroleum not replaced whale oil, it is possible that coal gas would have. Firms in New York, Boston, New Orleans, and Philadelphia copied the British in the early 1800s and started making hydrogen gas and pumping it into homes for lighting. They made the hydrogen from coal. "Coal was cheaper than whale oil," the historians noted, but it had other problems that whale oil didn't have. Those might have been worked out over time, but then petroleum came along and provided a cheap and better alternative to both whale oil and coal gas in most places.

    For a few years, and in some places, camphene, a kind of redistilled turpentine, also competed with whale oil. In other places, lard oil seemed cheaper, and of equal quantity. For a brief period it replaced sperm (whale) oil in US lighthouses. But eventually it was replaced by petroleum, which came to dominate the market. The historians concluded that "On the eve of the discovery of petroleum at Drake's well, then, the whaling industry was already in a kind of re-treat." Even so, the big-picture conclusion of historians remains the same today as it was 160 years ago. As the historians noted, "the birth of a large-scale petro-leum industry signaled the eventual death of the American whaling industry."

Davis et al.'s is the consensus view. "For lubrication and lighting, whale oil was squeezed out of the market by mineral oil [petroleum] and for various other purposes it could be replaced by vegetable oils, in particular cottonseed and linseed oil."

21. "Grand Ball Given by Whales," *Vanity Fair*, April 20, 1861.

22. Davis et al., *In Pursuit of Leviathan*, 516.

23. Ibid., 514.

24. Ibid., 498–499, 522. "Lower wages, lower opportunity costs of capital, and a lack of entrepreneurial alternatives pushed the Norwegians into exploiting the whale stocks. Higher wages, higher opportunity costs of capital, and a pleth-ora of entrepreneurial alternatives turned Americans—even those from New Bedford—toward the domestic economy."

Johan Nicolay Tønnessen and Arne Odd Johnsen, *The History of Modern Whaling* (Berkeley: University of California Press, 1982), 12. Again, there is agreement on this point. "In the United States there was simply no induce-ment for a regeneration of whaling."

25. Ibid., 508.

26. Ibid., 11.

27. "New Butter Substitute," *New York Times*, August 4, 1918, https://www.ny times.com. By 1929, scientists had improved the process so much that marga-rine could be made from just whale oil. Davis et al., *In Pursuit of Leviathan*, 503.

28. Tønnessen and Johnsen, *The History of Modern Whaling*, 529.

29. Walter Sullivan, "Extinction of Blue Whale Feared," *New York Times*, May 18, 1959, https://www.nytimes.com.

30. Ibid.

31. "Whaler Tells of Industry Shift from Oil to Production of Food," *New York Times*, May 12, 1968, https://www.nytimes.com.

32. Viktoria Schneider and David Pearce, "What Saved the Whales? An Eco-nomic Analysis of 20th Century Whaling," *Biodiversity and Conservation* 13 (2004): 453–562, https://doi.org/10.1023/B:BIOC.0000009489.08502.1a.

33. Ibid.

34. Ibid.

35. Tønnessen and Johnsen, *The History of Modern Whaling*, 136.

36. Viktoria Schneider and David Pearce, "What Saved the Whales? An Economic Analysis of 20th Century Whaling."

37. Davis et al., *In Pursuit of Leviathan*, 512.

38. Linus Blomqvist, Ted Nordhaus, and Michael Shellenberger, *Nature Unbound: Decoupling for Conservation*, The Breakthrough Institute, 2015, accessed December 5, 2019, 29, https://s3.us-east-2.amazonaws.com/uploads.thebreak through.org/legacy/images/pdfs/Nature_Unbound.pdf.

39. Schneider and Pearce, "What Saved the Whales? An Economic Analysis of 20th Century Whaling." The economists stress that there can still be a role for environmental regulation. "While international agreements may have limited effectiveness, there is plenty of evidence to suggest that domestic policies are potent ways of 'tunnelling through' the Environmental Kuznets Curve."

40. J. C. Fisher and R. H. Pry, "A Simple Substitution Model of Technological Change," *Technological Forecasting and Social Change* 3 (1971–72): 75–88, https://doi.org/10.1016/S0040-1625(71)80005-7.

41. Cesare Marchetti, "A Personal Memoir: From Terawatts to Witches," *Technological Forecasting and Social Change* 37, no. 4 (1990): 409–414, https://doi .org/10.1016/0040-1625(90)90049-2.

42. Ibid.

43. C. Marchetti and N. Nakićenović, *The Dynamics of Energy Systems and the Logistic Substitution Model*, International Institute for Applied Systems Analysis, December 1979, http://pure.iiasa.ac.at/id/eprint/1024/1/RR-79-013.pdf.

44. Marchetti, "A Personal Memoir: From Terawatts to Witches."

45. Cesare Marchetti, "Primary Energy Substitution Models: On the Interaction Between Energy and Society," *Technological Forecasting and Social Change* 10, no. 4 (1977): 345–56, https://doi.org/10.1016/0040-1625(77)90031-2.

46. Marchetti and Nakićenović, *The Dynamics of Energy Systems and the Logistic Substitution Model*.

47. Michael Williams, *Deforesting the Earth: From Prehistory to Global Crisis, An Abridgment* (Chicago: University of Chicago Press, 2006), 162.

48. Marchetti, "Primary Energy Substitution Models."

49. Richard Rhodes, *Energy: A Human History* (New York: Simon & Schuster, 2018).

50. Cesare Marchetti, "Primary Energy Substitution Models."

51. In liquid form, natural gas has 54 MJ/kg, but liquefaction of natural gas requires significant amounts of energy. Smil, *Power Density*, 13.

52. Ibid., 13, 122. The original calculations of the carbon-to-hydrogen ratio for fossil fuels are from Nebojša Nakićeović, "Freeing Energy from Carbon," *Daedalus* 125, no. 3 (Summer 1996), 95–112, http://pure.iiasa.ac.at /id/eprint/4850/1/RR-97-04.pdf. Natural gas is typically made up of methane, ethane, and propane. Methane comprises 73–95 percent of the natural gas that comes out of the earth and produces 35 MJ/kg compared to coal's 25 MJ/kg.

53. Nebojša Nakićeović, "Freeing Energy from Carbon," 99.

54. Tyler Hodge, "EIA Forecasts Slower Growth in Natural Gas–Fired Generation While Renewable Energy Rises," Today in Energy, U.S. Energy Information Administration, January 16, 2020, https://www.eia.gov/todayinenergy /detail.php?id=42497.

55. *BP Statistical Review of World Energy 2019, 68th Edition*, BP, June 2019, https://www.bp.com/content/dam/bp/business-sites/en/global/cor porate/pdfs/energy-economics/statistical-review/bp-stats-review-2019 -full-report.pdf. "Access to Clean Cooking," SDG7: Data and Projections, International Energy Agency, November 2019, https://www.iea.org/re ports/sdg7-data-and-projections/access-to-clean-cooking. *Energy Access Outlook 2017*, International Energy Agency, October 2017, https://webstore .iea.org/download/direct/274. In 2018, coal provided 27 percent of primary energy, not the 4 percent Marchetti had predicted. Natural gas provided 24 percent, not the 65 percent he had projected. Nuclear power provided 4 percent, not the 10 percent he had forecast.

56. Ibid.

57. *Gasland*, written and directed by Josh Fox, HBO, 2010.

58. Ian Urbina, "Regulation Lax as Gas Wells' Tainted Water Hits Rivers," *New York Times*, February 26, 2011, https://www.nytimes.com.

59. State of Colorado Oil and Gas Conservation Commission, Department of Natural Resources, *Gasland* (Denver: Department of Natural Resources, June 2013), accessed January 15, 2020, https://www.energyindepth.org /wp-content/uploads/2018/02/GASLAND-DOC.pdf.

60. Ibid.

61. kwashnak, "The Famous Burning Well of Colfax—Louisiana Historical Markers," Waymarking, accessed November 8, 2019, https://www.waymarking .com/waymarks/WM8XQ8_The_Famous_Burning_Well_of_Colfax.

62. Phelim McAleer, Ann McElhinney, and Magdalena Segieda, dirs., *FrackNation* (Ann and Phelim Media, 2013), quoted in Forrest M. Mims III, "How

flaming tap water ignited the fracking controversy: Gasland vs. FrackNation," *The Quad*, September 18, 2018, https://thebestschools.org/magazine/flam ing-tap-water-fracking-controversy-gasland-fracknation.

63. Forrest M. Mims III, "How flaming tap water ignited the fracking controversy: Gasland vs. FrackNation."

64. Bill McKibben, "Bad News for Obama: Fracking May Be Worse Than Coal," *Mother Jones*, September 8, 2014, https://www.motherjones.com; Bill Mc-Kibben, "The Literal Gaslighting That Helps America Avoid Acting on the Climate Crisis," *New Yorker*, October 9, 2019, https://www.newyorker .com.

65. Paulina Jaramillo, "Landfill-Gas-to-Energy Projects? Analysis of Net Private and Social Benefits," *Environmental Science and Technology* 39, no. 19 (2005): 7365–7373, https://doi.org/10.1021/es050633j.

66. Anil Markandya and Paul Wilkinson, "Electricity Generation and Health," *The Lancet* 370, no. 9501 (September 15, 2007): 979–90, https://doi .org/10.1016/S0140-6736(07)61253-7.

67. Bridget R. Scanlon, Robert C. Reedy, Ian Duncan et al., "Controls on Water Use for Thermoelectric Generation: Case Study Texas, US," *Environmental Science & Technology* 47, no. 19 (2013): 11326–34, https://doi.org/10.1021 /es4029183.

68. BP Energy Economics, "BP Statistical Review of World Energy 2019, 68th Edition," BP, June 2019, accessed January 16, 2020, https://www.bp.com/con tent/dam/bp/business-sites/en/global/corporate/pdfs/energy-economics /statistical-review/bp-stats-review-2019-full-report.pdf; Perry Lindstrom, "Carbon dioxide emissions from the US power sector have declined 28% since 2005," EIA, October 29, 2018, https://www.eia.gov/todayinenergy/detail .php?id=37392.

69. Sophie Dejonckheere, Mari Aftret Mørtvedt, and Eilif Ursin Reed, "Methane: A Climate Blind Spot?," Center for International Climate Research, March 25, 2019, https://www.cicero.oslo.no/en/posts/klima/methane-a -climate-blind-spot. Zeke Hausfather, "Bounding the Climate Viability of Natural Gas as a Bridge Fuel to Displace Coal," *Energy Policy* 86 (November 2015): 286–94, https://doi.org/10.1016/j.enpol.2015.07.012. Adam Voiland, "Methane Matters," National Air and Space Association, https:// earthobservatory.nasa.gov/features/MethaneMatter. Gunnar Mhyre and Drew Shindell, "Anthropogenic and Natural Radiative Forcing," in *Climate Change 2013: The Physical Science Basis*, Intergovernmental Panel on Climate

Change, 2013, 659–740, http://www.climatechange2013.org/images/re
port/WG1AR5_Chapter08_FINAL.pdf.

Most of the methane molecules that leak into the atmosphere today won't
be there in ten years. By contrast, most carbon dioxide will remain in the atmosphere for centuries. As a result, even if methane were leaking at a higher
rate than the U.S. EPA estimates, as some claim it is, its impact on global
warming would still be relatively small compared to the benefits of reduced
carbon emissions compared to coal.

70. Kevin Begos, "EPA Methane Report Further Divides Fracking Camps,"
    Yahoo! News, April 28, 2013, https://news.yahoo.com.

71. JenAlyse Arena, "Coal production using mountaintop removal mining de-
    creases by 62% since 2008," U.S. Energy Information Association, July 7,
    2015, https://www.eia.gov/todayinenergy/detail.php?id=21952.

72. Appalachian Voices, "Ecological Impacts of Mountaintop Removal," ac-
    cessed January 16, 2020, http://appvoices.org/end-mountaintop-removal
    /ecology.

73. Editorial Board, "The dirty effects of mountaintop removal mining," *Wash-
    ington Post*, October 21, 2014, https://www.washingtonpost.com.

74. Richard Schiffman, "A Troubling Look at the Human Toll of Mountaintop
    Removal Mining," Yale Environment 360, November 21, 2017, https://e360
    .yale.edu/features/a-troubling-look-at-the-human-toll-of-mountaintop-re
    moval-mining.

75. On average oil power densities are far less. Iraq's oil fields have a power
    density of only 5,000 W/m^2. But that is still twice as high as Australia's
    coal mines. As always there is a large range, with some petroleum fields
    producing as little as 100 W/m^2. A typical natural gas well in Alberta,
    Canada, has a power density of 2,300 W/m^2 while the Netherlands' gas
    fields have a power density of 16,000 W/m^2. A Liquified Natural Gas
    (LNG) terminal has a power density of 4,600 W/m^2 while a regasification
    terminal has a power density of an astonishing 60,000 W/m^2. Smil, *Power
    Density*, 197.

76. Chris Mooney, "Why We're Still So Incredibly Confused About Methane's
    Role in Global Warming," *Washington Post*, May 2, 2016, https://www.wash
    ingtonpost.com.

77. "Our salmon," AquaBounty, accessed October 6, 2019, https://aquabounty
    .com/our-salmon.

78. "Sustainable," AquaBounty, accessed February 2, 2020, https://aquabounty .com/sustainable.

79. "FDA Has Determined That the AquAdvantage Salmon Is as Safe to Eat as Non-GE Salmon," U.S. Food and Drug Administration, January 19, 2016, https://www.fda.gov/consumers/consumer-updates/fda-has-determined -aquadvantage-salmon-safe-eat-non-ge-salmon.

80. FAO, "The State of World Fisheries and Aquaculture," FAO, 2016, accessed January 16, 2020, http://www.fao.org/3/a-i5555e.pdf.

81. UNEP-WCMC, IUCN, and NGS, *Protected Planet Report 2018* (Cambridge, UK; Gland, Switzerland; and Washington, D.C.: UNEP-WCMC, IUCN, and NGS, 2018), https://livereport.protectedplanet.net/chapter-2.

82. FAO, "The State of World Fisheries and Aquaculture," FAO, 2016, accessed January 16, 2020, http://www.fao.org/3/a-i5555e.pdf.

83. Richard Waite et al., "Improving Productivity and Environmental Performance of Aquaculture," World Resources Institute, June 2014, accessed January 16, 2020, https://www.wri.org/publication/improving-aquaculture.

84. FAO, "The State of World Fisheries and Aquaculture" FAO, 2016, accessed January 16, 2020, http://www.fao.org/3/a-i5555e.pdf/; FAO, "The State of World Fisheries and Aquaculture," FAO, 2014, accessed January 16, 2020, http://www.fao.org/3/a-i3720e.pdf.

85. U.S. Soybean Export Council, "Aquaculture Is Fastest Growing Food Production Sector, According to FAO Report," July 16, 2018, accessed January 15, 2020, https://ussec.org/aquaculture-fastest-growing-food-production-sec tor-fao-report.

86. Martin Heller and Center for Sustainable Systems, University of Michigan, *Food Product Environmental Footprint Literature Summary: Land-Based Aquaculture*, State of Oregon Department of Environmental Quality, September 2017, https://www.oregon.gov/deq/FilterDocs/PEF-Aquaculture-FullReport .pdf.

87. Fong Yang Looi et al., "Creating Disease Resistant Chickens: A Viable Solution to Avian Influenza?" *Viruses* 10, no. 10 (October 2018): 561, https://doi .org/10.3390/v10100561; Bruce Y. Lee, "Can Gene Editing Stop the Bird Flu? Here Is the Latest With Chickens," *Forbes,* June 5, 2019, https://www .forbes.com.

88. William Robert Irvin et al., letter to Margaret A. Hamburg, M.D., Commis-

sioner of Food and Drugs, FDA, "Re: Draft Environmental Assessment and Preliminary Finding of No Significant Impact Concerning a Genetically Engineered Atlantic Salmon," April 26, 2013, https://www.nrdc.org/sites/default/files/hea_13042901a.pdf.

89. Center for Food Safety, "FDA approves first genetically engineered animal for human consumption over the objections of millions," Center for Food Safety, November 19, 2015, https://www.centerforfoodsafety.org/press-releases/4131/fda-approves-first-genetically-engineered-animal-for-human-consumption-over-the-objections-of-millions.

90. Sustainable Business, "Whole Foods, Trader Joe's say no to genetically modified fish" GreenBiz, March 21, 2013, https://www.greenbiz.com.

91. Alister Doyle, "Mangroves under threat from shrimp farms: UN," Reuters, November 14, 2012, https://www.reuters.com.

92. Michael Shellenberger, "Victory for Frankenfish and for Mother Earth," *USA Today*, November 19, 2015, https://www.usatoday.com.

93. Ibid.

94. AquaBounty, "Ask your supermarket to stock GMO salmon," AquaBounty, March 20, 2019, accessed October 8, 2019, https://aquabounty.com/ask-your-supermarket-to-stock-gmo-salmon.

95. Jesse Ausubel (environmental scientist) in conversation with the author, November 6, 2019.

96. Richard Rhodes, "Yale Environment 360: Why nuclear power must be a part of the energy solution," *Anthropocene Institute*, July 19, 2018, https://www.anthropoceneinstitute.com/2018/07/yale-environment-360-why-nuclear-power-must-be-part-of-the-energy-solution.

97. R.B. Allen, "Backward into the future: The shift to coal and implications for the next energy transition," *Energy Policy* 50 (2012): 17–23, https://doi.org/10.1016/j.enpol.2012.03.020.

98. Jesse Ausubel (environmental scientist) in conversation with the author, November 6, 2019.

99. Robert C. Rocha, Jr., Phillip J. Clapham, and Yulia Ivaschenko, "Emptying the Oceans: A Summary of Industrial Whaling Catches in the 20th Century," *Marine Fisheries Review* 76 (March 2015): 46, https://dx.doi.org/10.7755/MFR.76.4.3.

100. Walter Sullivan, "Extinction of Blue Whale Feared," *New York Times*, May 18, 1959, https://www.nytimes.com.

7: Have Your Steak and Eat It, Too

1. Jonathan Safran Foer, *Eating Animals* (New York: Little, Brown, 2009), 3.
2. Isabelle Gerretsen, "Change food production and stop abusing land, major climate report warns," CNN, August 8, 2019, https://www.cnn.com.
3. Jen Christensen, "To help save the planet, cut back to a hamburger and a half per week," CNN, July 17, 2019, https://www.cnn.com.
4. Quirin Schiermeier, "Eat Less Meat: UN climate-change report calls for change to human diet," *Nature*, August 8, 2019, https://doi.org/10.1038 /d41586-019-02409-7.
5. Ibid.
6. Cheikh Mbow et al., "Food Security," in Valerie Masson-Delmont et al. (eds.), *Climate Change and Land: An IPCC special report on climate change, desertification, land degradation, sustainable land management, food security, and greenhouse gas fluxes in terrestrial ecosystems*, IPCC, 2019, accessed January 21, 2020, https://www.ipcc.ch/srccl.
7. Abigail Abrams, "How Eating Less Meat Could Help Protect the Planet From Climate Change," *Time*, August 8, 2019, https://time.com.
8. Katia Dmitrieva, "The Green New Deal Progressives Really Are Coming For Your Beef," Bloomberg, March 13, 2019, https://www.bloomberg .com.
9. Livestock, Dairy, and Poultry Outlook, LDP-M-299, U.S. Department of Agriculture, Economic Research Service, May 16, 2019.
10. Jen Christensen, "To help save the planet, cut back to a hamburger and a half per week," CNN, July 17, 2019, https://www.cnn.com.
11. Foer, *Eating Animals*; "Why Go Organic, Grass-Fed and Pasture-Raised?" Meat Eater's Guide to Climate Change Health, Environmental Working Group, 2011, accessed January 21, 2020, www.ewg.org/meateatersguide.
12. Cheikh Mbow et al., "Food Security," in *Climate Change and Land: An IPCC special report on climate change, desertification, land degradation, sustainable land management, food security, and greenhouse gas fluxes in terrestrial ecosystems* (IPCC, 2019), 487.
13. Janina Grabs, "The rebound effects of switching to vegetarianism. A micro-economic analysis of Swedish consumption behavior," *Ecological Economics* 116 (2015): 270–279, https://doi.org/10.1016/j.ecolecon.2015.04.030.
14. Cheikh Mbow et al., "Food Security," in *Climate Change and Land: An IPCC*

special report on climate change, desertification, land degradation, sustainable land management, food security, and greenhouse gas fluxes in terrestrial ecosystems (IPCC, 2019), 487. In "business-as-usual," global greenhouse emissions will rise to 86 gigatons/year by 2050, and emissions from agriculture will rise to 11.6 gigatons/year. The "upper-bound" scenario of 100 percent veganism would reduce emissions by 8.1 gigatons/year from this baseline.

15. Gidon Eshel, "Environmentally Optimal, Nutritionally Sound, Protein and Energy Conserving Plant Based Alternatives to US Meat," *Nature: Scientific Reports* 9, no. 10345 (August 8, 2019), https://doi.org/10.1038/s41598-019-46590-1.

16. Elinor Hallström et al., "Environmental impact of dietary change: A systematic review," *Journal of Cleaner Production* 91 (March 15, 2015), https://doi.org/10.1016/j.jclepro.201y4.12.008. The best estimate of emissions reductions of going vegetarian was 540kg, while average developed nation CO_2 (Annex I) is 12.44t CO_2.

17. Robin R. White and Mary Beth Hall, "Nutritional and greenhouse gas impacts of removing animals from US agriculture," *Proceedings of the National Academy of Sciences* 114, no. 48 (2017), https://doi.org/10.1073/pnas.1707322114.

18. Janina Grabs, "The rebound effects of switching to vegetarianism. A microeconomic analysis of Swedish consumption behavior," *Ecological Economics* 116 (2015): 270–279, https://doi.org/10.1016/j.ecolecon.2015.04.030.

19. FAO, "Livestock Primary," FAOSTAT, http://www.fao.org/faostat/en/#data/QL.

20. Foer, *Eating Animals*, 129.

21. FAO, "Land Use," FAOSTAT, accessed January 27, 2020, http://www.fao.org/faostat/en. To be exact, 1.42 million km².

22. FAO, *World Livestock: Transforming the livestock sector through the Sustainable Development Goals* (Rome: FAO, 2018), Licence: CC BY-NC-SA 3.0 IGO, http://www.fao.org/3/CA1201EN/ca1201en.pdf.

23. Charles Stahler, "How many people are vegan?" Vegetarian Resource Group, based on March 7–11, 2019, Harris poll, accessed December 31, 2019, https://www.vrg.org/nutshell/Polls/2019_adults_veg.htm; Kathryn Asher et al., "Study of Current and Former Vegetarians and Vegans: Initial Findings, December 2014," Humane Research Council and Harris International, accessed

October 30, 2019, https://faunalytics.org/wp-content/uploads/2015/06/Faunalytics_Current-Former-Vegetarians_Full-Report.pdf.

24. FAO, "Land Use," FAOSTAT, accessed January 27, 2020, http://www.fao.org/faostat/en.

25. A. Shepon et al., "Energy and Protein Feed-to-Food Conversion Efficiencies in the US and Potential Food Security Gains from Dietary Changes," *Environmental Research Letters* 11, no. 10 (2016): 105002, https://doi.org/10.1088/1748-9326/11/10/105002. Beef has a protein conversion efficiency of 2.5 percent, pork of 9 percent, and poultry of 21 percent.

26. Vaclav Smil, *Should We Eat Meat? Evolution and Consequences of Modern Carnivory* (Oxford: John Wiley & Sons, Ltd., 2013), 92. When the U.S. started producing chickens indoors, in 1925, it took 112 days for one to reach maturity. By 1960 it took half as long and chickens gained one-third more weight. By 2017, chickens grown indoors only required just 48 days to reach maturity and they had more than doubled in weight since 1925.

27. FAO, "Data," FAOSTAT, http://www.fao.org/faostat/en/#data/RL, cited in Frank Mitloehner, "Testimony before the Committee on Agriculture, Nutrition and Forestry US Senate," May 21, 2019, accessed November 3, 2019, https://www.agriculture.senate.gov/imo/media/doc/Testimony_Mitloehner_05.21.2019.pdf.

28. Foer, *Eating Animals*, 196.

29. Ibid.

30. Durk Nijdam, Geertruida Rood, and Henk Westhoek, "The price of protein: Review of land use and carbon footprints from life cycle assessments of animal food products and their substitutes," *Food Policy* 37 (2012): 760–770, https://doi.org/10.1016/j.foodpol.2012.08.002.

31. David Gustafson et al., "Climate adaptation imperatives: Global sustainability trends and eco-efficiency metrics in four major crops—canola, cotton, maize, and soybeans," *International Journal of Agricultural Sustainability* 12 (2014): 146–163, https://doi.org/10.1080/14735903.2013.846017.

32. Durk Nijdam, Geertruida Rood, and Henk Westhoek, "The price of protein: Review of land use and carbon footprints from life cycle assessments of animal food products and their substitutes," *Food Policy* 37 (2012): 760–770, https://doi.org/10.1016/j.foodpol.2012.08.002.

"Production of 1 kg of extensively farmed beef results in roughly three to four times as many greenhouse gas emissions as the equivalent amount of intensively farmed beef."

33. C.D. Lupo, D. E. Clay, J. L. Benning, and J. J. Stone, "Life-Cycle Assessment of the Beef Cattle Production System for the Northern Great Plains, USA," *Journal of Environment Quality*, 42(5), 1386, doi:10.2134/jeq2013.03.0101.

34. M. de Vries, C. E. van Middelaar, and I. J. M. de Boer, "Comparing environmental impacts of beef production systems: A review of life cycle assessments," *Livestock Science* 178 (August 2015): 279–288, https://doi.org/10.1016/j.livsci.2015.06.020.

35. Foer, *Eating Animals*, 95–96.

36. Nina Teicholz, *The Big Fat Surprise: Why Butter, Meat and Cheese Belong in a Healthy Diet* (New York: Simon & Schuster, 2014), 1.

37. Ibid.

38. Ibid., 3.

39. Ibid., 2.

40. William R. Leonard, J. Josh Snodgrass, and Marcia L. Robertson, "Effects of brain evolution on human nutrition and metabolism," *Annual Review of Nutrition* 27 (2007): 311–327, https://doi.org/10.1146/annurev.nutr.27.061406.093659.

41. Teicholz, *The Big Fat Surprise: Why Butter, Meat and Cheese Belong in a Healthy Diet*, 1.

42. Marshall E. McCollough, *Optimum Feeding of Dairy Animals: For Milk and Meat* (Athens: University of Georgia Press, 1973), 64–92.

43. Temple Grandin, "Avoid Being Abstract When Making Policies on the Welfare of Animals," in *Species Matters: Humane Advocacy and Cultural Theory*, edited by Marianne DeKoven and Michael Lundblad (New York: Columbia University Press, 2010), 206.

44. Roman Pawlak, Scott James Parrott, Sudha Raj et al., "How Prevalent Is Vitamin B_{12} Deficiency Among Vegetarians?," *Nutrition Reviews* 71, no. 2 (February 2013): 110–17, https://doi.org/10.1111/nure.12001.

45. Matthew B. Ruby, "Vegetarianism. A blossoming field of study," *Appetite* 58 (2012): 144, https://doi.org/10.1016/j.appet.2011.09.019.

46. "What Percentage of Americans Are Vegetarian?" Gallup, 2018; Ruby, "Vegetarianism. A blossoming field of study," 144; Maria Chiorando, " 'Mind-Blowing' Rates of Vegans and Vegetarians Are Young People, Says New Poll," *Plant-Based News*, March 14, 2018

47. Ruby, "Vegetarianism. A blossoming field of study," 144.

48. Ines Testoni, "Representations of Death Among Italian Vegetarians: Eth-nographic Research on Environment, Disgust and Transcendence," *European Journal of Psychology* 13, no. 3 (August 2017): 378–395, https://doi.org/10.5964/ejop.v13i3.1301.

49. Foer, *Eating Animals*, 143.

50. Ibid.

51. Nathanael Johnson, "Temple Grandin Digs In on the Practical Side of What Animals Want," *Grist*, July 22, 2015, https://grist.org.

52. Ibid.

53. David Bianculli, "Temple Grandin: The Woman Who Talks to Animals," NPR, February 5, 2010, https://www.npr.org.

54. Foer, *Eating Animals*, 196.

55. Johnson, "Temple Grandin digs in on the practical side of what animals want."

56. Bianculli, "Temple Grandin: The Woman Who Talks to Animals."

57. Temple Grandin, "Behavioral Principles of Livestock Handling," *Applied Animal Science* 5, no. 2 (December 1989): 1–11, https://doi.org/10.15232/S1080-7446(15)32304-4. Foer, *Eating Animals*, 158.

58. Temple Grandin, "Avoid Being Abstract When Making Policies on the Welfare of Animals," in *Species Matters: Humane Advocacy and Cultural Theory*, edited by Marianne DeKoven and Michael Lundblad (New York: Columbia University Press, 2010), 195–217.

59. Foer, *Eating Animals*, 230, 256.

60. Johnson, "Temple Grandin digs in on the practical side of what animals want."

61. Foer, *Eating Animals*, 213.

62. Ibid.

63. Temple Grandin and Catherine Johnson, *Animals Make Us Human: Creating the Best Life for Animals* (Boston: Houghton-Mifflin Harcourt, 2009), 297.

64. Tom Levitt, "Jonathan Safran Foer: If You Care About Climate Change, Cut Out Meat," *Huffington Post*, October 3, 2019, https://www.huffpost.com.

65. Foer, *Eating Animals*, 189–96.

66. Gary Taubes in discussion with the author, November 1, 2019.

67. Foer, *Eating Animals*, 211.

68. Ibid., 192.

69. Ibid., 189–190.

70. Michael Pollan, *The Omnivore's Dilemma: A Natural History of Four Meals* (New York: Penguin, 2007), 361–362.

71. Foer, *Eating Animals*, 97.

72. Robert DuBroff and Michel de Lorgeril, "Fat or fiction: The diet-heart hypothesis," *British Medical Journal: Evidence-Based Medicine*, May 29, 2019, https://doi.org/10.1136/bmjebm-2019-111180.

73. Mi Ah Han, Dena Zeraatkar, Gordon H. Guyatt et al., "Reduction of Red and Processed Meat Intake and Cancer Mortality and Incidence: A Systematic Review and Meta-analysis of Cohort Studies," *Annals of Internal Medicine* 171, no. 10 (October 2019): 711–20, https://doi.org/10.7326/M19-0699. Dena Zeraatkar, Mi Ah Han, Gordon H. Guyatt et al., "Red and Processed Meat Consumption and Risk for All-Cause Mortality and Cardiometabolic Outcomes: A Systematic Review and Meta-analysis of Cohort Studies," *Annals of Internal Medicine* 171, no. 10 (October 2019): 703–10, https://doi.org/10.7326/M19-0655.

74. Aaron E. Carroll, "The Real Problem with Beef," *New York Times*, October 1, 2019, https://www.nytimes.com.

75. Smil, *Should We Eat Meat? Evolution and Consequences of Modern Carnivory*, 147. The difference is significant: "conversion of metabolizable energy to protein in pork peaks at about 45 percent, while the conversion of feed energy to fat can have efficiencies in excess of 70 percent."

76. Ibid., 132; U.S. Food and Drug Administration, "Steroid Hormone Implants Used for Growth in Food-Producing Animals," FDA, July 23, 2019, https://www.fda.gov/animal-veterinary/product-safety-information/steroid-hormone-implants-used-growth-food-producing-animals.

77. David S. Wilkie et al., "The Empty Forest Revisited," *Annals of the New York Academy of Sciences* 1223, no. 1 (2011): 120–128, https://doi.org/10.1111/j.1749-6632.2010.05908.x.

78. John E. Fa, Carlos A. Peres, and Jessica J. Meeuwig, "Bushmeat Exploitation in Tropical Forests: An Intercontinental Comparison," *Conservation Biology* 16 (2002): 232–237, https://doi.org/10.1046/j.1523-1739.2002.00275.x.

79. World Health Organization, "Availability and changes in consumption of meat products," accessed January 23, 2020, https://www.who.int/nutrition/topics/3_foodconsumption/en/index4.html.

80. John E. Fa, Carlos A. Peres, and Jessica J. Meeuwig, "Bushmeat Exploitation in Tropical Forests: An Intercontinental Comparison," *Conservation Biology* 16, no. 1 (January 2002): 232–37, https://doi.org/10.1046/j.1523 -1739.2002.00275.x.

81. Annette Lanjouw in discussion with the author, November 21, 2014.

82. The world average yield for chicken is 14 m²/kg, for pork 17 m²/kg, and for beef 43 m²/kg. Fa, et al., "Bushmeat Exploitation in Tropical Forests"; 5 million tons of bushmeat extracted in the Congo and Amazon basins: Emiel V. Elferink and Sanderine Nonhebel, "Variations in Land Requirements for Meat Production," *Journal of Cleaner Production* 15, no. 18 (2007): 1778–86, https://doi.org/10.1016/j.jclepro.2006.04.003.

83. Ricardo Grau, Nestor Gasparri, and T. Mitchell Aide, "Balancing Food Production and Nature Conservation in the Subtropical Dry Forests of NW Argentina," *Global Change Biology* 14, no. 5 (2008): 985–97, https://doi .org/10.1111/j.1365-2486.2008.01554.x.

84. Foer, *Eating Animals*, 263.

85. Ibid., 197.

86. Matthew B. Ruby, "Vegetarianism. A blossoming field of study," 142.

87. Foer, *Eating Animals*, 9.

88. Ibid., 198.

89. Ingrid Newkirk, "Temple Grandin: Helping the Animals We Can't Save," PETA, February 10, 2010, accessed October 12, 2019, https://www.peta.org /blog/temple-grandin-helping-animals-cant-save.

8: Saving Nature Is Bomb

1. Atsushi Komori, "Think Tank Puts Cost to Address Nuke Disaster up to 81 Trillion Yen," *Asahi Shimbun*, March 10, 2019, http://www.asahi.com/ajw /articles/AJ201903100044.html.

2. David Keohane and Nathalie Thomas, "EDF Increases Hinkley Point C Nuclear Plant Costs," *Financial Times*, September 25, 2019, http://www .ft.com.

3. Darrell Proctor, "Georgia PSC Backs Additional Costs for Vogtle Nuclear Project," *Power Magazine*, February 19, 2019, https://www.powermag.com /georgia-psc-backs-additional-costs-for-vogtle-nuclear-project.

4. "Olkiluoto 3 Reactor Delayed Yet Again, Now 12 Years Behind Schedule,"

Uutiset, December 23, 2019, https://yle.fi; Jussi Rosendahl and Tuomas For-
sell, "Areva's Finland Reactor to Start in 2019 After Another Delay," Reuters,
October 9, 2017, https://www.reuters.com.

5. Ran Fu, David Feldman, and Robert Margolis, "US Solar Photovoltaic System Cost Benchmark: Q1 2018," National Renewable Energy Laboratory,
September 2018, https://www.nrel.gov/docs/fy19osti/72399.pdf. Ryan
Wiser and Mark Bolinger, *2018 Wind Technologies Market Report,* US Department of Energy, 2018, https://www.energy.gov/sites/prod/files/2019/08
/f65/2018%20Wind%20Technologies%20Market%20Report%20Summary
.pdf.

6. IAEA, *Power Reactor Information System [PRIS],* IAEA, accessed January 31,
2020, https://www.iaea.org/resources/databases/power-reactor-informa
tion-system-pris. Prior to 1979, average construction time was 5.8 years in the
U.S., and 6.7 years in France. Reactor constructions after 1979 took on average 12.5 years and 14 years, respectively.

7. Gerry Thomas (professor of clinical pathology at Imperial College, London,
and founder of the Chernobyl Tissue Bank) in discussion with the author,
July 31, 2017.

8. United Nations Scientific Committee on the Effects of Atomic Radiation,
Sources and Effects of Ionizing Radiation, Vol. II (New York: United Nations,
2011), 149, https://www.unscear.org/unscear/publications.html.

9. "Firefighter fatalities in the United States," U.S. Fire Administration, accessed January 16, 2020, https://apps.usfa.fema.gov/firefighter-fatalities.

10. "Evaluation of Data on Thyroid Cancer in Regions Affected by the Chernobyl Accident," United Nations Scientific Committee on the Effects of
Atomic Radiation, https://www.unscear.org/unscear/publications.html, v.

11. Elisabeth Cardis and Maureen Hatch, "The Chernobyl Accident—an Epi-
demiological Perspective," *Clinical Oncology* 23, no. 4 (May 2011): 251–60,
https://doi.org/10.1016/j.clon.2011.01.510.

12. *Chernobyl,* episode 1, "1:23:45," directed by Johan Renck, HBO, May 6, 2019.

13. "Health Effects of the Chernobyl Accident: An Overview," World Health
Organization, April 2006, https://www.who.int/ionizing_radiation/cher
nobyl/backgrounder/en.

14. "Fact Sheet: Biological Effects of Radiation," Office of Public Affairs, United
States Nuclear Regulatory Commission, 2003, http://akorn.com/docu
ments/pentatate/about_radiation/02_Fact_Sheet_on_Biological_Effects
_of_Radiation_NRC.pdf.

15. Richard Pallardy, "Deepwater Horizon Oil Spill," *Encyclopaedia Britannica*, https://www.britannica.com/event/Deepwater-Horizon-oil-spill.

16. Kate Larsen, "PG&E Receives Maximum Sentence for 2010 San Bruno Explosion," ABC 7 News, January 26, 2017, https://abc7news.com.

17. Laura Smith, "The deadliest structural failure in history killed 170,000—and China tried to cover it up," Timeline, August 2, 2017, https://timeline.com. Justin Higginbottom, "230,000 Died in a Dam Collapse That China Kept Secret for Years," Ozy, February 17, 2019, https://www.ozy.com.

18. "Pedestrian Safety: A Road Safety Manual for Decision-Makers and Practitioners," United Nations Road Safety Collaboration, https://www.who.int/roadsafety/projects/manuals/pedestrian. "World Statistic," International Labour Organization, https://www.ilo.org/moscow/areas-of-work/occupational-safety-and-health/WCMS_249278/lang—en/index.htm. "Ambient Air Pollution: Health Impacts," World Health Organization, https://www.who.int/airpollution/ambient/health-impacts. "Global Status Report on Road Safety 2018," World Health Organization, https://www.who.int/violence_injury_prevention/road_safety_status/2018.

19. Jemin Desai et al., "Nuclear Deaths," Environmental Progress, February 2020, https://environmentalprogress.org/nuclear-deaths.

20. Vaclav Smil, *Power Density: A Key to Understanding Energy Sources and Uses* (New York: The MIT Press, 2015), 149.

21. "Ambient (outdoor) air pollution," World Health Organization, May 2, 2018, https://www.who.int/news-room/fact-sheets/detail/ambient-(outdoor)-air-quality-and-health.

22. Anil Markandya and Paul Wilkinson, "Electricity Generation and Health," *Lancet* 370, no. 9591 (2007): 979–990, https://doi.org/10.1016/S0140-6736(07)61253-7.

23. Pushker Kharecha, "Prevented Mortality and Greenhouse Gas Emissions from Historical and Projected Nuclear Power," *Environmental Science & Technology* 47, no. 9 (2013): 4889–4895, https://doi.org/10.1021/es3051197.

24. Stephen Jarvis, Olivier Deschenes, and Akshaya Jha, "The Private and External Costs of Germany's Nuclear Phase-Out" Working Paper 26598, National Bureau of Economic Research (NBER), Cambridge, MA, December 2019, https://doi.org/10.3386/w26598.

25. "GDP (Current US$)," World Bank Group, accessed October 15, 2019, https://data.worldbank.org/indicator/NY.GDP.MKTP.CD.

26. Electricity price data for industrial and residential consumers are from Eu-

rostat, 2019. France's residential price for the first half of 2018 was $.1746 per KWh in comparison to Germany's residential price of $.2987 per KWh. France's industrial price for the first half of 2018 was $.1174 per KWh, in comparison to Germany's, which was $.1967 per KWh. For a comparison of French and German carbon intensities of electricity, see Mark Nelson, "German Electricity Was Nearly 10 Times Dirtier than France's in 2016," Environmental Progress, http://environmentalprogress.org/big-news/2017/2/11/german-electricity-was-nearly-10-times-dirtier-than-frances-in-2016.

27. Mark Nelson and Madison Czerwinski, "With Nuclear Instead of Renewables, California and Germany Would Already Have 100% Clean Electricity," Environmental Progress, September 11, 2018, http://environmentalprogress.org.

28. *Projected Costs of Generating Electricity, 2015 Edition*, International Energy Agency, Nuclear Energy Agency, and Organisation for Economic Co-operation and Development, 2015, https://www.oecd-nea.org/ndd/pubs/2015/7057-proj-costs-electricity-2015.pdf. The cost of generating electricity is designated as the sum of operations and maintenance, fuel, waste, and carbon costs.

29. Mark Nelson et al., "Power to Decarbonize," Environmental Progress, 2017, last updated 2019, accessed October 24, 2019, http://environmentalprogress.org/the-complete-case-for-nuclear. Updated in 2019, in both cases using BP Energy Data for electricity production and Bloomberg New Energy Finance data for wind and solar investment volume.

30. U.S. Government Accountability Office, "Disposal of High-Level Nuclear Waste," GAO, accessed January 31, 2020, https://www.gao.gov/key_issues/disposal_of_highlevel_nuclear_waste/issue_summary.

31. BP Energy Economics, "BP Statistical Review of World Energy 2019, 68th Edition," BP, June 2019, accessed January 16, 2020, https://www.bp.com/content/dam/bp/business-sites/en/global/corporate/pdfs/energy-economics/statistical-review/bp-stats-review-2019-full-report.pdf. I consider nuclear, hydroelectricity, and renewables minus geothermal/biomass as zero-emission energy sources. In 1995, these consisted of 13 percent of the world's primary energy consumption; in 2018, they were 15 percent.

32. "Climate Change in Vermont: Emissions and Goals," State of Vermont, https://climatechange.vermont.gov/climate-pollution-goals. "Title 10: Conservation and Development, Chapter 023, Air Pollution Control," 10 V.S.A.

§ 578, passed 2005, amended 2007, Vermont General Assembly, https://leg islature.vermont.gov/statutes/section/10/023/00578.

33. Kristin Carlson, "Green Mountain Power Is First Utility to Help Customers Go Off-Grid with New Product Offering," Green Mountain Power, December 20, 2016, https://greenmountainpower.com/news/green-mountain -power-first-utility-help-customers-go-off-grid-new-product-offering.

34. "ACEEE 2018 State Energy Efficiency Scorecard," American Council for an Energy-Efficient Economy, October 4, 2018, https://aceee.org /press/2018/10/aceee-2018-state-energy-efficiency.

35. "Vermont Greenhouse Gas Emissions Inventory Update: Brief, 1990–2015," Vermont Department of Environmental Conservation Air Quality and Climate Division, June 2018, https://dec.vermont.gov/sites/dec/files/aqc /climate-change/documents/_Vermont_Greenhouse_Gas_Emissions_In ventory_Update_1990-2015.pdf, 3.

36. Bill McKibben, July 12, 2012 (2:53 p.m.), comment on Yes Vermont Yankee, "Carbon Dioxide and Nuclear Energy: The Great Divide and How to Cross It," July 11, 2012, https://yesvy.blogspot.com/2012/07/carbon-di oxide-and-nuclear-energy-great.html?showComment=134211923468 5#c9098175907964102387.

37. Bill McKibben, email correspondence with the author, January 31, 2019. Nadja Popovich, "How Does Your State Make Electricity?," *New York Times*, December 24, 2018, https://www.nytimes.com.

38. Robert Walton, "New England CO_2 Emissions Spike After Vermont Yankee Nuclear Closure," Utility Dive, February 6, 2017, https://www.utilitydive .com.

39. "Green New Deal Overview" (draft), Office of Rep. Alexandria Ocasio-Cortez, https://assets.documentcloud.org/documents/5729035/Green-New -Deal-FAQ.pdf. This document is a draft version of a "Green New Deal FAQ" that later appeared on AOC's website: "Green New Deal FAQ," Office of Rep. Alexandria Ocasio-Cortez, February 5, 2019, Internet Archive, https://web.archive.org/web/20190207191119/https://ocasio-cortez.house .gov/media/blog-posts/green-new-deal-faq. For more details, see Jordan Weissman, "Why the Green New Deal Rollout Was Kind of a Mess," Slate, February 8, 2019, https://slate.com.

40. Greta Thunberg, "On Friday March 15th 2019 well over 1,5 million students school striked for the climate in 2083 places in 125 countries on

all continents," Facebook post, March 17, 2019, https://www.facebook
.com/732846497083173/posts/on-friday-march-15th-2019-well-over-15
-million-students-school-striked-for-the-c/793441724356983.

41. Kerry Emanuel (climate scientist, MIT) in discussion with the author, November 2019.

42. Anne Van Tyne, interview with Kathleen Goddard Jones, "Defender of California's Nipomo Dunes, Steadfast Sierra Club Volunteer," in *Sierra Club Nationwide II*, 1984, https://digitalassets.lib.berkeley.edu/roho/ucb/text /sierra_club_nationwide2.pdf, 27.

43. Ann Lage, interview with William E. Siri, "Chapter XIII—Science and Mountaineering: Global Adventures, Interview 9, May 1, 1977," in *Reflections on the Sierra Club, the Environment and Mountaineering, 1950s–1970s* (Berkeley, CA: Regional Oral History Office, Bancroft Library, University of California, 1979), 228–250, accessed January 17, 2020, https://archive.org/details /reflectsierraclub00siririch.

44. Ibid., 25.

45. Ibid. Disney is discussed on pp. 80–91; the Grand Canyon situation is discussed on pp. 52–69; redwood forests are discussed on pp. 40–48.

46. Ibid., 93.

47. Ibid., 94.

48. Ibid., 97.

49. William Siri, quoted in Thomas Raymond Wellock, *Critical Masses: Opposition to Nuclear Power in California, 1958–1978* (Madison: University of Wisconsin Press, 1998), 74.

50. Lage, interview with William E. Siri, 105–106.

51. Wellock, *Critical Masses: Opposition to Nuclear Power in California, 1958–1978*, 81.

52. J. Robert Oppenheimer, "Atomic Weapons and American Policy," *Bulletin of the Atomic Scientists* 9, no. 6 (July 1953): 203.

53. Stephen Ambrose, *Eisenhower Volume II: The President* (New York: Simon & Schuster, 2014), 132.

54. Ibid., 132–133.

55. Ibid., 133.

56. Ibid., 133.

57. Ibid., 135.

58. Ibid., 135.

59. Ibid., 135.

60. Isaiah 2:4, *The Bible: New Revised Standard Version*, National Council of Churches, 1989, 550.

61. Ambrose, *Eisenhower Volume II: The President*, 147.

62. Ibid., 147.

63. Dwight D. Eisenhower, "United Nations Address," December 8, 1953.

64. Ibid.

65. Ibid. The word "Powers" is capitalized in the original.

66. Ibid.

67. Thomas Raymond Wellock, *Critical Masses: Opposition to Nuclear Power in California, 1958–1978* (Madison: The University of Wisconsin Press, 1998), 37.

68. Joseph J. Mangano, Jay M. Gould, Ernest J. Sternglass et al., "An Unexpected Rise in Strontium-90 in US Deciduous Teeth in the 1990s," *Science of The Total Environment* 317, nos. 1–3 (December 2003): 37–51, https://doi.org/10.1016/s0048-9697(03)00439-x.

69. Emma Brown, "A Mother Who Took a Stand for Peace," *Washington Post*, January 24, 2011, https://www.washingtonpost.com.

70. Linus Pauling et al., "An Appeal by American Scientists to the Governments and People of the World" (petition to the United Nations), January 15, 1958, https://profiles.nlm.nih.gov/spotlight/mm/catalog/nlm:nlmuid-101584639X78-doc.

71. Sheldon Novick, *The Electric War: The Fight over Nuclear Power* (San Francisco: Sierra Club Books, 1976), 22.

72. Tom Turner, *David Brower: The Making of the Environmental Movement* (Berkeley, CA: University of California Press, 2015), 109.

73. Wellock, *Critical Masses: Opposition to Nuclear Power in California, 1958–1978*, 37.

74. Ibid., 47.

75. Joel W. Hedgpeth, "Bodega Head—a Partisan View," *Bulletin of the Atomic Scientists* 21, no. 3 (March 1965): 2–7, https://doi.org/10.1080/00963402.1965.11454771.

76. Wellock, *Critical Masses: Opposition to Nuclear Power in California, 1958–1978*, 47.

77. Spencer R. Weart, *The Rise of Nuclear Fear* (Cambridge, MA: Harvard University Press, 2012), 192.

78. Ibid., 190.

79. Madison Czerwinski, Environmental Progress, "Nuclear Power Archive," 2020, http://environmentalprogress.org/nuclear-power-archive.

80. "Ohio EPA Celebrating 40 Years: Air Quality," Ohio Environmental Protection Agency, accessed January 17, 2020, https://epa.ohio.gov/40-Years-and -Moving-Forward/Air-Pollutants.

81. Czerwinski, "Nuclear Power Archive."

82. Wyndle Watson, "Shippingport Plant Gets Up-and-Atom Backing," *Pittsburgh Press*, May 26, 1970.

83. Michael Shellenberger, "Why the War on Nuclear Threatens Us All," Environmental Progress, March 28, 2017, http://environmentalprogress.org/big -news/2017/3/28/why-the-war-on-nuclear-threatens-us-all.

84. "AEC Board Gives Okay to Building Power Plant," *The Evening Review*, March 25, 1971, accessed January 17, 2020, https://www.newspapers.com /clip/39460528/the_evening_review.

85. Pat Norman, "Nader Forecasts Rocky Rejection," *Akron Beacon Journal*, October 15, 1974.

86. Wellock, *Critical Masses: Opposition to Nuclear Power in California, 1958–1978*, 105. "Fallout Fatal to 400,000 Babies, Physicist Says," *Greenville* [South Carolina] *News*, July 28, 1969, https://www.newspapers.com/image/189098668.

87. *The China Syndrome*, directed by James Bridges, featuring Jane Fonda, Jack Lemmon, and Michael Douglas (Culver City, CA: Paramount Pictures, 1979).

88. *The Simpsons*, season 4, episode 12, "Marge and the Monorail," directed by Rich Moore, written by Conan O'Brien, aired January 14, 1993, on Fox; *The Simpsons*, season 2, episode 4, "Two Cars in Every Garage and Three Eyes on Every Fish," directed by Wesley Archer, written by Sam Simon and John Swarzwelder, aired November 1, 1990, on Fox.

89. Sheldon Novick, *The Electric War: The Fight over Nuclear Power* (San Francisco: Sierra Club Books, 1976), 194.

90. "Electric Power Consumption (kWh per Capita)—United States, 1960–2014," IEA/OECD, 2014, https://data.worldbank.org/indicator/EG.USE .ELEC.KH.PC?locations=US. "Population and Housing Unit Estimate Tables," United States Census Bureau, https://www.census.gov.

91. Novick, *The Electric War*, 315.

92. Amory B. Lovins, "Energy Strategy: The Road Not Taken?," *Foreign Affairs*, October 20, 1976, 86, https://www.foreignaffairs.com.

93. Wellock, *Critical Masses: Opposition to Nuclear Power in California, 1958–1978*, 112.

94. Richard Halloran, "Carter Has Long Focused on Need to Curb Use of Energy," *New York Times*, July 14, 1979.

95. Richard D. Lyons, "Public Fears Over Nuclear Hazards Are Increasing," *New York Times,* July 1, 1979, https://www.nytimes.com.

96. Bruce Larrick, "Nuclear Rules Bother Edison's Rogers," *Akron Beacon Journal,* October 22, 1979.

97. James F. McCarty, "Promises of 'Clean Smoke' Assuage the Fears," *Cincinnati Enquirer,* December 8, 1985.

98. Czerwinski, "Nuclear Power Archive."

99. Wellock, *Critical Masses: Opposition to Nuclear Power in California, 1958–1978,* 83.

100. Ibid., 41.

101. Ibid., 46.

102. Ruth Teiser and Catherine Harroun, *Conversations with Ansel Adams* (Berkeley: University of California Press, 1978), https://archive.org/details/convan seladams00adamrich, 616.

103. Brad Lemley, "Loving on Nuclear Power," *Washington Post,* June 29, 1986, https://www.washingtonpost.com/archive.

104. Wellock, *Critical Masses: Opposition to Nuclear Power in California, 1958–1978,* 136.

105. "Nuclear Energy: Nuclear Power Is Dirty, Dangerous and Expensive. Say No to New Nukes," Greenpeace, accessed January 17, 2020, https://www.green peace.org/usa/global-warming/issues/nuclear; "Nuclear Free Future," Sierra Club, accessed January 17, 2020, https://www.sierraclub.org/nuclear -free; "Nuclear Power," Union of Concerned Scientists, accessed January 2020, https://www.ucsusa.org/energy/nuclear-power.

106. "Sharp World Wide Drop in Support for Nuclear Energy as 26% of New Opponents Say Fukushima Drove Their Decision," Ipsos Public Affairs, June 20, 2011, https://www.ipsos.com/sites/default/files/news_and _polls/2011–06/5265.pdf; "Nine in Ten Adults Agree That Solar Energy Should Be a Bigger Part of America's Energy Supply in the Future," Ipsos Public Affairs, September 28, 2012, https://www.ipsos.com/sites/default /files/news_and_polls/2012-09/5795.pdf.

107. Evelyn J. Bromet, "Mental Health Consequences of the Chernobyl Disaster," *Journal of Radiological Protection* 32, no. 1 (2012): N71–N75. https:// doi.org/10.1088/0952-4746/32/1/n71. Menachem Ben-Ezra, Yuval Palgi, Yechiel Soffer, and Amit Shrira, "Mental Health Consequences of the 2011 Fukushima Nuclear Disaster: Are the Grandchildren of People Living in Hiroshima and Nagasaki During the Drop of the Atomic Bomb More Vul-

nerable?," *World Psychiatry* 11, no. 2 (June 2012), https://doi.org/10.1016/j
.wpsyc.2012.05.011.

108. Linda E. Ketchum, "Lessons of Chernobyl: SNM Members Try to Decon-
taminate World Threatened by Fallout," *Journal of Nuclear Medicine* 28, no. 6
(1987): 933–942, http://jnm.snmjournals.org/content/28/6/933.citation.

109. Bromet, "Mental health consequences of the Chernobyl disaster."

110. Matthew J. Neidell, Shinsuke Uchida, and Marcella Veronesi, "Be Cautious
with the Precautionary Principle: Evidence from Fukushima Daiichi Nuclear
Accident" Working Paper 26395, National Bureau of Economic Research
(NBER), Cambridge, MA, October 2019, https://doi.org/10.3386/w26395.

111. "Stress-Induced Deaths in Fukushima Top Those from 2011 Natural Disasters," *The Mainichi*, September 9, 2013, Internet Archive, https://web
.archive.org/web/20130913092840/http://mainichi.jp/english/english
/newsselect/news/20130909p2a00m0na009000c.html. Molly K. Schnell and
David E. Weinstein, "Evaluating the Economic Response to Japan's Earthquake," Policy Discussion Paper 12003, Research Institute of Economy,
Trade and Industry, February 2012, https://www.rieti.go.jp/jp/publica
tions/pdp/12p003.pdf.

112. James Conca, "Shutting Down All of Japan's Nuclear Plants After Fukushima
Was a Bad Idea," *Forbes*, October 31, 2019, https://www.forbes.com; Justin
McCurry, "Fukushima disaster: First residents return to town next to nuclear
plant," *The Guardian*, April 10, 2019, https://www.theguardian.com; WHO,
*Health risk assessment from the nuclear accident after the 2011 Great East Japan
Earthquake and Tsunami* (Geneva, Switzerland: WHO, 2013), https://apps
.who.int/iris/bitstream/handle/10665/78218/9789241505130_eng.pdf.

113. Robin Harding, "Fukushima Nuclear Disaster: Did the Evacuation Raise the
Death Toll?," *Financial Times*, March 10, 2018, https://www.ft.com.

114. "Relative Radioactivity Levels," Berkeley RadWatch, 2014, accessed
January 17, 2020, https://radwatch.berkeley.edu/dosenet/levels; "Safecast Tile Map," Safecast, accessed January 17, 2020, http://safecast.org
/tilemap/?y=37.449&x=141.002&z=10&l=0&m=0.

115. Masaharu Tsubokura, Shuhei Nomura, Kikugoro Sakaihara et al., "Estimated Association Between Dwelling Soil Contamination and Internal Radiation Contamination Levels After the 2011 Fukushima Daiichi Nuclear
Accident in Japan," *BMJ Open* 6, no. 6 (June 29, 2016), https://bmjopen.bmj
.com/content/bmjopen/6/6/e010970.full.pdf.

116. Michel Berthélemy and Lina Escobar Rangel, "Nuclear Reactors' Construc-

tion Costs: The Role of Lead-Time, Standardization and Technological Progress," *Energy Policy* 82 (July 2015): 118–30, https://doi.org/10.1016/j.enpol.2015.03.015.

117. Jessica R. Lovering, Arthur Yip, and Ted Nordhaus, "Historical Construction Costs of Global Nuclear Power Reactors," *Energy Policy* 91 (April 2016): 371–81, https://doi.org/10.1016/j.enpol.2016.01.011. When the reactors in this study are grouped by reactor type, the average overnight capital costs are lowest for pressurized and boiling-water reactors and greater for heavy-water reactors and gas-cooled reactors. Sodium fast reactors are the most expensive. These cost trends correspond with both total deployment for each reactor type (more deployment with lower cost, less deployment with higher) and with total energy produced (more energy produced from lower-cost reactors, less energy from higher-cost reactors).

118. Berthélemy and Rangel, "Nuclear Reactors' Construction Costs."

119. Adriana Brasileiro, "Turkey Point Nuclear Reactors Get OK to Run Until 2053 in Unprecedented NRC Approval," *Miami Herald*, December 5, 2019, https://www.miamiherald.com.

120. *Chernobyl*, episode 5, "Vichnaya Pamyat," directed by Johan Renck, HBO, June 3, 2019.

121. Michael Dobbs, *One Minute to Midnight: Kennedy, Khrushchev, and Castro on the Brink of Nuclear War* (New York: Vintage Books, 2009), 312.

122. Ibid., 344.

123. John Lewis Gaddis, *The Long Peace: Inquiries into the History of the Cold War* (New York: Oxford University Press, 1987), 230. Gaddis first published his speech as a 1986 journal article, "The Long Peace: Elements of Stability in the Postwar International System," *International Security*, Spring 1986, 99–142.

124. Our World in Data, "Battle Related Deaths in State-Based Conflicts Since 1946," Our World in Data, https://ourworldindata.org/grapher/battle-related-deaths-in-state-based-conflicts-since-1946-by-world-region.

125. Devin T. Hagerty, *Nuclear Weapons and Deterrence: Stability in South Asia* (London: Palgrave Pivot, Cham, 2019), 94.

126. Dylan Matthews, "Meet the Political Scientist Who Thinks the Spread of Nuclear Weapons Prevents War," *Vox*, August 21, 2014, https://www.vox.com.

127. David Brunnstrom, "North Korea May Have Made More Nuclear Bombs, but Threat Reduced: Study," Reuters, February 11, 2019, https://www.reuters.com.

128. Matthew Kroenig, *A Time to Attack: The Looming Iranian Nuclear Threat* (New York: Palgrave Macmillan, 2014), 125–126.

129. Scott D. Sagan and Kenneth N. Waltz, "Is Nuclear Zero the Best Option?," *The National Interest*, no. 109 (September–October 2010): 88–96, https://fsi -live.s3.us-west-1.amazonaws.com/s3fs-public/Sagan_Waltz_-_National _Interest_-_The_Great_Debate.pdf.

130. Vannevar Bush, John S. Dickey, Allen W. Dulles et al., "Report by the Panel of Consultants of the Department of State to the Secretary of State: Armaments and American Policy," *Foreign Relations of the United States, 1952–1954*, vol. 2, part 2, *National Security Affairs*, document 67, January 1953, https://history.state.gov/historicaldocuments/frus1952-54v02p2/d67. Richard Rhodes, *Dark Sun: The Making of the Hydrogen Bomb* (New York: Simon & Schuster, 2012), 588.

131. Max Born, Percy W. Bridgman, Albert Einstein, Bertrand Russell et al., "Statement: The Russell-Einstein Manifesto," July 9, 1955, presented at the 1st Pugwash Conference on Science and World Affairs, Pugwash, Nova Scotia, 1957, https://pugwash.org/1955/07/09/statement-manifesto.

132. "CNN Poll: Public Divided on Eliminating All Nuclear Weapons," CNN, April 12, 2010, http://www.cnn.com.

133. Kai Bird and Martin J. Sherwin, *American Prometheus: The Triumph and Tragedy of J. Robert Oppenheimer* (New York: Vintage Books, 2006), 309.

134. Ibid., 317.

9: Destroying the Environment to Save It

1. Bryan Bishop and Josh Dzieza, "Tesla Energy Is Elon Musk's Battery System That Can Power Homes, Businesses, and the World," *The Verge*, May 1, 2015, https://www.theverge.com.

2. Tesla, "Tesla introduces Tesla Energy," YouTube, May 2, 2015, https://www.youtube.com/watch?time_continue=82&v=NvCIhn7_FXI&feature=emb_logo.

3. H. J. Mai, "Tesla Powerwall, Powerpack deployment grows 81% to 415 MWh in Q2," *Utility Dive*, July 30, 2019, https://www.utilitydive.com.

4. Andy Sendy, "Pegging the All-in, Installed Cost of a Tesla Powerwall 2," *Solar Reviews*, October 3, 2017, https://www.solarreviews.com; Sean

O'Kane, "Tesla Launches a Rental Plan to Help Its Slumping Home Solar Panel Business," *The Verge*, August 19, 2019, https://www.theverge.com.

5. U.S. Energy Information Administration, *International Energy Outlook 2019 With Projections to 2050* (Washington, D.C.: EIA, September 2019), https://www.eia.gov/outlooks/ieo/pdf/ieo2019.pdf; U.S. Energy Information Administration, *Annual Energy Outlook 2020* (Washington, D.C.: EIA, January 2020), https://www.eia.gov/outlooks/aeo.

6. BP Energy Economics, "BP Statistical Review of World Energy 2019, 68th Edition," BP, June 2019, accessed January 16, 2020, https://www.bp.com/content/dam/bp/business-sites/en/global/corporate/pdfs/energy-eco nomics/statistical-review/bp-stats-review-2019-full-report.pdf. In the U.S. and Europe in 2018, renewables generated 7 and 16 percent of total primary energy and 17 and 34 percent of total electricity, respectively. Of that, hydroelectricity generated 2.8 and 7.1 percent of total primary energy and 6.5 and 15.7 percent of total electricity. Biomass generated 0.7 and 2.3 percent of total primary energy and 1.5 and 5.0 percent of total electricity. Biofuels generated 1.7 and 0.8 percent of total primary energy.

7. BP Energy Economics, *BP Statistical Review of World Energy 2019, 68th Edition* (London: BP, June 2019), accessed January 16, 2020, https://www.bp.com/content/dam/bp/business-sites/en/global/corporate/pdfs/ener gy-economics/statistical-review/bp-stats-review-2019-full-report.pdf. Solar energy figure from BP Statistical Review of World Energy 2019. Geothermal electricity production in 2018 is estimated by the International Energy Agency to be 90 terawatt-hours, which, using BP's method for conversion to primary energy equivalent, would be 20.3 MTOE, or 0.1 percent of 2018 global primary energy.

8. "Tesla Mega-battery in Australia Activated," BBC News, December 1, 2017, https://www.bbc.com.

9. Julian Spector, "California's Big Battery Experiment: A Turning Point for Energy Storage?," *The Guardian*, September 15, 2017, https://www.theguardian.com. In 2018, according to the U.S. Energy Information Agency, 134 million residential electricity accounts used about 11,000 KWh per year, or 1.25 KW on average throughout the year.

10. The U.S. Department of Energy finds that four-hour-capacity lithium battery facilities cost $1,876 per kilowatt of power in 2018, with prices expected to fall 23 percent to $1,446 per kilowatt by 2025. Escondido Energy Storage

is a 30 megawatt, four-hour facility. The U.S. Energy Information Agency reported 4,174 terawatt-hours of electricity use in 2018, giving an average hourly power of 476,600 megawatts.

11. Falko Ueckerdt, Lion Hirth, Gunnar Luderer, and Ottmar Edenhofer, "System LCOE: What Are the Costs of Variable Renewables?," *Energy* 63 (2013): 61–75, https://doi.org/10.1016/j.energy.2013.10.072.

12. Matthew Shaner, "Geophysical Constraints on the Reliability of Solar and Wind Power in the United States," *Energy and Environmental Science* 11, no. 4 (2018): 914–25, https://doi.org/10.1039/c7ee03029k.

13. "New Poll: Nearly Half of Americans Are More Convinced than They Were Five Years Ago That Climate Change Is Happening, with Extreme Weather Driving Their Views," Energy Policy Institute at the University of Chicago, January 22, 2019, https://epic.uchicago.edu/news/new-poll-nearly-half -of-americans-are-more-convinced-than-they-were-five-years-ago-that-cli mate-change-is-happening-with-extreme-weather-driving-their-views.

14. Ivan Penn, "The $3 Billion Plan to Turn Hoover Dam into a Giant Battery," *New York Times*, July 24, 2018, https://www.nytimes.com. It describes a specific pumped-hydroelectric proposal of the type that would be repeated on a massive scale in Jacobson's plan.

15. In a *Guardian* column in 2017, Professor Mark Jacobson and U.S. senator Bernie Sanders announced new energy legislation based on Jacobson's work. Bernie Sanders and Mark Jacobson, "The American People—Not Big Oil— Must Decide Our Climate Future," *The Guardian*, April 29, 2017, https:// www.theguardian.com.

16. Christopher T. M. Clack, Steffan A. Qvist, Jay Apt et al., "Evaluation of a Proposal for Reliable Low-Cost Grid Power with 100% Wind, Water, and Solar," *Proceedings of the National Academy of Sciences of the United States of America* 114, no. 26 (June 27, 2017): 6722–27, https://doi.org/10.1073 /pnas.1610381114. Chris Mooney, "A Bitter Scientific Debate Just Erupted over the Future of America's Power Grid," *Washington Post*, June 19, 2017, https://www.washingtonpost.com.

17. Ivan Penn, "California Invested Heavily in Solar Power. Now There's So Much That Other States Are Sometimes Paid to Take It," *Los Angeles Times*, June 22, 2017, https://www.latimes.com.

18. Vanessa Dezem, "Germany Plans Incentives to Boost Hydrogen in Energy Mix," Bloomberg, November 20, 2019, https://www.bloomberg.com.

19. Frank Dohmen, Alexander Jung, Stefan Schultz, and Gerald Traufetter,

"German Failure on the Road to a Renewable Future," *Spiegel International*, May 13, 2019, https://www.spiegel.de.

20. "Annexe 6: Historique des charges de service public de l'électricité et de la contribution unitaire," Commission de Régulation de l'Énergie, http://www.cre.fr/documents/deliberations/proposition/cspe-2014/consulter-l-annexe-6-historique-des-charges-de-service-public-de-l-electricite-et-de-la-contribution-unitaire.

21. The French grid operator RTE France publishes hourly historical data for electricity production from 2012 onward, including an hourly carbon intensity rate useful for calculating annual carbon intensity averages. Since carbon intensity hit a minimum of 41 grams of CO_2 in 2014, much higher power production from natural gas, wind, and solar electricity has accompanied declining nuclear power production.

22. "Big Oil's Real Agenda on Climate Change," InfluenceMap, March 2019, https://influencemap.org/report/How-Big-Oil-Continues-to-Oppose-the-Paris-Agreement-38212275958aa21196dae3b76220bddc.

23. Image of gas company Statoil's ad campaign at Brussels airport, Bellona, February 5, 2016, https://network.bellona.org/content/uploads/sites/3/2016/02/WP_20151022_008.jpg.

24. BP (@BP_plc), "Our natural gas is a smart partner to renewable energy: http://on.bp.com/possibilitieseverywhere . . . #PossibilitiesEverywhere #NatGas," Twitter, January 27, 2019, 1:38 a.m., https://twitter.com/bp_plc/status/1089457584694685696.

25. Mike Shellenberger (@ShellenbergerMD), "No sooner had I landed in Germany, for 2017 U.N. climate talks, when I was confronted by airport ads paid for by Total, the French oil and gas company reading, 'Committed to Solar' and 'Committed to Natural Gas,' " Twitter (text/image), March 28, 2019, 8:22 a.m., https://twitter.com/ShellenbergerMD/status/1111287183497789440.

26. Lisa Linowes (executive director, Wind Action Group) in conversation with the author, November 1, 2019.

27. Ibid.

28. Ibid.

29. John van Zalk and Paul Behrens, "The Spatial Extent of Renewable and Non-renewable Power Generation: A Review and Meta-analysis of Power Densities and Their Application in the U.S.," *Energy Policy* 123 (December 2018): 83–91, https://doi.org/10.1016/j.enpol.2018.08.023.

30. Lisa Linowes (executive director, Wind Action Group) in conversation with the author, November 1, 2019.

31. Ibid.

32. Paul M. Cryan, "Wind Turbines as Landscape Impediments to the Migratory Connectivity of Bats," *Journal of Environmental Law* 41 (May 2011): 355–70, https://www.lclark.edu/live/files/8520-412cryan.

33. Ana T. Marques, Carlos D. Santos, Frank Hanssen et al., "Wind Turbines Cause Functional Habitat Loss for Migratory Soaring Birds," *Journal of Animal Ecology* 89, no. 1 (2019): 1–11, https://doi.org/10.1111/1365-2656 .12961.

34. Michael W. Collopy, Brian Woodridge, and Jessi L. Brown, "Golden Eagles in a Changing World," *Journal of Raptor Research* 51, no. 3 (2017): 193–96, https://doi.org/10.3356/0892-1016-51.3.193.

35. "Domestic Cat Predation on Birds and Other Wildlife," (Washington, D.C.: American Bird Conservancy), accessed January 27, 2020, http://trnerr.org /wp-content/uploads/2013/06/ABCBirds_predation.pdf; Shawn Small-wood and Carl Thelander, "Bird Mortality in the Altamont Pass Wind Resource Area, California," *Journal of Wildlife Management* 72, no. 1 (2008): 215–223, https://doi.org/10.2193/2007-032.

36. Adam Welz, "How Kenya's Push for Development is Threatening its famed wild lands," Yale School of Forestry & Environmental Studies, April 24, 2019, https://e360.yale.edu/features/how-kenyas-push-for-development-is -threatening-its-prized-wild-lands.

37. Ibid.

38. William Wilkes, Hayley Warren, and Brian Parkin, "Germany's Failed Climate Goals: A Wake-up Call for Governments Everywhere," Bloomberg, August 15, 2018, https://www.bloomberg.com.

39. "Annual Electricity Generation in Germany," Fraunhofer ISE, accessed January 10, 2020, https://www.energy-charts.de/energy.htm.

40. Frank Dohmen, "German Failure on the road to a renewable future," *Der Spiegel,* May 13, 2019, https://www.spiegel.de; "Annual Electricity Generation in Germany," Fraunhofer ISE, accessed January 10, 2020, https://www .energy-charts.de/energy.htm. GDP conversion between Germany and USA made using OECD data for Purchasing Power Parity.

41. Fridolin Pflugmann, Ingmar Ritzenhofen, Fabian Stockhausen, and Thomas Vahlenkamp, "Germany's Energy Transition at a Crossroads," McKinsey & Company, November 2019, https://www.mckinsey.com/industries

/electric-power-and-natural-gas/our-insights/germanys-energy-transition
-at-a-crossroads.

42. "Electricity Prices for Household Consumers—Bi-annual Data (from 2007
Onwards)," Eurostat, December 1, 2019, https://appsso.eurostat.ec.eu
ropa.eu/nui/show.do?dataset=nrg_pc_204&lang=en. "Electricity Data
Browser: Retail Sales of Electricity Annual," U.S. Energy Information Administration, https://www.eia.gov/electricity/data/browser.

43. Michael Greenstone and Ishan Nath, "Do Renewable Portfolio Standards Deliver?," Working Paper no. 2019-62, Energy Policy Institute at the University
of Chicago, 2019, https://epic.uchicago.edu/wp-content/uploads/2019/07
/Do-Renewable-Portfolio-Standards-Deliver.pdf. "Renewable Portfolio
Standards Reduce Carbon Emissions—But, at a High Cost," Energy Policy
Institute at the University of Chicago, April 22, 2019, https://epic.uchicago
.edu/insights/renewable-portfolio-standards-reduce-carbon-emissions-but
-at-a-high-cost.

44. "Electricity Data Browser: Retail Sales of Electricity," and "Electricity Data
Browser: Revenue from Retail Sales of Electricity," U.S. Energy Information
Administration, https://www.eia.gov/electricity/data/browser.

45. Dohmen, "German Failure on the road to a renewable future."

46. International Energy Agency, "Renewable capacity growth worldwide stalled
in 2018 after two decades of strong expansion," May 6, 2019, https://www
.iea.org/news/renewable-capacity-growth-worldwide-stalled-in-2018-after
-two-decades-of-strong-expansion.

47. Dohmen, "German Failure on the road to a renewable future."

48. John Etzler, *The Paradise Within the Reach of All Men, Without Labor, by Powers of Nature and Machinery* (London: John Brooks, 1833), 1.

49. Ibid., 36.

50. Henry David Thoreau, "Paradise (to Be) Regained," 1843, in *The Writings
of Henry David Thoreau*, vol. 4 (Cambridge, MA: Riverside Press, 1906),
283.

51. Steven Stoll, *The Great Delusion: A Mad Inventor, Death in the Tropics, and
the Utopian Origins of Economic Growth* (New York: Farrar, Straus & Giroux,
2008), 14–15.

52. Alexis Madrigal, *Powering the Dream: The History and Promise of Green Technology* (Cambridge, MA: Da Capo Press, 2011), 124–125.

53. Ibid., 125.

54. Ibid., 87.

55. Ibid., 92–126.

56. Martin Heidegger, *The Question Concerning Technology and Other Essays* (New York & London: Garland Publishing Inc., 1954), 14.

57. Murray Bookchin, *The Murray Bookchin Reader* (Montreal: Black Rose Books, 1999), 18–19, 91.

58. Barry Commoner, *Making Peace with the Planet* (New York: New Press, 1992), 193.

59. Barry Commoner, *Crossroads: Environmental Priorities for the Future* (Washington, DC: Island Press, 1988), 146.

60. Ibid., 182.

61. Amory B. Lovins, "Energy Strategy: The Road Not Taken?," *Foreign Affairs*, October 20, 1976, 65–96, https://www.foreignaffairs.com.

62. Diane Tedeschi, "Solar Impulse Crosses America: Who Needs Fuel When the Sun Can Keep You Aloft?," *Air & Space*, June 19, 2013, https://www.airspacemag.com.

63. Ibid.

64. "Around the World in a Solar Airplane," Solar Impulse Foundation, https://aroundtheworld.solarimpulse.com/adventure. Solar Impulse 2 in 2015 was little improvement. It flew at 75 km/h and had the same wingspan as the Boeing 747, a jumbo jet, and similarly could not hold passengers other than the pilot.

65. "California," *Environmental Progress*, last updated August 12, 2019, http://environmentalprogress.org/california.

66. Gilbert Masters, "Wind Power Systems," in Masters, *Renewable and Efficient Electric Power Systems* (Hoboken, NJ: Wiley, 2004): 307–83, https://nature.berkeley.edu/er100/readings/Masters_2004_Wind.pdf.

67. Vaclav Smil, *Power Density: A Key to Understanding Energy Sources and Uses* (Cambridge, MA: The MIT Press, 2015), 52–57, 199.

68. Peter Laufer, "The Tortoise Is Collateral Damage in the Mojave Desert," *High Country News*, March 19, 2014, https://www.hcn.org/wotr/the-tortoise-is-collateral-damage-in-the-mojave-desert.

69. *Quadrennial Technology Review: An Assessment of Energy Technologies and Research Opportunities*, U.S. Department of Energy, 2015, https://www.energy.gov/sites/prod/files/2017/03/f34/quadrennial-technology-review-2015_1.pdf, 402.

70. Jemin Desai and Mark Nelson, "Are We Headed for a Solar Waste Crisis?," Environmental Progress, June 21, 2017, http://environmentalprogress.org/big-news/2017/6/21/are-we-headed-for-a-solar-waste-crisis.

71. Kelly Pickerel, "It's Time to Plan for Solar Panel Recycling in the United States," Solar Power World, April 2, 2018, https://www.solarpowerworldonline.com/2018/04/its-time-to-plan-for-solar-panel-recycling-in-the-united-states.

72. Nadav Enbar, "PV Life Cycle Analysis Managing PV Assets over an Uncertain Lifetime," slide presentation, Solar Power International, September 14, 2016, https://www.solarpowerinternational.com/wp-content/uploads/2016/09/N253_9-14-1530.pdf, 20.

73. Idiano D'Adamo, Michela Miliacca, and Paolo Rosa, "Economic Feasibility for Recycling of Waste Crystalline Silicon Photovoltaic Modules," *International Journal of Photoenergy*, June 27, 2017, article no. 4184676, https://doi.org/10.1155/2017/4184676.

74. Stephen Cheng, "China's Ageing Solar Panels Are Going to Be a Big Environmental Problem," *South China Morning Post*, July 30, 2017, https://www.scmp.com.

75. Travis Hoium, "Solar Shake-up: Why More Bankruptcies Are Coming in 2017," The Motley Fool, April 18, 2017, https://www.fool.com. Travis Hoium, "Bankruptcies Continue in Solar Industry," The Motley Fool, May 19, 2017, https://www.fool.com.

76. As an example, Solyndra's bankruptcy led to its abandonment of a cadmium waste storage facility in Milpitas, California, due to the facts that the facility was not financed by Solyndra's federal loan guarantees and that Solyndra had not made much money prior to its bankruptcy. Dustin Mulvaney, *Solar Power: Innovation, Sustainability, and Environmental Justice* (Oakland: University of California Press, 2019), 83–84.

77. Daniel Wetzel, "Study Warns of Environmental Risks from Solar Modules," *Die Welt*, May 13, 2018, https://cloudup.com/c1xW_Mk$f85.

78. Cheng, "China's Ageing Solar Panels Are Going to Be a Big Environmental Problem."

79. "Illegally Traded and Dumped E-Waste Worth up to $19 Billion Annually Poses Risks to Health, Deprives Countries of Resources, Says UNEP Report," United Nations Environment Programme, May 12, 2015, https://www.unenvironment.org/news-and-stories/press-release/illegally-traded-and-dumped-e-waste-worth-19-billion-annually-poses.

80. Amy Yee, "Electronic Marvels Turn into Dangerous Trash in East Africa," *New York Times*, May 12, 2019, https://www.nytimes.com.

81. Smil, *Power Density: A Key to Understanding Energy Sources and Uses*, 205, 208.

82. We can check this claim by assuming that his plant would be similar to Mesquite Solar, one of the largest solar plants in the world, located in the Sonoran Desert. In 2018, Mesquite Solar 1, 2, and 3 generated 1.15 TWh, according to the U.S. Energy Information Administration, on a land area of 10 km². In 2018, the United States generated 4,174.4 TWh of energy. Replacing this electricity would require 3,636 Mesquites with an area of 36,422.3 km². In comparison, the state of Maryland has an area of 32,131 km².

83. Using monthly generation data for Mesquite Solar plants from the U.S. Energy Information Administration, we can see the lowest and highest six-month periods of generation at Mesquite, and compare them to the same lowest and highest half years of generation of electricity in the entire United States. If 3,642 Mesquites provided all electricity, the worst six-month energy deficit, from October 2018 to March 2019, would have been 444 TWh. Overbuilding solar in this example by 30 percent would reduce this deficit to 16 TWh.

84. The U.S. Department of Energy found that in 2018, four-hour-capacity lithium ion battery facilities cost $469 per kWh of energy storage capacity. K. Mongird, V. Foledar, V. Viswanathan et al., *Energy Storage Technology and Cost Characterization Report*, U.S. Department of Energy, July 2019, https://www.energy.gov/sites/prod/files/2019/07/f65/Storage%20Cost%20and%20Performance%20Characterization%20Report_Final.pdf.

85. Tesla, "Tesla Introduces Tesla Energy," YouTube, May 2, 2015, https://www.youtube.com/watch?time_continue=82&v=NvCIhn7_FXI&feature=emb_logo.

86. Victor Tangermann, "Elon Musk Is Talking About Powering All of America with Solar," *Futurism*, December 12, 2019, https://futurism.com.

87. Smil, *Power Density: A Key to Understanding Energy Sources and Uses*, 247.

88. Ibid., 247.

89. D. Weißbach, G. Ruprecht, A. Huke et al., "Energy Intensities, EROIs (Energy Returned on Invested), and Energy Payback Times of Electricity Generating Power Plants," *Energy* 52 (April 2013): 210–21, https://doi.org/10.1016/j.energy.2013.01.029.

90. Ibid.

91. Dohmen et al., "German Failure on the Road to a Renewable Future."

92. The power density of wood-based biomass is from van Zalk and Behrens, "The Spatial Extent of Renewable and Non-renewable Power Generation." U.S. electricity generation data are from the U.S. Energy Information Agency.

93. Nana Yaw Amponsah, Mads Troldborg, Bethany Kington et al., "Greenhouse Gas Emissions from Renewable Energy Sources: A Review of Lifecycle Considerations," *Renewable and Sustainable Energy Reviews* 39 (November 2014), 461–75, https://doi.org/10.1016/j.rser.2014.07.087.

94. Timothy Searchinger, Ralph Heimlich, R. A. Houghton et al., "Use of U.S. Croplands for Biofuels Increases Greenhouse Gases Through Emissions from Land-Use Changes," *Science* 319, no. 5867 (2008): 1238–40, https://doi.org/10.1126/science.1151861. Tiziano Gomiero, "Are Biofuels an Effective and Viable Energy Strategy for Industrialized Societies? A Reasoned Overview of Potentials and Limits," *Sustainability* 7 (2015): 8491–521, https://doi.org/10.3390/su7078491.

95. David Schaper, "Record Number of Miles Driven in US Last Year," NPR, February 21, 2017, https://www.npr.org. "How Much Gasoline Does the United States Consume?," U.S. Energy Information Administration, https://www.eia.gov/tools/faqs/faq.php?id=23&t=10. "How Much Ethanol Is in Gasoline, and How Does It Affect Fuel Economy?," U.S. Energy Information Administration, https://www.eia.gov/tools/faqs/faq.php?id=27&t=4. "Data," FAO, http://www.fao.org/faostat/en/#data/RL.

In 2017, US drivers drove a total of 3.22 trillion miles. The fuel economy of U.S. cars averaged 24.9 miles per gallon. For every gallon of gasoline, 1.5 gallons of ethanol (E100) is needed. In 2017, U.S. cropland totaled 396 million acres. The amount of land needed to replace all gasoline with pure ethanol would be 653.3 million acres, an increase of 51 percent.

96. John van Zalk and Paul Behrens, "The spatial extent of renewable and non-renewable power generation: A review and meta-analysis of power densities and their application in the US," *Energy Policy* 123 (2018): 83–91, https://doi.org/10.1016/j.enpol.2018.08.023.

97. Ibid., 83–91.

98. Jesse Jenkins, Mark Moro, Ted Nordhaus et al., *Beyond Boom & Bust: Putting Clean Tech on a Path to Subsidy Independence*, Breakthrough Institute, April 2012, https://s3.us-east-2.amazonaws.com/uploads.thebreakthrough.org/articles/beyond-boom-and-bust-report-overview/Beyond_Boom_and_Bust.pdf, 18.

99. Tom Jackson, "PacificCorp Sues to Block Release of Bird-Death Data at Wind Farms," *Sandusky Register*, November 17, 2014, http://www.sanduskyregister.com.

100. Shawn Smallwood, "Bird and Bat Impacts and Behaviors at Old Wind Turbines

at Foebay, Altamont Pass Wind Resource Area," California Energy Commission, November 2016, accessed January 17, 2020, https://ww2.energy.ca.gov /2016publications/CEC-500-2016-066/CEC-500-2016-066.pdf.

101. Shawn Smallwood, letter to Fish and Wildlife Service, "Comments on the Updated Collision Risk Model Priors for Estimating Eagle Fatalities at Wind Energy Facilities," Center for Biological Diversity, August 17, 2018.

102. Ibid.

103. Michael Hutchins, "To Protect Birds from Wind Turbines, Look to Hawai'i's Approach," American Bird Conservancy, June 21, 2016, https://ab cbirds.org/to-protect-birds-and-bats-from-wind-turbines-adopt-hawaiis -approach.

104. Joseph Goldstein, "A Climate Conundrum: The Wind Farm vs. the Eagle's Nest," *New York Times*, June 25, 2019, https://www.nytimes.com.

105. Clifford P. Schneider, *pro se*, "Motion for Dismissal for Fraud upon the Siting Board," *Application of Galloo Island Wind LLC for a Certificate of Environmental Compatibility and Public Need Pursuant to Article 10 to Construct a Wind Energy Project*, Case No. 15-F-0327, September 13, 2018, 2; Joseph Goldstein, "A Climate Conundrum: The Wind Farm vs. The Eagle's Nest," *New York Times*, June 25, 2019, https://www.nytimes.com.

106. Goldstein, "A Climate Conundrum."

107. Lisa Linowes (executive director, Wind Action Group) in discussion with author, November 6, 2019.

108. Scientists have found that curtailing the use of wind turbines when wind speed is low can reduce bat fatalities by 44 percent, to 93 percent. Edward B. Arnett, Manuela M. P. Huso, Michael R. Schirmacher, and John P. Hayes, "Altering Turbine Speed Reduces Bat Mortality at Wind-Energy Facilities," *Frontiers in Ecology and the Environment* 9, no. 4 (November 2010): 209–14, https:// doi.org/10.1890/100103. Lori Bird, Jaquelin Cochran, and Xi Wang, *Wind and Solar Energy Curtailment: Experience and Practices in the United States*, National Renewable Energy Laboratory, March 2014, https://www.nrel.gov /docs/fy14osti/60983.pdf.

109. Smallwood, letter to Fish and Wildlife Service.

110. Shawn Smallwood, "Estimating Wind Turbine–Caused Bird Mortality," *Journal of Wildlife Management* 71, no. 8 (2007): 2781–91, https://doi .org/10.2193/2007-006.

111. Franz Trieb, "Study Report: Interference of Flying Insects and Wind Parks," Deutsches Zentrum für Luft- und Raumfahrt, September 30, 2018, https://

www.dlr.de/tt/Portaldata/41/Resources/dokumente/st/FliWip-Final-Re
port.pdf. Caspar A. Hallmann, Martin Song, Eelke Jongejans et al., "More
than 75 Percent Decline over 27 Years in Total Flying Insect Biomass in Protected Areas," *PLOS ONE* 12, no. 10 (October 18, 2017): e0185809, https://
doi.org/10.1371/journal.pone.0185809.

112. Trieb, "Interference of Flying Insects and Wind Parks."

113. Ibid.

114. Robert F. Contreras and Stephen J. Frasier, "High-Resolution Observations
of Insects in the Atmospheric Boundary Layer," *Journal of Atmospheric and
Oceanic Technology* 25 (2008): 2176–87, https://doi.org/10.1175/2008JTE
CHA1059.1.

115. Trieb, "Interference of Flying Insects and Wind Parks."

116. U.S. Census of Agriculture, 2017. EU, "EU Agricultural Outlook 2017–
2030," www.ec.europa.eu.

117. Franz Trieb and Denise Nuessele (Institut für Technische Thermodynamik),
email correspondence with the author, November 16, 2019, and November 19,
2019.

118. "Wind turbines and birds and bats," Sierra Club, accessed June 26, 2019,
https://www.sierraclub.org/michigan/wind-turbines-and-birds-and-bats.

119. Susan Crosier, "A wind farm in the great lakes? Let's give it a twirl," NRDC,
February 21, 2017, https://www.nrdc.org/stories/wind-farm-great-lakes
-lets-give-it-twirl.

120. Stacy Small-Lorenz et al., "Can birds and wind energy co-exist?" *Energy
Exchange* (blog), EDF, December 5, 2014, http://blogs.edf.org/energyex
change/2014/12/05/can-birds-and-wind-energy-co-exist.

121. Joseph Goldstein, "A Climate Conundrum: The Wind Farm vs. The Eagle's
Nest," *New York Times*, June 25, 2019, https://www.nytimes.com.

122. Science Advisory Board, *Review of EPA's Accounting Framework for Biogenic
CO$_2$ Emissions from Stationary Sources* (Washington, D.C.: United States
Environmental Protection Agency, 2011), https://yosemite.epa.gov/sab
/sabproduct.nsf/0/57B7A4F1987D7F7385257A87007977F6/$File/EPA
-SAB-12-011-unsigned.pdf.

123. Louis Sahagun, "Companies won't face charges in condor deaths," *Los Angeles Times*, May 10, 2013, https://www.latimes.com.

124. Ibid.

125. Barry Brook, "An Open Letter to Environmentalists on Nuclear Energy," Brave New Climate, December 15, 2014, https://bravenewclimate

.com/2014/12/15/an-open-letter-to-environmentalists-on-nuclear-en
ergy.

126. Peter Laufer, "The Tortoise Is Collateral Damage in the Mojave Desert," *High Country News*, March 19, 2014, https://www.hcn.org.

127. Thomas Lovejoy, "A Mojave Solar Project in the Bighorn's Way," *New York Times*, September 11, 2015, https://www.nytimes.com.

128. Jonathan Franzen, "Carbon Capture," *The New Yorker*, March 30, 2015, https://www.newyorker.com.

129. Michael Casey, "30,000 Wind Turbines Located in Critical Bird Habitats," CBS News, May 20, 2015, https://www.cbsnews.com.

130. *Wind Power: Impacts on Wildlife and Government Responsibilities for Regulating Development and Protecting Wildlife*, GAO Report to Congressional Requesters, GAO-05-906 (Washington, D.C.: GAO, September 2005), https://www.gao.gov/new.items/d05906.pdf.

131. W. F. Frick, E. F. Baerwald, J. F. Pollock et al., "Fatalities at Wind Turbines May Threaten Population Viability of a Migratory Bat," *Biological Conservation* 209 (May 2017): 172–77, http://doi.org/10.1016/j.biocon.2017.02.023.

132. Fridolin Pflugmann et al., "Germany's energy transition at a crossroads," McKinsey & Company, November 2019, https://www.mckinsey.com/indus tries/electric-power-and-natural-gas/our-insights/germanys-energy-transi tion-at-a-crossroads.

133. Vermont Department of Environmental Conservation Air Quality and Climate Division, *Vermont Greenhouse Gas Emissions Inventory Update: Brief 1990–2015*, June 2018, 3, https://dec.vermont.gov/sites/dec/files/aqc/cli mate-change/documents/_Vermont_Greenhouse_Gas_Emissions_Inven tory_Update_1990–2015.pdf.

134. "Preliminary Monthly Electric Generator Inventory," U.S. Energy Information Agency, September 2019, https://www.eia.gov/electricity/data /eia860M.

135. Jim Therrien, "Long Awaited Deerfield Wind Turbines Go Online," *VT Digger*, December 30, 2017, https://vtdigger.org. Howard Weiss-Tisman, "After Agreement Over Bear Habitat, Deerfield Wind Project Will Move Forward," *VPR*, August 11, 2016, https://www.vpr.org.

136. Solar energy production has been increasing in Vermont by 50 GWh per year according to the U.S. Energy Information Agency's Electricity Data Browser. Together with Deerfield's 9 GWh per year effective increase in wind energy from 2009 to 2018, Vermont Yankee's 5000 GWh per year

will be replaced if these rates of solar and wind deployment continue for eighty-five years. However, almost no new commercial solar projects were planned as of November 2019, putting the current rate of solar increase in doubt.

137. Katharine Q. Seelye, "After 16 Years, Hopes for Cape Cod Wind Farm Float Away," *New York Times*, December 19, 2017, https://www.nytimes.com.

138. John Laumer, "US Wind Industry Follows 'Starbucks Rule' for Turbine Siting," *Treehugger*, October 15, 2009, https://www.treehugger.com.

10: All About the Green

1. Brady Dennis, "Trump: I'm Not a Big Believer in Man-Made Climate Change," *Washington Post*, March 22, 2016, https://www.washingtonpost.com.

2. Priyanka Boghani, "Meet Myron Ebell, the Climate Contrarian Leading Trump's EPA Transition," *Frontline*, PBS, November 14, 2016, https://www.pbs.org.

3. Tik Root, Lisa Friedman, and Hiroko Tabuchi, "Following the Money That Undermines Climate Science," *New York Times*, July 10, 2019, https://www.nytimes.com.

4. Ibid.

5. "ExxonMobil Foundation & Corporate Giving to Climate Change Denier & Obstructionist Organizations," Greenpeace and Union of Concerned Scientists, 2014, https://www.ucsusa.org/sites/default/files/attach/2015/07/ExxonMobil-Climate-Denial-Funding-1998-2014.pdf.

6. Lee Wasserman, "Did Exxon Deceive Its Investors on Climate Change?," *New York Times*, October 21, 2019, https://www.nytimes.com.

7. Jon Queally, " 'Pete Takes Money from Fossil Fuel Billionaires': Climate Activists Disrupt Buttigieg Rally in New Hampshire," Common Dreams, January 17, 2020, https://www.commondreams.org/news/2020/01/17/pete-takes-money-fossil-fuel-billionaires-climate-activists-disrupt-buttigieg-rally.

8. Griffin Sinclair-Wingate, LinkedIn, accessed February 1, 2020. https://www.linkedin.com/in/griffin-sinclair-wingate-6b1bba119.

9. Bill Allison and Tom Maloney, "Billionaire Candidate Steyer Admits to Carbon 'Dregs' from His Hedge Fund Days," Bloomberg, September 20, 2019,

https://www.bloomberg.com. Forbes estimated Tom Steyer's real-time net worth to be $1.6 billion on January 21, 2020. "Financial Data: 2018 Financials," 350.org, accessed January 30, 2020, https://350.org/2018-annual-report-financials.

10. Michael Barbaro and Coral Davenport, "Aims of Donor Are Shadowed by Past in Coal," *New York Times*, July 4, 2014, https://www.nytimes.com. Carol D. Leonnig, Tom Hamburger, and Rosalind S. Helderman, "Tom Steyer's Slow, and Ongoing, Conversion from Fossil-Fuels Investor to Climate Activist," *Washington Post*, June 9, 2014, https://www.washingtonpost.com.

11. Bill McKibben (@billmckibben), "Important, and unsurprising since @Tom Steyer has long been a climate champ. 21 of 24 Dems now signed on. Things are changing!," Twitter, July 17, 2019, 5:06 p.m., https://twitter.com/billmckibben/status/1151644321453674501; Bill McKibben (@billmckibben), "Since @TomSteyer has been laser-focused on climate policy for longer than just about any political player, it's not surprising that his just-released climate policy is damned good! #demunitytwitterproject," Twitter, July 25, 2019, 3:22 p.m., https://twitter.com/billmckibben/status/1154517342467969024.

12. Mike Brune (@bruneski), "@TomSteyer has been a climate leader for yrs & I'm glad to see yet another climate champ join the primary. Dem candidates' bold proposals to tackle the crisis continue to illustrate the divide between a field of candidates listening to the people vs. Trump who's endangering them," Twitter, July 9, 2019, 8:54 a.m., https://twitter.com/bruneski/status/1148621407233937408.

A few weeks after that, McKibben tweeted praise to Steyer for opposing the Keystone pipeline and for investing millions to impeach Donald Trump.

Bill McKibben (@billmckibben), "Worth noting that @TomSteyer was outpokenly opposed to #KXL pretty much from the jump," Twitter, August 20, 2019, 2:22 p.m., https://twitter.com/billmckibben/status/1163924331577184259; Bill McKibben (@billmckibben), "Credit where due: @TomSteyer worked long and hard to build the baseline support for impeachment," Twitter, September 24, 2019, 3:34 p.m., https://twitter.com/billmckibben/status/1176626062559760384.

13. Eric Orts (@EricOrts), "Please urge this guy to step down. Huge waste of money. And why say you nothing positive of @JayInslee? Too practical for

you?" Twitter, July 17, 2019, 8:51 p.m., https://twitter.com/EricOrts/sta tus/1151700931765055489.

14. Daniel Payettte (@danielpayette), "Bill, if you're seriously in cahoots with Steyer, you lose any and all credibility in this fight," Twitter, August 24, 2019, 7:37 p.m., https://twitter.com/danielpayette/status/1165453054637985795.

15. TomKat Charitable Trust, 990-PF IRS Form 990-PF, 2012, https://www .tomsteyer.com/wp-content/uploads/2019/08/2012_TFS_Tax_Returns _Public.pdf. TomKat Charitable Trust, IRS form 990-PF, 2014, https:// www.tomsteyer.com/wp-content/uploads/2019/08/2014_TFS_Tax_Re turns_Public.pdf. TomKat Charitable Trust, IRS form 990-PF, 2014, https://www.tomsteyer.com/wp-content/uploads/2019/08/2015_TFS _Tax_Returns_Public.pdf. "Financial Data: 2018 Financials," 350.org, https://350.org/2018-annual-report-financials.

16. "Financial Data: 2018 Financials," 350.org, https://350.org/2018-annual -report-financials.

17. TomKat Charitable Trust, IRS Form 990-PF, 2012, https://www.tomsteyer .com/wp-content/uploads/2019/08/2012_TFS_Tax_Returns_Public .pdf; TomKat Charitable Trust, IRS Form 990-PF, 2014, https://www .tomsteyer.com/wp-content/uploads/2019/08/2014_TFS_Tax_Returns_ Public.pdf. TomKat Charitable Trust, IRS Form 990-PF, 2015, https:// www.tomsteyer.com/wp-content/uploads/2019/08/2015_TFS_Tax_Re turns_Public.pdf. "Financial Data: 2018 Financials,"350.org, https://350 .org/2018-annual-report-financials.

18. Carol D. Leonnig et al., "Tom Steyer's slow, and ongoing, conversion from fossil-fuels investor to climate activist," *Washington Post*, June 9, 2014, https://www.washingtonpost.com.

19. Ibid.

20. Michael Barbaro and Coral Davenport, "Aims of Donor Are Shadowed by Past in Coal," *New York Times*, July 4, 2014, https://www.nytimes.com.

21. ABC News, " 'This Week' Transcript 7-14-19: Ken Cuccinelli, Sen. Amy Klobuchar, Tom Steyer," ABC News, July 14, 2019, https://abcnews.go .com.

22. Bill Allison and Tom Maloney, "Billionaire Candidate Steyer Admits to Carbon 'Dregs' from His Hedge Fund Days," Bloomberg, September 20, 2019, https://www.bloomberg.com.

23. TomKat Charitable Trust, IRS Form 990-PF, 2012, https://www.tomsteyer

.com/wp-content/uploads/2019/08/2012_TFS_Tax_Returns_Public
.pdf; TomKat Charitable Trust, IRS Form 990-PF, 2014, https://www
.tomsteyer.com/wp-content/uploads/2019/08/2014_TFS_Tax_Returns_
Public.pdf. TomKat Charitable Trust, IRS Form 990-PF, 2015, https://
www.tomsteyer.com/wp-content/uploads/2019/08/2015_TFS_Tax_Re
turns_Public.pdf. "Financial Data: 2018 Financials,"350.org, https://350
.org/2018-annual-report-financials.

Bloomberg Philanthropies gave EDF $13.8 million from 2012 to
2017. Bloomberg Philanthropies, IRS Form 990-PF, 2011–2015, https://
www.influencewatch.org/non-profit/bloomberg-family-founda
tion-bloomberg-philanthropies. Bloomberg Philanthropies, IRS Form 990-
PF, 2016–2017, https://projects.propublica.org/nonprofits/organizations
/205602483.

Bloomberg Philanthropies gave the Sierra Club $99 million from
2011 to 2017. Bloomberg Philanthropies, IRS Form 990-PF, 2011–2015,
https://www.influencewatch.org/non-profit/bloomberg-family-founda
tion-bloomberg-philanthropies. Bloomberg Philanthropies, IRS Form
990-PF, 2016–2017, https://projects.propublica.org/nonprofits/organizations
/205602483. "Bloomberg Philanthropies Boosts Investment in Sierra Club
to Retire U.S. Coal Fleet & Transition the Nation towards Clean Energy
Sources," Bloomberg Philanthropies, April 8, 2015, https://www.bloomberg
.org.

Bloomberg Philanthropies gave NRDC $5.6 million from 2015 to
2017. Bloomberg Philanthropies, IRS Form 990-PF, 2011–2015, https://
www.influencewatch.org/non-profit/bloomberg-family-founda
tion-bloomberg-philanthropies. Bloomberg Philanthropies, IRS Form
990-PF, 2016-2017, https://projects.propublica.org/nonprofits/organiza
tions/205602483.

24. A single nuclear plant such as Indian Point, in New York, which Sierra Club, NRDC, and EDF have all worked together to close, can provide electricity for two million residents. Indian Point's operator receives
about $50 per MWh of electricity, and an average New Yorker needs about
8 MWh per year. Indian Point produces more than 16 TWh per year according to data from IAEA, with New York's electricity consumption
from the U.S. Energy Information Association and electricity pricing data
from David B. Patton, Pallas LeeVanSchaick, Jie Chen, and Raghu Pa-

lavadi Naga, *2018 State of the Market Report for the New York ISO Markets*, New York Independent System Operator, May 2019, https://www.nyiso .com/documents/20142/2223763/2018-State-of-the-Market-Report.pdf /b5bd2213-9fe2-b0e7-a422-d4071b3d014b?t=1557775606716.

25. For a history of EDF's work, see David Roe, *Dynamos and Virgins* (New York: Random House, 1984).

26. "The Sierra Club Foundation Has a Bright Shining Conflict of Interest," Big Green Radicals, May 1, 2014, accessed February 2, 2020, https://www.big greenradicals.com/the-sierra-club-foundation-has-a-bright-shining-con flict-of-interest. As of early 2020, the composition of the board of directors of the Sierra Club foundation had changed.

27. Michael Shellenberger, "Environmental Defense Fund," Environmental Progress, 2018, http://environmentalprogress.org/edf.

28. Michael Shellenberger, "NRDC," Environmental Progress, April 4, 2018, http://environmentalprogress.org/nrdc.

29. Clayton Aldern, "Looking for a Fossil-Free Investment Fund? Check the Fine Print," *Grist*, December 8, 2015, https://grist.org.

30. Tom Turner, *David Brower: The Making of the Environmental Movement* (Berkeley, CA: University of California Press, 2015), 161.

31. Sharon Beder, "How Environmentalists Sold Out to Help Enron," PR Watch, 2003, accessed January 19, 2020, https://www.prwatch.org.

32. Peter Bondarenko, "Enron Scandal," *Encyclopaedia Britannica*, accessed February 1, 2020, https://www.britannica.com/event/Enron-scandal.

33. Matt Taibbi, "The Great American Bubble Machine," *Rolling Stone*, April 5, 2010, https://www.rollingstone.com.

34. *New Energy, New Solutions: Annual Report 2019*, Environmental Defense Fund, 2019, https://www.edf.org/sites/default/files/content/2019_EDF _Annual_Report.pdf. Natural Resources Defense Council, IRS Form 990, 2017, https://www.nrdc.org/sites/default/files/nrdc-2017-form-990.pdf. Sierra Club Foundation, IRS Form 990 , 2018, https://www.sierraclubfoun dation.org/sites/sierraclubfoundation.org/files/2018%20SCF%20990%20 final%20signed%20Pub%20Disc.pdf. Competitive Enterprise Institute, IRS Form 990, 2017, https://cei.org/sites/default/files/CEI%20990%20FY%20 ending%20Sept.%202018%20Public%20V.pdf. The Heartland Institute, IRS Form 990, 2018, https://www.heartland.org/_template-assets/documents /about-us/2017%20Form%20IRS%20990.pdf. The Environmental Defense

Fund (EDF) had revenues of $202 million, along with $215 million in assets, in 2019. The Natural Resources Defense Council (NRDC) raised $182 million and had $411 million in assets. The Competitive Enterprise Institute brought in just $7 million revenue in 2018 and had just shy of $1 million assets, while the Heartland Institute had revenues of $6 million ($1.5 million in assets).

Michael Shellenberger, "Exxon," Environmental Progress, April 2020, http://environmentalprogress.org/exxon.

35. The Heritage Foundation, "2018 Financial Statements," December 31, 2018, https://thf-reports.s3.amazonaws.com/Financial/2018_AnnualReport_Financials.pdf.

36. American Enterprise Institute, *2018 Annual Report*, 2018, https://www.aei.org/wp-content/uploads/2019/01/AEI-2018-Annual-Report.pdf.

37. Cato Institute, "Life Liberty the Pursuit of Happiness: 2018 Annual Report," https://www.cato.org/sites/cato.org/files/pubs/pdf/cato-annual-report-2018.pdf.

38. Steven F. Hayward, Mark Muro, Ted Nordhaus, and Michael Shellenberger, *Post-partisan Power*, AEI, Brookings, and Breakthrough Institute, 2010, https://www.politico.com/pdf/PPM170_post-partisan_power-1.pdf, 8. Alex Brill and Alex Flint, "Carbon Tax Most Efficient in Tackling Climate Change," American Enterprise Institute, March 1, 2019, https://www.aei.org.

39. Center for American Progress, IRS Form 990, 2018, https://projects.propublica.org/nonprofits/organizations/300126510/201912639349301106/full; The Nature Conservancy, "Consolidated Financial Statements," June 30, 2019, https://www.nature.org/content/dam/tnc/nature/en/documents/TNC-Financial-Statements-FY19.pdf.

40. "Howard Learner, President and Executive Director," Staff, Environmental Law and Policy Center, accessed October 2, 2019, http://elpc.org/staff/howard-a-learner.

41. Edith Brady-Lunny, "Absent Legislative Action, Clinton Power Station Faces Dark Future," *Herald & Review*, April 24, 2016, https://herald-review.com.

42. Bryan Walsh, "Exclusive: How the Sierra Club Took Millions From the Natural Gas Industry—and Why They Stopped," *Time*, February 2, 2012, https://time.com.

43. Alleen Brown, "The Noxious Legacy of Fracking King Aubrey McClendon," *Intercept*, March 7, 2016, https://theintercept.com.

44. Bryan Walsh, "Exclusive: How the Sierra Club Took Millions From the Natural Gas Industry—and Why They Stopped."

45. Ibid.

46. Russell Mokhiber, "Sierra Club Tells Members—We Don't Take Money from Chesapeake Energy—When in Fact They Took $25 Million," Corporate Crime Reporter, February 2, 2012, https://www.corporatecrimereporter.com.

47. Derek Seidman, "Climate Summit Co-chair Michael Bloomberg Backs Fracking and Invests in Fossil Fuels," *Truthout*, September 11, 2018, https://truthout.org; Timothy Gardner, "Bloomberg's Charity Donates $64 Million to 'War on Coal,' " Reuters, October 11, 2017, https://www.reuters.com.

48. Carl Nelburger, "Rally Spurs Brown to Oppose Diablo," San Luis Obispo County *Telegraph-Tribune*, July 1, 1979, https://static1.square space.com/static/56a45d683b0be33df885def6/t/5a81f84424a6941151341938/1518467142362/No-nukes-concert.pdf.

49. Ibid.

50. Thomas Wellock, *Critical Masses: Opposition to Nuclear Power in California, 1958–1978* (Madison: University of Wisconsin Press, 1992), 200.

51. Ibid., 200.

52. Ibid., 176.

53. Ibid., 176.

54. Madison Czerwinski, Environmental Progress, "Nuclear Power Archive," 2020, http://environmentalprogress.org/nuclear-power-archive.

55. Gayle Spinazze, "California Governor Jerry Brown to Join the Bulletin of the Atomic Scientists as Executive Chair," *Bulletin of the Atomic Scientists*, October 31, 2018, https://thebulletin.org/2018/10/california-governor-jerry-brown-to-join-the-bulletin-of-the-atomic-scientists-as-executive-chair.

56. Dan Walters, "The Brown Link to Indonesian Oil Firm," *Lodi News-Sentinel*, October 17, 1990, https://news.google.com/newspapers?nid=2245&dat=19901017&id=HhYzAAAAIBAJ&sjid=MTIHAAAAIBAJ&pg=4793%2C5589478.

57. Peter Bryne, "Bringing up Baby Gavin," *SF Weekly*, April 2, 2003, https://sfweekly.com.

58. Dan Walters, "The Brown Link to Indonesian Oil Firm."

59. Jeff Gerth, "Gov. Brown Supporting Projects That Aid a Mexican Contributor," *New York Times*, March 11, 1979, https://timesmachine.nytimes.com.

60. Ibid.

61. Ibid.

62. Mark A. Stein, "Utility to Ask Voters for 18-Month Rancho Seco Reprieve," *Los Angeles Times*, March 11, 1988, https://www.latimes.com.

63. Rob Nikolewski, "Regulators Vote to Shut Down Diablo Canyon, California's Last Nuclear Power Plant," *Los Angeles Times*, January 11, 2018, https://www.latimes.com.

64. Melinda Henneberger, "Al Gore's Petrodollars Once Again Make Him a Chip off the Old Block," *Washington Post*, January 8, 2013, https://www.washingtonpost.com.

65. "Al Gore Is Big Oil," *Washington Post*, September 29, 2000, https://www.washingtonpost.com.

66. Center for Public Integrity, "How the Gores, Father and Son, Helped Their Patron Occidental Petroleum," January 10, 2000, https://publicintegrity.org.

67. Ibid.

68. Brad Plumer, "Al Gore: 'The Message Still Has to Be About the Reality We're Facing,' " *Washington Post*, September 12, 2011, https://www.washingtonpost.com.

69. Emily Steel, "Al Gore Sues Al Jazeera over TV Deal," *New York Times*, August 15, 2014, https://www.nytimes.com.

70. Henneberger, "Al Gore's Petrodollars Once Again Make Him a Chip off the Old Block."

71. John Koblin, "Al Jazeera America to Shut Down by April," *New York Times*, January 13, 2016.

72. Alexei Koseff, " 'It's Literally Drill, Baby, Drill': Did Jerry Brown's Climate Crusade Give Big Oil a Pass?," *Sacramento Bee*, September 13, 2018, https://www.sacbee.com.

73. Liza Tucker, "Brown's Dirty Hands," *Consumer Watchdog*, September 2017, https://www.consumerwatchdog.org/resources/BrownsDirtyHands.pdf.

74. Associated Press, "Gov. Jerry Brown Had State Workers Research Oil on Family Ranch," *Los Angeles Times*, November 5, 2015, https://www.latimes.com.

75. Tucker, "Brown's Dirty Hands."

76. Ibid.

77. *Times* Editorial Board, "Editorial: Does California Now Have a Utilities Regulator It Can Trust?," *Los Angeles Times*, June 29, 2016, https://www.latimes.com.

78. Elizabeth Douglas, "Outage at San Onofre May Cost Hundreds of Millions," *San Diego Union-Tribune*, June 14, 2012, https://www.sandiegouniontribune .com. Morgan Lee, "Nuclear settlement gets mixed reception," *San Diego Union-Tribune*, June 16, 2014, https://www.sandiegouniontribune.com.

79. Liza Tucker, "Brown's Dirty Hands." Lucas Davis, "Too Big to Fail?," Energy Institute at Haas, March 31, 2014, https://energyathaas.wordpress .com/2014/03/31/too-big-to-fail.

80. Jeff McDonald, "Aguirre Pushing for Brown's Emails," *San Diego Union-Tribune*, November 13, 2015, https://www.sandiegouniontribune.com.

81. Tony Kovaleski, Liz Wagner, and Felipe Escamilla, "Attorneys Suggest Evidence Isn't Safe at CPUC amid Federal Investigation," NBC Bay Area, October 16, 2014, https://www.nbcbayarea.com.

82. Jaxon Van Derbeken, "Judge: Regulator Should Release Brown Emails on Nuclear Shutdown," SF Gate, November 28, 2018, https://www.sfgate.com.

83. Jeff McDonald, "San Onofre Plan Detail Under Scrutiny in Probe," *San Diego Union-Tribune*, March 14, 2015, https://www.sandiegouniontribune .com.

84. John Doerr, "Salvation (and Profit) in Greentech," TED2007, March 2007, https://www.ted.com/talks/john_doerr_salvation_and_profit_in_green tech/transcript?language=en.

85. Michael Shellenberger and Adam Werbach, "It's the Oil Economy, Stupid," SF Gate, March 10, 2003, https://www.sfgate.com.

86. Michael Shellenberger et al., "Beyond Boom and Bust: Putting Clean Tech On a Path To Subsidy Independence," Brookings, April 12, 2012, https:// bloomberg.com.

Ira Boudway, "The 5 Million Green Jobs That Weren't," *Bloomberg Businessweek*, October 12, 2012, https://www.bloomberg.com.

87. Peter Schweitzer, *Throw Them All Out: How Politicians and Their Friends Get Rich off Insider Stock Tips, Land Deals, and Cronyism That Would Send the Rest of Us to Prison* (New York: Mariner Books, 2012), 96.

88. Frank Rusco, *Recovery Act: Status of Department of Energy's Obligations and Spending, Testimony Before the Subcommittee on Oversight and Investigations, Committee on Energy and Commerce, House of Representatives*, GAO-11-483T, 10 (Washington, D.C.: Diane Publishing, March 17, 2011).

89. Carol D. Leonnig and Joe Stephens, "Energy Dept. E-mails on Solyndra Provide New Details on White House Involvement," *Washington Post*, August 9, 2012, https://www.washingtonpost.com. Schweitzer, *Throw Them All Out*, 96.

90. Ibid.

91. "Reply Comments of Tesla, Inc. on Proposed Decision Approving Retirement of Diablo Canyon Nuclear Power Plant," August 11, 2016, Decision, http://docs.cpuc.ca.gov/PublishedDocs/Efile/G000/M201/K923/201923970.PDF; "California home battery rebate: Self-Generation Incentive Program (SGIP) explained," Energy Sage, December 6, 2019; https://news.energysage.com. "Energy storage resources paired with solar generation can take advantage of federal tax credits," noted Tesla's attorney.

92. Lachlan Markay, "How Dem Moneyman Nat Simons Profits from Political Giving," *Washington Free Beacon*, November 4, 2014, https://freebeacon.com.

93. Ibid.

94. The World Watch Institute, *State of the World 2014: Governing Sustainability* (Washington, D.C.: Island Press, 2014), 121.

95. "LD-2 Disclosure Form," United States Senate, accessed January 3, 2020, https://soprweb.senate.gov/index.cfm?event=getFilingDetails&filingID=f87d42e3–8272–4c5d-914d-85198279ac67&filingTypeID=71; "LD-2 Disclosure Form," United States Senate, accessed January 3, 2020, https://soprweb.senate.gov/index.cfm?event=getFilingDetails&filingID=ff992139–85el–48b5–86eb-c8ecfc42a083&filingTypeID=60.

96. David Colgan, "Renewable Energy Ignites Debate," UCLA Institute of the Environment and Sustainability, February 23, 2016, https://www.ioes.ucla.edu/article/renewable-energy-ignites-debate.

97. "Precourt Institute Energy Advisory Council," People, Precourt Institute, accessed January 6, 2020, https://energy.stanford.edu/people/precourt-institute-energy-advisory-council.

98. Howard Fischer, "Tom Steyer's PAC to Spend More Money Telling Arizona Voters About APS' Influence," *Arizona Daily Star*, May 14, 2019, https://tucson.com/tom-steyer-s-pac-to-spend-more-money-telling-arizona/article_12e594f0-73a1-5e67-9c1a-eb83a261a801.html.

99. Steyer's group NextGen Climate Action gave the Sierra Club $250,000 in 2016. NextGen Climate Action, IRS Form 990, 2017, https://projects.propublica.org/nonprofits/display_990/461957345/12_2017_prefixes_45-46%2F461957345_201612_990O_2017123015067289. Steyer and his wife, Kat Taylor, have also hosted a benefit for the Sierra Club. See: "2017 Guardians of Nature Benefit," Sierra Club Lome Prieta, https://www.sierraclub.org/loma-prieta/benefit17.

Todd Schifeling and Andrew Hoffman, "How Bill McKibben's radical idea of fossil fuel divestment transformed the climate debate," Conversation, December 11, 2017, https://theconversation.com/how-bill-mckibbens-radical -idea-of-fossil-fuel-divestment-transformed-the-climate-debate-87895.
Bloomberg Philanthropies gave EDF $13.8 million from 2012 to 2017. Bloomberg Philanthropies, IRS Form 990-PF, 2011–2015, https:// www.influencewatch.org/non-profit/bloomberg-family-founda tion-bloomberg-philanthropies. Bloomberg Philanthropies, IRS Form 990-PF, 2016–2017, https://projects.propublica.org/nonprofits/organizati ons/205602483.
Derek Seidman, "Climate Summit Co-Chair Michael Bloomberg Backs Fracking and Invests in Fossil Fuels," *Truthout*, September 11, 2018.
TomKat Charitable Trust, IRS Form 990-PF, 2012, https://www.tom steyer.com/wp-content/uploads/2019/08/2012_TFS_Tax_Returns_Public .pdf; TomKat Charitable Trust, IRS Form 990-PF, 2014, https://www.tom steyer.com/wp-content/uploads/2019/08/2014_TFS_Tax_Returns_Public .pdf. TomKat Charitable Trust, IRS Form 990-PF, 2015, https://www.tom steyer.com/wp-content/uploads/2019/08/2015_TFS_Tax_Returns_Public .pdf. "Financial Data: 2018 Financials," 350.org, https://350.org/2018-annu al-report-financials.
Bloomberg Philanthropies gave NRDC $5.6 million from 2015 to 2017. Bloomberg Philanthropies, IRS Form 990-PF, 2011–2015, https:// www.influencewatch.org/non-profit/bloomberg-family-founda tion-bloomberg-philanthropies. Bloomberg Philanthropies, IRS Form 990-PF, 2016–2017, https://projects.propublica.org/nonprofits/organiza tions/205602483.
Bloomberg Philanthropies gave the Sierra Club $99 million from 2011 to 2017. Bloomberg Philanthropies, IRS Form 990-PF, 2011–2015, https://www.influencewatch.org/non-profit/bloomberg-family-founda tion-bloomberg-philanthropies. Bloomberg Philanthropies, IRS Form 990-PF, 2016–2017, https://projects.propublica.org/nonprofits/organizations /205602483. "Bloomberg Philanthropies Boosts Investment in Sierra Club to Retire U.S. Coal Fleet & Transition the Nation towards Clean Energy Sources," Bloomberg Philanthropies, April 8, 2015; https://www .bloomberg.org.
Steyer's group NextGen Climate Action gave the EDF $54,000 in 2016. NextGen Climate Action, IRS Form 990, 2017, https://projects

.propublica.org/nonprofits/display_990/461957345/12_2017_prefixes
_45-46%2F461957345_201612_990O_2017123015067289.

100. Bryan Anderson, "Did Tom Steyer Buy His Way into the Democratic Debate? How He Worked His way to the Stage," *Sacramento Bee*, October 15, 2019, https://www.sacbee.com.

11: The Denial of Power

1. Neal Baker, Emily Smith, Ebony Bowden, and Grant Rollings, "HAZ-GREEN: Prince Harry 'Gives Barefoot Speech' at Google Climate Change Retreat to Celebs Who Flocked In on 114 Gas-Guzzling Jets and Superyachts," *The Sun*, August 2, 2019, https://www.thesun.co.uk. "Climate change: Why have Prince Harry and Meghan been criticised for using a private jet?" BBC, August 20, 2019, https://www.bbc.co.uk.

2. Neal Baker, Emily Smith, Ebony Bowden, and Grant Rollings, "HAZ-GREEN: Prince Harry 'Gives Barefoot Speech' at Google Climate Change Retreat to Celebs Who Flocked In on 114 Gas-Guzzling Jets and Superyachts," *The Sun*, August 2, 2019, https://www.thesun.co.uk.

3. Ibid.

4. "Climate Change: Why Have Prince Harry and Meghan Been Criticised for Using a Private Jet?," BBC, August 20, 2019, https://www.bbc.com.

5. Michael Hamilton, "DUMBO JET: 'Eco-warriors' Meghan Markle and Prince Harry Fly on Private Jet Again to France After Gas-Guzzling Ibiza Trip," *The Sun*, August 17, 2019, https://www.thesun.co.uk.

6. Ibid.

7. Reality Check Team, "Prince Harry and Private Jets: What's the Carbon Footprint?," BBC, August 20, 2019, https://www.bbc.com.

8. Suzy Byrne, "Ellen DeGeneres, Pink Join Elton John in Defending Prince Harry and Meghan Markle over Private Jet Drama: 'Imagine Being Attacked for Everything You Do,' " Yahoo! Entertainment, August 20, 2019, https://www.yahoo.com/entertainment.

9. Lynette (@Lynette55), "So stop lecturing us on how we live our lives and live by example. No private jets every couple of weeks etc etc. You only have what you have because of us, the people, both celebrities and the Royals," Twitter, August 19, 2019, 2:09 p.m., https://twitter.com/Lynette55/status/1163558755583373312.

10. Associated Press, "Gore gets green kudos for home renovation," *NBC News*, December 13, 2007, http://www.nbcnews.com. Al Gore, *An Inconvenient Truth: The Planetary Emergency of Global Warming and What We Can Do About It* (Emmaus, PA: Rodale, 2006), 286.

11. Malena Ernman, "Malena Ernman on daughter Greta Thunberg: 'She was slowly disappearing into some kind of darkness,' " *The Guardian*, February 23, 2020, https://www.theguardian.com.

12. Agence-France Presse, "CO_2 Row over Climate Activist Thunberg's Yacht trip to New York," France 24, August 18, 2019, https://www.france24.com /en/20190818-co2-row-over-climate-activist-thunberg-s-yacht-trip-to -new-york.

13. "Energy Use vs. GDP per Capita, 2015," Our World in Data, https://our worldindata.org/grapher/energy-use-per-capita-vs-gdp-per-capita.

14. Hannah Ritchie and Max Roser, "Primary Energy Consumption by World Region" (table), "Energy," Our World in Data, July 2018, https://ourworld indata.org/energy.

15. A 2009 study found that 12–17 percent of China's energy consumption was embodied in products for export. Ming Xu, Braden Allenby, and Weiqiang Chen, "Energy and Air Emissions Embodied in China-US Trade: Eastbound Assessment Using Adjusted Bilateral Trade Data," *Environmental Science and Technology* 43, no. 9 (2009): 3378–84, https://doi.org/10.1021/es803142v. A 1994 study found that 13 percent of the carbon emissions of the world's six largest economies (the United States, Great Britain, Japan, Germany, France, and Canada) was "embodied" in manufacturing imports. Andrew W. Wyckoff and Joseph M. Roop, "The Embodiment of Carbon in Imports of Manufactured Products: Implications for International Agreements on Greenhouse Gas Emissions," *Energy Policy* 22, no. 3 (1994): 187–94, https://doi .org/10.1016/0301-4215(94)90158-9. Per capita energy consumption in the U.S. rose from 227 million BTUs in 1950 to its peak of 350 million BTUs in 2000 and declined to 309 million BTUs in 2018. "Primary Energy Consumption," U.S. Energy Information Administration, https://www.eia.gov/total energy/data/monthly/pdf/sec1_17.pdf.

16. Zeke Hausfather, "Mapped: The World's Largest CO_2 Importers and Exporters," *Carbon Brief*, July 5, 2017, https://www.carbonbrief.org.

17. Jillian Ambrose and Jon Henley, "European Investment Bank to Phase Out Fossil Fuel Financing," *The Guardian*, November 15, 2019, https://www .theguardian.com.

18. Sebastian Mallaby, *The World's Banker: A Story of Failed States, Financial Crises, and the Wealth and Poverty of Nations* (New York: Penguin, 2004), 336.

19. John Briscoe, "Invited Opinion Interview: Two Decades at the Center for World Water Policy," *Water Policy* 13, no. 2 (2011): 147–60, https://doi .org/10.2166/wp.2010.000.

20. Ibid.

21. Ibid.

22. John Briscoe, "Infrastructure First? Water Policy, Wealth, and Well-Being," Belfer Center, January 28, 2012, https://www.belfercenter.org/publication /infrastructure-first-water-policy-wealth-and-well-being.

23. "Report of the World Commission on Environment and Development: Our Common Future," Annex to U.N. General Assembly document A/42/427, ¶ 63 (1987), United Nations, http://www.un-documents.net/wced-ocf.htm.

24. Ibid.

25. Charles Recknagel, "What Can Norway Teach Other Oil-Rich Countries?," Radio Free Europe/Radio Liberty, November 27, 2014, https://www.rferl .org/a/what-can-norway-teach-other-oil-rich-countries/26713453.html.

26. José Goldemberg, "Leapfrog Energy Technologies," *Energy Policy* 26, no. 10 (1998): 729–41, https://sites.hks.harvard.edu/sed/docs/k4dev/goldemberg _energypolicy1998.pdf.

27. Mark Malloch Brown, Nitin Desai, Gerald Doucet et al., *World Energy Assessment: Energy and the Challenge of Sustainability*, United Nations Development Programme, 2000, https://www.undp.org/content/dam/aplaws /publication/en/publications/environment-energy/www-ee-library /sustainable-energy/world-energy-assessment-energy-and-the-chal lenge-of-sustainability/World%20Energy%20Assessment-2000.pdf, 22.

28. Ibid., 55.

29. Hinh Dinh (former World Bank economist) in discussion with the author, February 21, 2016.

30. John Briscoe, "Hydropower for Me but Not for Thee—with Two Postscripts," Center for Global Development, March 6, 2014, https://www.cgdev .org/blog/hydropower-me-not-thee.

31. Justin Gillis, "Sun and Wind Alter Global Landscape, Leaving Utilities Behind," *New York Times*, September 13, 2014, https://www.nytimes.com.

32. Ricci Shryock, "Rwanda's Prison System Innovates Energy from Human Waste," *VOA*, April 1, 2012, https://www.voanews.com.

33. Eva Müller, "Forests and Energy: Using Wood to Fuel a Sustainable, Green

Economy," International Institute for Sustainable Development, M
2017, http://sdg.iisd.org/commentary/guest-articles/forests-and-e
ing-wood-to-fuel-a-sustainable-green-economy.

34. Joyashree Roy et al., "Chapter Five: Sustainable Developmen
Eradication, and Reducing Inequalities," in *Global Warming*
An IPCC Special Report on the impacts of global warming of 1.5°
industrial levels and related global greenhouse gas emission pathway
text of strengthening the global response to the threat of climate chang
development, and efforts to eradicate poverty, V. Masson-Delmon
(IPCC, 2018), 4–29, 4–59, 4–114, 4–117, https://www.ipcc.ch

35. Heike Mainnhardt, "World Bank Group Financial Flows Unde
Climate Agreement," Urgewald, March 2019, https://urgewa
fault/files/World_Bank_Fossil_Projects_WEB.pdf.

36. Briscoe, "Hydropower for Me but Not for Thee—with Two

37. Robert J. Mayhew, *Malthus: The Life and Legacies of an*
(Cambridge, MA: Harvard University Press, 2014), 43–45
troduction to William Godwin, *An Enquiry Concerning P*
ford, UK: Oxford University Press, 2013), x–xxii.

38. "Each century will add new enlightenment to that of the
ceded it," Condorcet wrote in 1782, "and this progress
henceforth halt or delay, will have no other limits than t
the universe." Ibid.

39. Ibid., 41.

40. Thomas Robert Malthus, *An Essay on the Principle of F*
ford University Press, 1993), 61.

41. Thomas Malthus, *Essay on the Principle of Population*
Haven, CT: Yale University Press, 2018), 417.

42. Robert J. Mayhew, *Malthus: The Life and Legaci*
(Cambridge, MA: Harvard University, 2014), 45.

43. Godwin, *An Enquiry Concerning Political Justice,* 45

44. Malthus, *An Essay on the Principle of Population,* 6

45. Thomas Robert Malthus, *Observations on the Eff*
a Rise or Fall in the Price of Corn on the Agricult
Country (London: John Murray, 1915), 30.

46. Mayhew, *Malthus: The Life and Legacies of an U*

47. Ibid., 17, 18.

48. Christine Kinealy, *The Great Irish Famine:*

(London: Palgrave, 2002), 105–111. In 1846, Ireland exported three million quarts of grain and corn flour to Britain, and 730,000 cattle and livestock.

49. Quoted in Fred Pearce, *The Coming Population Crash: And Our Planet's Surprising Future* (Boston: Beacon Press, 2010), 18.

50. Malthus, *An Essay on the Principle of Population: The 1803 Edition*, 265.

51. Thomas Malthus, letter to David Ricardo, 1817, in *Thomas Robert Malthus: Critical Assessments*, John Cunningham Wood, ed. (London: Routledge, 1994), 262.

52. John and Richard Strachey, *The Finances and Public Works of India* (London: K. Paul Trench & Company, 1882), 172.

53. The House of Commons of the United Kingdom, "Copy of Correspondence Between the Secretary of State for India and the Government of India, on the Subject of the Famine in Western and Southern India," in *Parliamentary Papers*, vol. 59, H.M. Stationery Office, 1878, 14.

54. Mayhew, *Malthus: The Life and Legacies of an Untimely Prophet*, 174.

55. Ibid., 177.

56. Ibid., 179–181.

57. "Valley of the Dams: The Impact & Legacy of the Tennessee Valley Authority," National Archives at Atlanta, accessed January 27, 2020, https://www.archives.gov/atlanta/exhibits/tva-displacement.html.

John Crowe Ransom, "Reconstructed but Unregenerate," in *I'll Take My Stand: The South and the Agrarian Tradition*, edited by Louis D. Rubin, Jr. (Baton Rouge: Louisiana State University Press, 1977), 1–27.

Mayhew, *Malthus: The Life and Legacies of an Untimely Prophet*, 141–143.

Henry George, *Progress and Poverty: An Inquiry Into the Cause of Industrial Depressions and of Increase of Want with Increase of Wealth. The Remedy.* (New York: D. Appleton and Company, 1881), 304.

William Tucker, *Progress and Privilege: America in the Age of Environmentalism* (New York: Anchor Press/Doubleday, 1982), 37.

Mayhew, *Malthus: The Life and Legacies of an Untimely Prophet*, 206, 208.

Ibid., 190.

William Vogt, *Road to Survival* (New York: William Sloane, 1948), 226–27.

Ibid., 48.

Ibid.

Ibid., 58.

68. Ibid., 278.

69. Lyndon Baines Johnson, "State of the Union," January 4, 1965, https://millercenter.org/the-presidency/presidential-speeches/january-4-1965-state-union.

70. The New York Times Editorial Board, "Johnson vs. Malthus," *New York Times*, January 24, 1966, https://www.nytimes.com.

71. Mayhew, *Malthus: The Life and Legacies of an Untimely Prophet*, 199.

72. Paul Ehrlich, *The Population Bomb* (New York: Ballantine Books, 1969), xi.

73. Ibid., 15–16.

74. Paul Ehrlich and Anne Ehrlich, "The Population Bomb Revisited," *The Electronic Journal of Sustainable Development* 1, no. 3 (2009), https://www.populationmedia.org/wp-content/uploads/2009/07/Population-Bomb-Revisited-Paul-Ehrlich-20096.pdf.

75. Garrett Hardin, "Lifeboat Ethics: The Case Against Helping the Poor," *Psychology Today*, September 1974.

76. Peter Passell, Marc Roberts, and Leonard Ross, "The Limits to Growth," [review] *New York Times*, April 2, 1972, https://www.nytimes.com.

77. Barry Commoner, *Crossroads: Environmental Priorities for the Future* (Washington, D.C.: Island Press, 1988), 146.

78. William Tucker, *Progress and Privilege* (New York: Anchor Press/Doubleday, 1982), 108, 109.

79. Ibid., 108, 109.

80. Amory B. Lovins, "Energy Strategy: The Road Not Taken?," *Foreign Affairs*, October 1976, 65–96, https://www.foreignaffairs.com/articles/united-states/1976-10-01/energy-strategy-road-not-taken.

81. Ibid.

82. Ibid.

83. The Mother Earth News Editors, "Amory Lovins: Energy Analyst and Environmentalist," *Mother Earth News*, November–December 1977, https://www.motherearthnews.com/renewable-energy/amory-lovins-energy-analyst-zmaz77ndzgoe.

84. Paul Ehrlich, "An Ecologists's Perspective on Nuclear Power," *F.A.S. Public Interest Report*, May–June 1975, https://fas.org/faspir/archive/1970-1981/May-June1975.pdf, 3–6.

85. Paul Ehrlich, John P. Holdren, and Anne Ehrlich, *Ecoscience: Population, Resources, Environment* (San Francisco, CA: W. H. Freeman and Company, 1977), 923.

86. Ibid., 350, 401.

87. Ibid., 350.

88. Ibid., 350.

89. Ibid., 350–351.

90. Ibid., 435.

91. Gerald O. Barney, ed., *Entering the Twenty-First Century*, vol. 1 of *The Global 2000 Report to the President of the United States* (Washington, D.C.: Government Printing Office, 1980), 1.

92. David Stradling, ed., *The Environmental Moment: 1968–1972* (Seattle, WA: University of Washington Press, 2012), 131.

93. Ben Wattenberg, "The Nonsense Explosion," in *Image & Event: American Now*, ed. David L. Bicknell and Richard L. Brengle (New York: Meredith Corporation, 1971), 262.

94. Mayhew, *Malthus: The Life and Legacies of an Untimely Prophet*, 203–204.

95. Amartya Sen, *Poverty and Famines: An Essay on Entitlement and Deprivation* (Oxford: Oxford University Press, 1981).

96. Mayhew, *Malthus: The Life and Legacies of an Untimely Prophet*, 214.

97. Harold J. Barnett and Chandler Morse, *Scarcity and Growth* (London: Routledge, 1963), 7.

98. Frederick Soddy, *The Interpretation of Radium* (New York: G.P. Putnam's Sons, 1922).

99. Thomas R. Wellock, *Critical Masses: Opposition to Nuclear Power in California, 1958–1978* (Madison, WI: University of Wisconsin Press, 1992), 85.

100. Bertrand Russell, *Detente or Destruction, 1955–57* (London: Routledge, 2005), 328.

101. Ehrlich, Holdren, and Ehrlich, *Ecoscience: Population, Resources, Environment*, 445.

102. United Nations World Commission on Environment and Development, *Our Common Future, from One Earth to One World* [*Brundtland Report*] (Oxford and New York: Oxford University Press, 1987), https://sustainabledevelopment.un.org/content/documents/5987our-common-future.pdf.

103. Stephen H. Schneider, *Science As a Contact Sport* (Washington, D.C.: National Geographic, 2009), 39. Schneider writes, "Opponents called them neo-Malthusians," as though it were some kind of a smear when in fact the Ehrlichs and Holdren explicitly endorse Malthusian ideas in their 1977 textbook.

104. Ibid., 65.

105. Ibid., 72.

106. Mark Sagoff, "The Rise and Fall of Ecological Economics," *Breakthrough Journal*, Winter 2012, https://thebreakthrough.org/.

107. Ibid.

108. Bill McKibben, *The Bill McKibben Reader: Pieces from an Active Life* (New York: Holt Paperbacks, 2008), 189.

109. Bill McKibben, "A Special Moment in History," *The Atlantic Monthly*, May 1998.

110. Schneider, *Science As a Contact Sport*, 193.

111. John Briscoe, "Invited Opinion Interview: Two decades at the center for world water policy," *Water Policy* 13, no. 2 (February 2011): 156, https://doi.org/10.2166/wp.2010.000.

112. John Briscoe, "Invited Opinion Interview: Two decades at the center for world water policy," *Water Policy* 13, no. 2 (February 2011): 150, https://doi.org/10.2166/wp.2010.000.

113. Ibid., 151.

114. International Rivers Network, "History and Accomplishments," International Rivers, accessed November 20, 2019, https://www.internationalrivers.org/history-accomplishments.

115. Sebastian Mallaby, *The World's Banker* (New York: Penguin, 2004), 7–8.

116. Yolanda Machena and Sibonginkosi Maposa, "Zambezi basin dam boom threatens delta," *World Rivers Review*, June 10, 2013, https://www.internationalrivers.org/world-rivers-review.

117. Ranjit Deshmukh, Ana Mileva, and Grace C. Wu, "Renewable energy alternatives to mega hydropower: A case study of Inga 3 for Southern Africa," *Environmental Research Letters* 13, no. 6 (June 5, 2018), https://iopscience.iop.org/article/10.1088/1748–9326/aabf60. International Rivers Network, "Inga 3: An Exclusive Development Deal for Chinese and European companies," International Rivers, October 17, 2018, https://www.internationalrivers.org/resources/press-statement-inga-3-an-exclusive-development-deal-for-chinese-and-european-companies; Ranjit Deshmukh, Ana Mileva, and Grace C. Wu, *Renewable Riches: How Wind and Solar Can Power DRC and South Africa* (Hatfield, South Africa: International Rivers, 2017), https://www.internationalrivers.org/sites/default/files/attached-files/ir_inga_re_report_2017_fa_v2_email_1.pdf.

118. "Hydropower," International Energy Agency, accessed November 20, 2019, https://www.iea.org/topics/renewables/hydropower. "Hydroelectricity's many advantages include reliability, proven technology, large storage capac-

ity, and very low operating and maintenance costs. . . . Many hydropower plants also provide flood control, irrigation, navigation and freshwater supply."

119. The power densities for hydroelectric dams range from as little as 0.4 watts per square meter (We/m^2) on the flat Egypt desert to 56 We/m^2 for tall dams in places like Switzerland. Larger dams are nearly ten times more land-efficient than smaller dams, and dams that require larger reservoirs are less land-efficient than ones that require smaller ones. Vaclav Smil, *Power Density: A Key to Understanding Energy Sources and Uses* (Cambridge, MA: MIT Press, 2016), 73.

120. Opponents at IRN claim that the Inga Dam's reservoir would occupy an area of 22,000 hectares (220 km^2). With a power of 39 GWe, this would mean a power density of 177 We/m^2. "Grand Inga Hydroelectric Project: An Overview," International Rivers, accessed February 1, 2020, https://www.internationalrivers.org/resources/grand-inga-hydroelectric-project-an-overview-3356.

121. Berkeley and much of the Bay Area received power from the O'Shaughnessy Dam, which is part of the Hetch Hetchy Project and was built between 1919 and 1923. *Hetch Hetchy Reservoir Site: Hearing on H.R. 7207, before the Committee on Public Lands*, 63rd Cong. (1913).

122. Alice Scarsi, "Prince Harry under fire for attending eco event packed with private jets," *Daily Express*, August 1, 2019, https://www.express.co.uk; Baker et al., "HAZ-GREEN: Prince Harry 'gives barefoot speech' at Google climate change retreat to celebs who flocked in on 114 gas-guzzling jets and superyachts."

123. Jost Maurin, "Thunbergs Segelreise in die USA Gretas Törn schädlicher als Flug," *Taz*, August 15, 2019, https://taz.de/Thunbergs-Segelreise-in-die-USA-!5615733.

124. Greta Thunberg, *No One Is Too Small to Make a Difference* (New York: Penguin, 2019), 96–99.

125. Ben Geman, "Obama: Power Africa Energy Plan a 'Win-Win,'" *The Hill*, July 2, 2013. Preeti Aroon, "Sorry, Obama, Soccer Balls Won't Bring Progress to Africa," *Foreign Policy*, July 2, 2013, https://foreignpolicy.com.

126. Gayathri Vaidyanathan, "Coal Trumps Solar in India," ClimateWire, *Scientific American*, October 19, 2015, https://www.scientificamerican.com/article/coal-trumps-solar-in-india.

127. Joyashree Roy et al., "Chapter 5: Sustainable Development, Poverty Eradication 3 and Reducing Inequalities," in V. Masson-Delmotte et al., eds., *Global Warming of 1.5°C. An IPCC Special Report on the impacts of global warming of 1.5°C above pre-industrial levels and related global greenhouse gas emission pathways, in the context of strengthening the global response to the threat of climate change, sustainable development, and efforts to eradicate poverty* (IPCC, 2018), https://www.ipcc.ch/sr15.

128. Joyashree Roy (IPCC author) in discussion with the author, November 11, 2019.

12: False Gods for Lost Souls

1. Cristina Mittermeier, "Starving-Polar-Bear Photographer Recalls What Went Wrong," *National Geographic*, July 26, 2018, https://www.nationalgeographic.com.

2. Jonathan Watts, "Greta Thunberg, Schoolgirl Climate Change Warrior: Some People Can Let Things Go. I Can't," *The Guardian*, March 11, 2019, https://www.theguardian.com.

3. Erica Goode, "Climate Change Denialists Say Polar Bears Are Fine. Scientists Are Pushing Back," *New York Times*, April 10, 2018, https://www.nytimes.com. Roz Pidcock, "Polar Bears and Climate Change: What Does the Science Say?," *Carbon Brief*, July 18, 2017, https://www.carbonbrief.org.

4. Naomi Oreskes and Erik Conway, *Merchants of Doubt: How a Handful of Scientists Obscured the Truth on Issues from Tobacco Smoke to Global Warming* (New York: Bloomsbury, 2010). Claudia Dreyfus, "Naomi Oreskes Imagines the Future History of Climate Change," *New York Times*, October 12, 2014, https://www.nytimes.com. Melissa Block, " 'Merchants of Doubt' Explores Work of Climate Change Deniers," NPR, March 6, 2014, https://www.npr.org. Justin Gillis, "Naomi Oreskes, a Lightning Rod in a Changing Climate," *New York Times*, June 15, 2015, https://www.nytimes.com. Timothy P. Scanlon, "A Climate Change 'Critic' with Ties to the Fossil Fuel Industry," *Washington Post*, December 24, 2015, https://www.washingtonpost.com.

5. Oreskes and Conway, *Merchants of Doubt*, 177.

6. Ibid., 175, 182, 176.

7. William Grimes, "Thomas C. Schelling, Master Theorist of Nuclear Strategy, Dies at 95," *New York Times*, December 14, 2016, https://www.nytimes.com; "The Sveriges Riksbank Prize in Economic Sciences in Memory of Alfred Nobel 2018," NobelPrize.org, https://www.nobelprize.org/prizes/economic-sciences/2018/nordhaus/facts. "William D. Nordhaus: Facts," The Nobel Prize, https://www.nobelprize.org/prizes/economic-sciences/2018/nordhaus/facts.

8. Nicolas Nierenberg, Walter R. Tschinkel, and Victoria J. Tschinkel, "Early Climate Change Consensus at the National Academy: The Origins and Making of *Changing Climate*," *Historical Studies in the Natural Sciences*, 40, no. 3 (2010), https://www.bio.fsu.edu/~tschink/publications/HSNS4003_021.pdf, 318–49. Jesse H. Ausubel, *La liberazione dell'ambiente* (Roma: Di Renzo Editore, 2014), 19. "Excerpts from the Climate Report," *New York Times*, October 21, 1983, https://timesmachine.nytimes.com.

9. Oreskes and Conway, *Merchants of Doubt*, 348, 349.

10. Ibid., 341, 348, 349.

11. Mittermeier, "Starving-Polar-Bear Photographer Recalls What Went Wrong."

12. Polar Bear Study Group, "Summary of Polar Bear Population Status per 2019," 2019, http://pbsg.npolar.no/en/status/status-table.html.

13. Rachel Fobar, "Should Polar Bear Hunting Be Legal? It's Complicated," *National Geographic*, May 28, 2019, https://www.nationalgeographic.com.

14. Susan Crockford (zoologist) in discussion with the author, January 31, 2020.

15. "International Poll: Most Expect to Feel Impact of Climate Change, Many Think It Will Make Us Extinct," YouGov, September 14, 2019, https://yougov.co.uk.

16. Richard Tol, "IPCC Again," Richard Tol, April 24, 2014, http://richardtol.blogspot.com/2014/04/ipcc-again.html.

17. Ibid.

18. Ibid.

19. Richard Tol, "Why I resigned from the IPCC WGII," The Global Warming Policy Forum, April 26, 2014, https://www.thegwpf.com/richard-tol-why-i-resigned-from-the-ipcc-wgii.

20. Richard Tol, interview with Roger Harrabin, "In Conversation: Roger Harrabin and Richard Tol," *The Changing Climate*, BBC-TV, November 16, 2015.

21. Tol, "IPCC again."

22. J. Graham Cogley, Jeffrey S. Kargel, G. Kaser, and C. J. van der Veen,

"Tracking the Source of Glacier Misinformation," *Science* 327, no. 5965 (2010): 522, https://doi.org/10.1126/science.327.5965.522-a. Lauren Morello of the *New York Times* wrote, "Pielke said his concern is heightened because he believes Working Group II also misrepresented his research about the link between climate change and monetary damages of natural disasters, highlighting a white paper produced for a conference he organized—when ultimately, attendees at the conference 'came up with a contrary conclusion to what the background paper said.' " Lauren Morello, "Climate Science Panel Apologizes for Himalayan Error," *New York Times*, January 21, 2010, https://www.nytimes.com.

23. Roger Pielke, Jr., *The Climate Fix: What Scientists and Politicians Won't Tell You About Global Warming* (New York: Basic Books, 2010), 176–78, 182. *Topics 2000: Natural Catastrophes—the Current Position*, Munich Re Group, https://www.imia.com/wp-content/uploads/2013/05/EP17-2003-Loss-Potential-of-Natural-Hazzards-sm.pdf.

24. Harold T. Shapiro, Roseanne Diab, Carlos Henrique de Brito Cruz, et al., *Climate Change Assessments: Review of the Processes and Procedures of the IPCC*, Committee to Review the Intergovernmental Panel on Climate Change, October 2010, http://reviewipcc.interacademycouncil.net/report/Climate%20Change%20Assessments,%20Review%20of%20the%20Processes%20&%20Procedures%20of%20the%20IPCC.pdf.

25. Christopher Flavelle, "Climate Change Threatens the World's Food Supply, United Nations Warns," *New York Times*, August 8, 2019, https://www.nytimes.com.

26. Nicholas Wade, "A Passion for Nature, and Really Long Lists," *New York Times*, April 25, 2011, https://www.nytimes.com.

27. Richard Tol, interview with Roger Harrabin, "In Conversation: Roger Harrabin and Richard Tol," *The Changing Climate*, BBC-TV, November 16, 2015.

28. Tol, "IPCC Again."

29. Raúl M. Grijalva, Ranking Member, House Committee on Natural Resources, letter to Bruce Benson, President, University of Colorado, February 24, 2015, https://naturalresources.house.gov/imo/media/doc/2015-02-24_RG%20to%20UoC%20re_climate%20research.pdf.

30. John Schwartz, "Lawmakers Seek Information on Funding for Climate Change Critics," *New York Times*, February 25, 2015, https://www.nytimes.com.

31. Roger Pielke, Jr., "Wikileaks and Me," Roger Pielke, Jr., November 14, 2016,

https://rogerpielkejr.com/2016/11/14/wikileaks-and-me. Joe Romm, "Obama Science Advisor John Holdren Schools Political Scientist Roger Pielke on Climate and Drought," ThinkProgress, March 3, 2019, https://thinkprogress.org. Lindsay Abrams, "FiveThirtyEight's Science Writer Accused of Misrepresenting the Data on Climate Change," *The New Republic*, March 19, 2014, https://www.salon.com.

32. Joe Romm, "Why Do Disinformers like Pielke Shout Down Any Talk of a Link Between Climate Change and Extreme Weather?," ThinkProgress, June 23, 2009, https://thinkprogress.org/. Joe Romm, "Foreign Policy's 'Guide to Climate Skeptics' includes Roger Pielke," ThinkProgress, February 28, 2010, https://thinkprogress.org.

33. Christina Larson and Joshua Keating, "The FP Guide to Climate Skeptics," *Foreign Policy*, February 26, 2010, https://foreignpolicy.com.

34. Roger Pielke, Jr., "I Am Under Investigation," *The Climate Fix* (blog), February 25, 2015, https://theclimatefix.wordpress.com/2015/02/25/i-am-under-investigation.

35. John Holdren, email to John Podesta, January 5, 2014, WikiLeaks, https://wikileaks.org/podesta-emails/emailid/12098.

36. Congressional Record, Proceedings and Debates of the 113rd Congress, Second Session, vol. 160, part 3, 3977. John Holdren, "Drought and Global Climate Change: An Analysis of Statements by Roger Pielke Jr.," February 28, 2014, https://www.whistleblower.org.

37. "Gone Fishing," *Nature*, March 5, 2015, https://www.nature.com.

38. American Meteorological Society, letter to Raúl Grijalva, February 27, 2015.

39. Ben Geman and *National Journal*, "Grijalva: Climate Letters Went Too Far in Seeking Correspondence," *The Atlantic*, March 2, 2015, https://www.theatlantic.com.

40. Roger Pielke, Jr., (climate scientist) in discussion with the author, December 2019.

41. Dave Davies, "Climate Change Is 'Greatest Challenge Humans Have Ever Faced,' Author Says," NPR, April 16, 2019, https://www.npr.org.

42. "Financial Data: 2018 Financials," 350.org, accessed January 30, 2020, https://350.org/2018-annual-report-financials.

43. Giuseppe Formetta and Luc Feyen, "Empirical Evidence of Declining Global Vulnerability to Climate-Related Hazards," *Global Environmental Change* 57 (July 2019): 101920, https://doi.org/10.1016/j.gloenvcha.2019.05.004. Hannah Ritchie and Max Roser, "Global Deaths from Natural Disasters," Our World in Data, https://ourworldindata.org/natural-disasters. Data pub-

lished by EMDAT (2019): OFDA/CRED International Disaster Database, Université Catholique de Louvain–Brussels–Belgium. Data for individual years are summed over ten-year intervals from first to last year of each calendar decade.

44. Matthew C. Nisbet, "How Bill McKibben Changed Environmental Politics and Took on the Oil Patch," *Policy Options*, May 1, 2013, https://policyoptions.irpp.org.

45. McKibben, *The Bill McKibben Reader: Pieces from an Active Life*, 183.

46. Bill McKibben, *The End of Nature* (New York: Random House Trade Paperbacks, 2006), 71.

47. William James, *The Varieties of Religious Experience: A Study in Human Nature* (New York: Random House, 1994), 661.

48. Alston Chase, *A Mind for Murder: The Education of the Unabomber and the Origins of Modern Terrorism* (New York: W. W. Norton & Company, Inc., 2003), 320–322.

49. Mark Sagoff, "The Rise and Fall of Ecological Economics," *Breakthrough Journal* 2 (Fall 2011), https://thebreakthrough.org/journal/issue-2/the-rise-and-fall-of-ecological-economics.

50. Clarence J. Glacken, *Traces on the Rhodian Shore: Nature and Culture in Western Thought from Ancient Times to the End of the Eighteenth Century* (Berkeley: University of California Press, 1967), 423.

51. Environmentalism is the dominant religion in many nations outside of the "West," as well. Both South Korea and Taiwan are led by ostensibly secular anti-nuclear, pro-renewables environmentalists who make apocalyptic statements about climate change nearly identical to those made by the European Greens and U.S. Democrats such as Rep. Alexandria Ocasio-Cortez. Robert H. Nelson, *The New Holy Wars: Economic Religion Versus Environmental Religion in Contemporary America* (Philadelphia: Pennsylvania State University Press, 2010), xv.

52. Carl M. Hand and Kent D. Van Liere, "Religion, Mastery-over-Nature, and Environmental Concern," *Social Forces* 63, no. 2 (December 1984): 555–70, https://doi.org/10.2307/2579062. Mark Morrison, Roderick Duncan, and Kevin Parton, "Religion Does Matter for Climate Change Attitudes and Behavior," *PLOS ONE* 10, no. 8 (2015), https://doi.org/10.1371/journal.pone.0134868. James L. Guth, John C. Green, Lyman A. Kellstedt, and Corwin E. Smidt, "Faith and the Environment: Religious Beliefs and Attitudes on Environmental Policy," *American Journal of Political Science* 39, no. 2

(1995): 364–82, https://doi.org/10.2307/2111617. Paul Wesley Schultz and Lynette Zelezny, "Values as Predictors of Environmental Attitudes: Evidence for Consistency Across 14 Countries," *Journal of Environmental Psychology* 19, no. 3 (1999): 255–65, https://doi.org/10.1006/jevp.1999.0129. Bron Taylor, Gretel Van Wieren, and Bernard Zaleha, "The Greening of Religion Hypothesis (Part Two): Assessing the Data from Lynn White, Jr., to Pope Francis," *Journal for the Study of Religion, Nature and Culture* 10, no. 3 (2016), https://doi.org/10.1558/jsrnc.v10i3.29011. Paul Wesley Schultz, Lynnette Zelezny, and Nancy Dalrymple, "A Multinational Perspective on the Relation Between Judeo-Christian Religious Beliefs and Attitudes of Environmental Concern," *Environment and Behavior* 32, no. 4 (2000): 576–91, https://doi .org/10.1177/00139160021972676.

53. Oliver Milman and David Smith, " 'Listen to the Scientists': Greta Thunberg Urges Congress to Take Action," *The Guardian*, September 18, 2019, https:// www.theguardian.com.

54. Robert H. Nelson, *The New Holy Wars* (Philadelphia: Pennsylvania State University Press, 2010), xv.

55. Ibid.

56. Jonathan Haidt, "Moral Psychology and the Misunderstanding of Religion," *Edge*, September 21, 2007, https://www.edge.org.

57. Michael Barkun, "Divided Apocalypse: Thinking About the End in Contemporary America," *Soundings: An Interdisciplinary Journal* 6, no.3 (1983): 258, https://www.jstor.org/stable/41178260.

58. Friedrich Nietzsche, *Beyond Good and Evil: Prelude to a Philosophy of the Future* (Cambridge, UK: Cambridge University Press, 2002). Friedrich Nietzsche, *On the Genealogy of Morality* (Cambridge, UK: Cambridge University Press, 2007). Chase, *A Mind for Murder: The Education of the Unabomber and the Origins of Modern Terrorism*, 203.

59. Chase, *A Mind for Murder*, 201.

60. Ibid., 205–206.

61. Pascal Bruckner, *The Fanaticism of the Apocalypse: Save the Earth, Punish Human Beings* (Boston: Polity Press, 2013), 19.

62. Ernest Becker, *The Denial of Death* (New York: Free Press, 1973).

63. Ibid., 5.

64. Zion Lights (Extinction Rebellion spokesperson), email correspondence with the author, January 4, 2020. Leslie Hook, "Greta Thunberg: 'All My Life I've

Been the Invisible Girl,' " *Financial Times*, February 22, 2019, https://www
.ft.com. Sarah Lunnon (Extinction Rebellion spokesperson) in discussion
with the author, November 26, 2019.

65. Ines Testoni, Tommaso Gheller, Maddalena Rodelli et al., "Representations
of Death Among Italian Vegetarians: Ethnographic Research on Environment, Disgust and Transcendence," *Europe's Journal of Psychology* 13, no. 3
(August 2017): 378–95, https://doi.org/10.5964/ejop.v13i3.1301.

66. Ibid., 378–395.

67. Rachel N. Lipari and Eunice Park-Lee, *Key Substance Use and Mental Health
Indicators in the United States: Results from the 2018 National Survey on Drug
Use and Health*, Substance Abuse and Mental Health Services Administration,
U.S. Department of Health and Human Services, 2019, http://www.samhsa
.gov/data/sites/default/files/cbhsq-reports/NSDUHNationalFindings
Report2018/NSDUHNationalFindingsReport2018.pdf. "Adolescent Mental
Health in the European Union," World Health Organization Regional Of-
fice for Europe, Copenhagen, Denmark, http://www.euro.who.int/__data
/assets/pdf_file/0005/383891/adolescent-mh-fs-eng.pdf?ua=1.

68. Jean M. Twenge, A. Bell Cooper, Thomas E. Joiner et al., "Age, Period, and
Cohort Trends in Mood Disorder Indicators and Suicide-Related Outcomes in
a Nationally Representative Dataset, 2005–2017," *Journal of Abnormal Psychology* 128, no. 3 (2019): 185–99, https://doi.org/10.1037/abn0000410. Juliana
Menasce Horowitz and Nikki Graf, "Most US Teens See Anxiety and Depression as a Major Problem Among Their Peers," Pew Research Center, February
20, 2019, https://www.pewsocialtrends.org/2019/02/20/most-u-s-teens-see
-anxiety-and-depression-as-a-major-problem-among-their-peers.

69. Richard Rhodes, *A Hole in the World: An American Boyhood* (New York:
Simon & Schuster, 1990).

70. Richard Rhodes (historian) in discussion with the author, November 12,
2019.

71. Brendan O'Neill, "The Madness of Extinction Rebellion," *Spiked*, October 7,
2019, https://www.spiked-online.com/2019/10/07/the-madness-of-extinc
tion-rebellion.

72. Greta Thunberg, " 'Our House Is on Fire': Greta Thunberg, 16, Urges
Leaders to Act on Climate," *The Guardian*, January 25, 2019, https://www
.theguardian.com.

73. Greta Thunberg, "If standing up against the climate and ecological break-

down and for humanity is against the rules then the rules must be broken. #ExtinctionRebellion," Twitter, October 15, 2019, 12:24 p.m., https://twitter .com/gretathunberg/status/1184188303295336448.

74. Sarah Lunnon, interviewed by Phillip Schofield and Holly Willoughby, *This Morning*, ITV, October 17, 2019, https://www.youtube.com/watch?v=ACL mQsPocNs.

75. Myles Allen and Sarah Lunnon, interviewed by Emma Barrett, *Newsnight*, BBC, October 10, 2019.

76. Thunberg, *No One Is Too Small to Make a Difference*, 96.

77. Roger Scruton, *Fools, Frauds and Firebrands: Thinkers of the New Left* (London: Bloomsbury Continuum, 2019), 15.

78. Ibid., 15.

79. Savannah Lovelock and Sarah Lunnon, interviewed by Sophy Ridge, *Sophy Ridge on Sunday*, Sky News, October 6, 2019, YouTube, https://www.you tube.com/watch?v=ArO_-xH5Vm8.

80. Ibid.

81. Lauren Jeffrey (British YouTuber) in discussion with the author, December 3, 2019.

82. Greta Thunberg, "School Strike for Climate—Save the World by Changing the Rules," TEDxStockholm, January 27, 2019, https://www.ted.com. Malena Ernman, "Malena Ernman on daughter Greta Thunberg: 'She was slowly disappearing into some kind of darkness,' " *The Guardian*, February 23, 2020.

83. Scruton, *Fools, Frauds and Firebrands*, 277, 283.

84. Ibid., 288.

85. Sir Francis Bacon, *The Advancement of Learning* (London, Paris, and Melbourne: Cassell & Sons, Ltd., 1893), https://www.gutenberg.org/files /5500/5500-h/5500-h.htm.

86. Jean Delumeau, *Sin and Fear: The Emergence of a Western Guilt Culture, 13th– 18th Centuries*, translated by Eric Nicholson (New York: St. Martin's Press, 1990).

87. Brian L. Burke, "Two Decades of Terror Management Theory: A Meta-analysis of Mortality Salience Research," *Personality and Social Psychology Review* 14, no. 2 (2010): 155–195, https://doi.org/10.1177/1088868309352321.

88. "World's Fairs, 1933–1939," Encyclopedia.com, December 9, 2019, https:// www.encyclopedia.com/education/news-and-education-magazines/worlds -fairs-1933-1939.

89. In truth, gorillas can walk upright but most of the time walk on their hands as well as their feet. But the impression of the gorilla standing upright has been a potent one, in part because it is what humans see when gorillas charge. George Schaller, *The Year of the Gorilla* (Chicago: Chicago University Press, 1988), 4.

90. I am not the first person to make this point. "Our living standards are largely—or completely—unaffected by the presence or absence of spotted owls," wrote Amartya Sen, "but I strongly believe that we should not let them become extinct, for reasons that have nothing much to do with human living standards." Quoted in Mark Sagoff, *The Economy of the Earth: Philosophy, Law, and the Environment* (Cambridge, UK: Cambridge University Press, 2008), 2.

Epilogue

1. Christine Figgener (sea turtle biologist) in conversation with the author, November 6, 2019. Zion Lights (Extinction Rebellion spokesperson) in discussion with the author, December 5, 2019.

2. David Wallace-Wells, "We're Getting a Clearer Picture of the Climate Future—and It's Not as Bad as It Once Looked," *New York*, December 20, 2019, http://nymag.com.

3. Roger Pielke, Jr., "In 2020 Climate Science Needs to Hit the Reset Button, Part One," *Forbes*, December 22, 2019, https://www.forbes.com.

INDEX

ABOUT THE AUTHOR

Michael Shellenberger is a *Time* magazine "Hero of the Environment"; the winner of the 2008 Green Book Award from the Stevens Institute of Technology's Center for Science Writings; and an invited expert reviewer of the next Assessment Report for the Intergovernmental Panel on Climate Change (IPCC). He has written on energy and the environment for the *New York Times*, the *Washington Post*, the *Wall Street Journal*, *Nature Energy*, and other publications for two decades. He is the founder and president of Environmental Progress, an independent, nonpartisan research organization based in Berkeley, California.